MACROERGONOMICS

THEORY, METHODS, AND APPLICATIONS

Human Factors and Ergonomics
Gavriel Salvendy, Series Editor

HCI 1999 Proceedings   2-Volume Set

—**Bullinger, H.-J., and Ziegler, J.** (Eds.): Human–Computer Interaction:
Ergonomics and User Interfaces, *Volume 1 of the Proceedings of HCI
International 1999, the 8th International Conference on Human-Computer
Interaction*
—**Bullinger, H.-J., and Ziegler, J.** (Eds.): Human–Computer Interaction:
Communication, Cooperation, and Application Design, *Volume 2 of the
Proceedings of HCI International 1999, the 8th International Conference on
Human-Computer Interaction*

Stanney, Kay M. (Ed.): Handbook of Virtual Environments: Design,
Implementation, and Applications

Meister, D., and Enderwick, T. (Eds.): Human Factors in System Design,
Development, and Testing

Stephanidis, C. (Ed.): User Interfaces for All: Concepts, Methods, and Tools

HCI 2001 Proceedings   3-Volume Set

—**Smith / Salvendy / Harris / Koubek**
Usability Evaluation and Interface Design: Cognitive Engineering, Intelligent
Agents, and Virtual Reality (Volume 1 of set of 3-volume set)
—**Smith / Salvendy**
Systems, Social, and Internationalization Design Aspects of Human–Computer
Interactions (Volume 2 of 3-volume set)
—**Stephanidis**
Universal Access in HCI: Towards an Information Society for All
(Volume 3 of 3-volume set)

For more information on LEA titles, please contact Lawrence Erlbaum
Associates, Publishers, at www.erlbaum.com.

MACROERGONOMICS
Theory, Methods, and Applications

Edited by

Hal W. Hendrick
University of Southern California

Brian M. Kleiner
*Virginia Polytechnic Institute
and State University*

CRC Press
Taylor & Francis Group
Boca Raton London New York

CRC Press is an imprint of the
Taylor & Francis Group, an informa business

Senior Acquisitions Editor:	Anne Duffy
Editorial Assistant:	Kristin Duch
Cover Design:	Kathryn Houghtaling Lacey
Textbook Production Manager:	Paul Smolenski
Full-Service Compositor:	UG / GGS Information Services, Inc.
Text and Cover Printer:	Hamilton Printing Company

Reprinted 2009 by CRC Press

Library of Congress Cataloging-in-Publication Data

Macroergonomics: theory, methods, and applications / edited by Hal W. Hendrick, Brian M. Kleiner.
    p. cm.—(Human factors and ergonomics)
Includes bibliographical references and index.
ISBN 0-8058-3191-6 (casebound)
1. Human engineering. I. Hendrick, Hal W. II. Kleiner, Brian M., 1959- III. Series.

TA166.M29 2002
620.8′2–dc21   2001058626

10 9 8 7 6 5 4 3 2 1

Contents

Preface

This volume represents a major milestone in the development and dissemination of knowledge in the comparatively new sub-discipline of human factors or ergonomics that is known as *macroergonomics*. Macroergonomics is concerned with the analysis, design and evaluation of work systems. The editors have assembled an international array of academicians in this field—persons who have been contributing knowledge through research and applying macroergonomics to a broad spectrum of organizations. This comprehensive, advanced set of works follows the release of an introductory volume, *Macroergonomics: An Introduction to Work System Design*, published by the Human Factors and Ergonomics Society (HFES) and authored by the editors of this volume. With the introduction to the field established by HFES, the editors now seek to advance the understanding and application of macroergonomics by assembling a group of their distinguished colleagues from the U.S., Europe, and Japan.

The volume is divided into two major Parts. In Part I, macroergonomics is introduced. Chapter 1, by Hendrick, gives an overview of the basic concepts and theoretical constructs of macroergonomics. Chapter 2, presented by Brown, covers the method of participation as applied in macroergonomics interventions. The method of assessing work system structure (Chapter 3) is presented by Hendrick. Chapter 4 addresses the assessment of work system processes and is presented by Robertson, Kleiner, and O'Neill. Hendrick then presents additional methods in Chapter 5. In Chapter 6, Nagamachi, using his own research as a basis, discusses the relationship among job design, macroergonomics and productivity. Part I concludes with a chapter by Kleiner that illustrates how research methods can be used to empirically test and evaluate macroergonomic constructs and factors for increased work system understanding.

In Part II, chapters are presented that focus on macroergonomics applications. These application areas include reducing work related injuries (Chapter 8) presented by Imada; information and communications technology (Chapter 9) presented by Bradley; hazards (Chapter 10) presented by T. Smith; manufacturing (Chapter 11) presented by Karwowski and Salvendy, et al.; training system development (Chapter 12) presented by Robertson; large-scale organizational change (Chapter 13) presented by Kleiner; community planning and development (Chapter 14) presented by H.J. Smith and Carayon et al.; technology transfer (Chapter 15) presented by Shahnavaz; and aviation (Chapter 16) presented by Meshkati. The need for macroergonomic analysis and design is realized with a discussion by Meshkati of macroergonomic aspects of major disasters (Chapter 17). In the final chapter (Chapter 18) Zink offers a vision of macroergonomics for the future.

This book can be used as a text in an academic setting or can be a vital reference resource for the macroergonomics specialist or ergonomics generalist. Virtually anyone with a role of organizational intervention at the individual, system or organizational level can benefit from a working knowledge of macroergonomics. It is the hope of the editors that through this text, work systems will be optimized, yielding organizations with improved worker well being and productivity.

Series Foreword

With the rapid introduction of highly sophisticated computers, (tele) communication, service, and manufacturing systems, a major shift has occurred in the way people use technology and work with it. The objective of this book series on Human Factors and Ergonomics is to provide researchers and practitioners a platform where important issues related to these changes can be discussed, and methods and recommendations can be presented for ensuring that emerging technologies provide increased productivity, quality, satisfaction, safety, and health in the new workplace and Information Society.

This edited book by Hal W. Hendrick, the father of macroergonomics, and Brian M. Kleiner cover most effectively, in 18 chapters, the depth and breadth of macroergonomics. While ergonomics traditionally refers to the design of systems for human use, macroergonomics extends this notion by including industrial and organizational psychology approaches in combination with ergonomics approaches. By so doing, productivity, efficiency and satisfaction should increase in the enterprise.

By utilizing concepts and methods of macroergonomics as a supplement to, rather than as a substitute for the use of traditional ergonomics theories and methods, a more comprehensive approach to the design of human work as it impacts enterprise effectiveness may be derived than it was possible to do so before the mergence of macroergonomics.

This book effectively presents this road map with concepts, theories and methods of macroergonomics. Thus, this book could be used both as a textbook and as a guide to effectively implement macroergonomics in the workplace.

—**Gavriel Salvendy**
Series Editor

About the Editors

Dr. Hal W. Hendrick is Emeritus Professor of Human Factors at the University of Southern California (USC) and Principal of Hendrick and Associates, an ergonomics and I/O psychology consulting firm. He is a Certified Professional Ergonomist and holds a Ph.D. in Industrial Psychology and MS in Human Factors from Purdue University, with a minor in Industrial Engineering. He is a Past President of the Human Factors and Ergonomics Society (HFES), the International Ergonomics Association, and the Board of Certification in Professional Ergonomics. Hal is the recipient of USC's highest award for teaching excellence and both the HFES Jack A. Kraft Innovator Award and Alexander C. Williams, Jr. Design Award. He is the author of over 150 professional publications, including two textbooks, and the editor of ten books. Hal conceptualized and initiated the sub-discipline of macroergonomics.

Dr. Brian M. Kleiner is associate professor, coordinator of the Human Factors Engineering and Ergonomics Center, and director of the Macroergonomics and Group Decision Systems Laboratory in the Grado Department of Industrial and Systems Engineering, Virginia Polytechnic Institute and State University (Virginia Tech). He received his Ph.D. in 1990 in Industrial Engineering (human factors concentration) from the State University of New York at Buffalo, his MS in 1983 in Industrial Engineering (human factors concentration) from the State University of New York at Buffalo and his BA in 1981 in Psychology from SUNY.

MACROERGONOMICS

THEORY, METHODS, AND APPLICATIONS

1

An Overview of Macroergonomics

Hal W. Hendrick
University of Southern California

INTRODUCTION

This chapter provides a comprehensive overview of the subdiscipline of human factors or ergonomics that has come to be known as *macroergonomics*. Macroergonomics is concerned with the analysis, design, and evaluation of *work systems*. The term *work* is used herein to refer to any form of human effort or activity, including recreation and leisure pursuits. As used herein, *system* refers to *sociotechnical* systems. These systems may be as simple as a single individual using a hand tool or as complex as a multinational organization. A *work system* consists of two or more persons interacting with some form of (1) job design, (2) hardware and/or software, (3) internal environment, (4) external environment, and (5) an organizational design (i.e., the work system's structure and processes). *Job design* includes work modules, tasks, knowledge and skill requirements, and such factors as the degree of autonomy, identity, variety, meaningfulness, feedback, and opportunity for social interaction. The *hardware* typically consists of machines or tools. The *internal environment* consists of various physical parameters, such as temperature, humidity, illumination, noise, temperature, humidity, air quality, and vibration. It also includes psychosocial factors (see Chapter 3). The *external environment* consists of those elements that permeate the organization to which the organization must be responsive to be successful. Included are political, cultural, and economic factors (e.g., materials and parts resources, customers, available labor pool, and educational resources). Of particular importance is the degree of stability or change of these external environment factors and, taken together for a given work system, the degree of environmental complexity they present to the organization. The *organizational design* of a work system consists of its organizational structure and the processes by which the work system accomplishes its functions.

THE CONCEPT OF MACROERGONOMICS

The Technological Context of Macroergonomics

Any scientific discipline can most readily and distinctly be defined by the nature of its unique technology. Based on its survey of human factors/ergonomics internationally, the Strategic Planning Committee of the Human Factors and Ergonomics Society (HFES) identified the unique technology of human factors/ergonomics (HF/E) as *human–system interface technology*. Included are the interfaces between the people portion of systems and the other sociotechnical system components. As noted above, these components include jobs, hardware, software, internal and external environments, and work system structures and processes. As a science, HF/E involves the study of human performance capabilities, limitations, and other characteristics. These data then are used to develop human–system interface (HSI) technology. HSI technology takes the form of interface design principles, guidelines, specifications, methods, and tools. As a practice, HF/E professionals apply HSI technology to the design, analysis, test and evaluation, standardization, and control of systems. The overall goal of the discipline is to improve the human condition, including health, safety, comfort, productivity, and quality of life (Human Factors and Ergonomics Society, 1998).

HSI technology has at least five clearly identifiable subparts, each with a related design focus (Hendrick, 1998; Hendrick & Kleiner, 2001). These are as follows:

1. *Human–machine* interface technology or hardware ergonomics
2. *Human–environment* interface technology, or environmental ergonomics
3. *Human–software* interface technology, or cognitive ergonomics
4. *Human–job* interface technology, or work design ergonomics
5. *Human–organization* interface technology, or macroergonomics

The first four of these technologies primarily are focused at the individual or, at best, subsystem level. They thus constitute the technologies of what often is referred to in the literature as *micro-ergonomics*. The fifth is focused at the overall work system level and, accordingly, is the primary technology of *macroergonomics*.

Macroergonomics

Macroergonomics is a perspective, a methodology, and a recognized subdiscipline of ergonomics/human factors. It is backed by empirical science. From its foundational research roots in the sociotechnical systems tradition to modern laboratory and field investigations of the sociotechnical system elements and their relationships, new scientific knowledge about work systems has emerged. Systematic macroergonomic methodologies for the analysis, design, and evaluation of work systems also have emerged from this research base. As a perspective, macroergonomics provides the ergonomist with an appreciation for the larger system. It is a perspective that increases the likelihood of micro-ergonomic interventions having a relatively greater effectiveness than otherwise might be the case.

As a subdiscipline, macroergonomics is concerned with human–organization interface (HOI) technology. The empirical science supporting this subdiscipline is concerned with factors in the technological subsystem, personnel subsystem, external environment, and their interactions as they impact on work system design. This will be described and explained further in Chapter 3. As a perspective, certain guiding principles enlighten the ergonomist. These include participation, flexibility, joint optimization, joint design, system harmonization, and continuous improvement of processes (a principle shared with total quality management).

Conceptually, macroergonomics may be defined as a top-down sociotechnical systems approach to the design of work systems, and the carry-through of the overall work system design to the design of the human–job, human–machine, and human–software interfaces (Hendrick, 1997; Hendrick & Kleiner, 2001).

The goal of macroergonomics is to optimize the work system's design in terms of its sociotechnical system characteristics, and then carry the characteristics of the overall work system design down through to the design of individual jobs and human–machine and human–software interfaces to ensure a fully harmonized work system. As explained later in this chapter, when this goal is achieved, the result should be dramatic improvements in various aspects of organizational performance and effectiveness.

Implementing Macroergonomics

Although conceptually, macroergonomics is a top-down approach, in actual practice, it is a top-down, bottom-up, and middle out analysis, design, and evaluation process. *Top down*, an overall general work system structure may be prescribed to match the organization's sociotechnical characteristics, as described in Chapter 3. *Middle out*, an analysis of subsystems and work process can be assessed both up and down the hierarchy from intermediate levels, and changes made to ensure a harmonized work system design. *Bottom up* most often involves identification of problems by employees and lower-level supervisors that result from higher-level work system structural or processes design. A true macroergonomic intervention effort most often requires employee participation at all levels of the organization. Rather than being a sequential linear process, macroergonomic interventions often are nonlinear, stochastic, in that inferences and decisions often have to be derived from incomplete data, and iterative (i.e., design, try, evaluate, redesign, etc.).

In actual practice, macroergonomic design rarely is a pure process; for example, one usually has to live with a number of existing subsystems and components. Union–management contracts, aspects of corporate culture, ongoing projects, and financial considerations all may serve to inhibit carrying out what appears to be optimally desirable from a macroergonomic design standpoint. In some cases, these factors may serve simply to delay implementation of some macroergonomic improvements. In other cases, the desired changes may never be possible.

A full macroergonomic effort is most feasible under four conditions. First, when developing a new work system from scratch, such as when a new organization is being formed. Second, and more often the case, when a major work system change already is to take place; for example, when updating equipment or changing over to a new technology, or moving to a new facility. A third opportunity is when a major change occurs in the goals, scope, or direction of the organization. A fourth situation is when the organization has a costly chronic problem that has not proven correctable with a purely micro-ergonomic effort, or via other intervention strategies. Under these conditions, management is likely to be receptive to a true macroergonomic intervention effort if it can be shown to have good promise for success. Recently, the desire to reduce lost-time accidents and injuries, and related costs, has led senior managers in some organizations to support a true macroergonomic effort. As illustrated in Chapters 7 and 8, many of these efforts have achieved dramatic results.

Frequently, a true macroergonomic intervention is not possible initially. Rather, the ergonomist or ergonomic team has to begin by making micro-ergonomic improvements that yield positive results within a relatively short period of time (often called the "picking the low hanging fruit" strategy). When managers see these positive results, they become interested in supporting further ergonomic interventions. In the course of this process, the ergonomist has the opportunity to raise the consciousness level of the decision-makers about the full scope of ergonomics and its potential value to the organization. Over time, senior management comes to support progressively larger ergonomic projects that actually change the nature of the work system as a whole. Based on the author's intervention experiences, this process typically takes about two years from the time one first has completed several successful micro-ergonomic projects, and thus gained the confidence of the key decision-makers.

Macroergonomics Versus Industrial and Organizational Psychology

Industrial and organizational (I/O) psychology can be viewed conceptually as the flip side of the coin from ergonomics. Whereas ergonomics focuses on designing work systems to fit people, I/O psychology primarily is concerned with selecting people to fit work systems. This is especially true of classical industrial psychology—the opposite side of the coin from micro-ergonomics. Micro-ergonomics focuses on designing jobs, work environments, hardware, and software to fit individuals. In contrast, classical industrial psychology focuses on identifying and selecting individuals to fit jobs and work environments.

Organizational psychology can be viewed as the opposite side of the coin from macroergonomics. Here, however, there is a greater overlap. Both organizational psychology and macroergonomics are concerned with the design of organizational structures and processes, but the focus is somewhat different. In the case of organizational psychology, improving motivation and job satisfaction, developing effective incentive systems, enhancing leadership and organizational climate, and fostering teamwork are common objectives. While these objectives also are important to macroergonomics, the primary focus of macroergonomics is to design work systems that are compatible with an organization's sociotechnical system characteristics; and then to ensure that the micro-ergonomic elements are designed to harmonize with the overall work system structure and processes.

Because there is an overlap of macroergonomics with organizational psychology, many of the empirically developed methods and tools of organizational psychology are adaptable for use in implementing macroergonomics. Adaptations of some of these classical methods are described in Chapter 5.

HISTORICAL DEVELOPMENT

The design of sociotechnical systems to optimize human–system interfaces is not new to human factors/ergonomics. In fact, it has been going on since the formal inception of human factors/ergonomics in the late 1940s. During the first three decades of the discipline, the focus of this design concern primarily was on optimizing the interface between individual operators and their immediate work environment, or what today often is referred to as micro-ergonomics.

Initially this focus was labeled as *man–machine* interface design. As our culture became more sensitive to gender issues, this focus came to be known as *human–machine* interface design. Scientific knowledge about human capabilities, limitations, and other characteristics was developed and applied to the design of operator controls, displays, tools, workspace arrangements, and physical environments to enhance health, safety, comfort, and productivity, and to minimize human error via design.

The advent of the silicon chip led to the subsequent rapid development of computers and the automation of sociotechnical systems. With this development, a new subdiscipline of human factors/ergonomics emerged. It was centered on software design and became known as *cognitive ergonomics* (Dray, 1985, 1986). Although the primary focus still was on the individual operator, the emphasis was on how humans think and process information, and how to design software to dialog in the same manner. One impact of this new subdiscipline has been to increase the number of ergonomic positions in the industrialized world by roughly 25%.

Select Committee on Human Factors Futures, 1980–2000

Although various ergonomists had written about the importance of a true systems approach and considering organizational and management factors in ergonomics, the formal beginning of macroergonomics as an identifiable subdiscipline had its roots in the work of the Human Factors Society's Select Committee on Human Factors Futures, 1980–2000. In 1978, a distinguished former president of the Human Factors Society (HFS), Arnold Small, noted that there were many dramatic changes occurring in all aspects of our society and its built environments. Arnold had become convinced that, in the coming decades, traditional human factors/ergonomics would not be adequate to respond to these trends effectively. At his urging, the HFS (now the Human Factors and Ergonomics Society) formed a "Select Committee on Human Factors Futures, 1980–2000." The purpose of this committee was to study these trends and determine their implications for the human factors discipline for the next two decades and beyond. Arnold Small was appointed chair of this committee. I was appointed to the committee and specifically charged to research trends related to the management and organization of work systems.

In October 1980, at the HFS Annual Meeting in Los Angeles, California, the committee members reported on their findings. Among other things, I noted the following six major trends as part of my report (Hendrick, 1980).

1. *Technology.* Recent breakthroughs in the development of new materials, microminiaturization of components, and the rapid development of new technology in the computer and telecommunications industries will fundamentally alter the nature of work in offices and factories during the 1980–2000 time frame. In

general, we were entering a true information age of automation that would profoundly affect work organization and related human–machine interfaces.

2. *Demographic shifts.* The average age of the work populations in the industrialized countries of the world will increase by approximately six months for each passing year during the 1980s and most of the 1990s. Two major factors account for this "graying" of the workforce: first, the aging of the post–World War II "baby boom" demographic bulge that now has entered the work force; second, the lengthening of the average productive life span of workers because of better nutrition and health care. In short, during the next two decades, the workforce would become progressively more mature, experienced, and professionalized.

As the organizational literature has shown (e.g., see Robbins, 1983 for a review published after my 1980 report), as the level of professionalism (i.e., education, training, and experience) increases, it becomes important for work systems to become less formalized (i.e., less controlled by standardized procedures, rules, and detailed job descriptions), tactical decisionmaking to become decentralized (i.e., delegated to the lower-level supervisors and workers), and management systems to accommodate similarly.

These requirements represent profound changes to traditional bureaucratic work systems and related human–system interfaces.

3. *Value changes.* Beginning in the mid 1960s and progressing into the 1970s, a fundamental shift occurred in the value systems of work forces in the United States and Western Europe. These value system changes and their implications for work systems design were noted by a number of prominent organizational behavior researchers, and were summarized by Argyris (1971). In particular, Argyris noted that workers now both value and expect to have greater control over the planning and pacing of their work, greater decision-making responsibility, and more broadly defined jobs that enable a greater sense of both responsibility and accomplishment. Argyis further noted that, to the extent organizations and work system designs do not accommodate to these values, organizational efficiency and quality of performance will deteriorate.

These value changes were further validated in the 1970s by Yankelovich (1979), based on extensive longitudinal studies of workforce attitudes and values in the United States. Yankelovich found these changes to be particularly dramatic and strong among those workers born after the World War II. Of particular note from his findings was the insistence that jobs become less depersonalized and more meaningful.

4. *World competition.* Progressively, U.S. industry is being forced to compete with high-quality products from Europe and Japan; and other countries, such as Taiwan and Korea, soon will follow. Put simply, the post–World War II dominance by U.S. industry is gone. In light of this increasingly competitive world market, the future survival of most companies will depend on their efficiency of operation and production of state-of-the-art products of high quality. In the final analysis, the primary difference between successful and unsuccessful

competitors will be the quality of the ergonomic design of their products and of their total work organization, and the two are likely to be interrelated.

5. *Ergonomics-based litigation.* In the United States, litigation based on the lack of ergonomics safety design of both consumer products and the workplace is increasing, and awards of juries often have been high. The message from this litigation is clear: Managers are responsible for ensuring that adequate attention is given to the ergonomic design of both their products and their employees' work environments to ensure safety.

One impact of this message, as well as from the competition issue, noted earlier, is that ergonomists are likely to find themselves functioning as true management consultants. A related implication of equal importance is that ergonomic education programs will need to provide academic courses in organizational theory, behavior, and management to prepare their students for this consultant role.

6. *Failure of traditional (micro-) ergonomics.* Early attempts to incorporate ergonomics into the design of computer work stations and software resulted in improvement, but have been disappointing in terms of (a) reducing the work system productivity costs of white-collar jobs (b) improving intrinsic job satisfaction, and (c) reducing symptoms of high job stress.

As I noted several years later, we had begun to realize that it was entirely possible to do an outstanding job of ergonomically designing a system's components, modules, and subsystems, yet fail to reach relevant system effectiveness goals because of inattention to the macroergonomic design of the overall work system (Hendrick 1984, 1986a, 1986b). Investigations by Meshkati (1986) and Meshkati & Robertson (1986) of failed technology transfer projects, by Meshkati (1991) of major system disasters (e.g., Three Mile Island and Chernobyl nuclear power plants, and the Bhopal chemical plant), and by Munipov (1990) of the Chernobyl accident, all have resulted in similar conclusions.

Integrating Organizational Design and Management Factors Into Ergonomics

Based on the aforementioned observations, I concluded in my 1980 report that, for the human factors/ergonomic discipline to be truly effective, and responsive to the foreseeable requirements of the next two decades and beyond, there was a strong need to integrate organizational design and management (ODAM) factors into our research and practice.

As we since have seen, these predictions from 1980 have come to pass, and are continuing. I believe this need to integrate ODAM factors into our research and practice largely accounts for the rapid growth and development of macroergonomics that has occurred since. In 1984, as a direct response to the Select Committee's report, an ODAM technical group was formed within the Human

Factors Society. Similar groups also were developed in both the Japan Ergonomics Research Society and the Hungarian society, and less formal ODAM groups were formed in other ergonomics societies around the world. In 1985, based on a survey of the IEA Federated Societies, the International Ergonomics Association (IEA) formed a Science and Technology Committee comprised of eight technical groups specifically requested in the survey responses. One of these eight was an ODAM Technical Group. This technical group since has been one of the IEA's most active; for example, it has helped organize six highly successful IEA International Symposia on Human Factors in ODAM, and plans to continue this activity indefinitely on a biennial basis.

In 1988, ODAM was one of the five major themes of the 10th IEA Triennial Congress in Sidney, Australia, and 1 of 12 themes for the 11th Triennial Congress in Paris, France, in 1990. At the 12th Triennial Congress in Toronto, Canada, in 1994, the 13th in Tempare, Finland, in 1997, and the 14th in San Diego, California, in 2000, a major multisession symposium on Human Factors in ODAM was organized. For the 12th and 13th Congresses, more paper proposals were received on macroergonomics and ODAM than on any other topic, and it was one of the three largest topics in terms of papers presented at the 14th Congress.

By 1986, conceptualization of the ergonomics of work systems had been developed to the point of identifying it as a separate subdiscipline. At that time, it became formally identified as macroergonomics (Hendrick, 1986a, 1986b). In 1998, in response to the considerable methodology, research findings, and practice experience that had developed internationally during the 1980s and 1990s, the Human Factors and Ergonomics Society ODAM Technical Group changed its name to the "Macroergonomics Technical Group" (ME TG).

THE STRUCTURAL DIMENSIONS OF WORK SYSTEMS

Earlier in this chapter, in defining macroergonomics, I noted that it is concerned with optimizing the structure and related processes of work systems. Accordingly, an understanding of macroergonomics first requires an understanding of the key dimensions of work system structure. As will be described in Chapter 3, knowledge of the specific sociotechnical characteristics of a given work system will guide us in macroergonomically optimizing these key dimensions of the work system's organizational structure.

Before describing the dimensions of organizational structure, a few basic concepts need to be clarified to provide a common framework. These are the terms, organization and organization design.

Organization. An organization can be defined as "the planned coordination of two or more people who, functioning on a relatively continuous basis and through division of labor and a hierarchy of authority, seek to achieve a common goal or set of goals" (Robbins, 1983, p. 5). Breaking down this definition we can note the following. First, the planned coordination of collective activities implies management. Second, because organizations are made up of more than one person, individual activities must be designed and functionally allocated so as to be complementary, balanced, harmonized, and integrated to ensure an effectively functioning work system. Third, because organizations accomplish their activities and functions through a division of labor and a hierarchy of authority, they have structure. Thus, how this structure is designed is critical to the organization's functioning and effectiveness. Fourth, because the collective activities and functions of an organization are oriented toward achieving a common goal, or set of goals, it implies that criteria for assessing an organization's design exist. These criteria should be identified, weighted, and utilized in evaluating alternative designs for the work system.

Organizational Design.   Organizational design specifically means the design of an organization's work system structure and related processes to achieve the organization's goals.

The structure of a work system often is conceptualized as having three core dimensions: complexity, formalization, and centralization (Stevenson, 1993; Bedeian & Zammuto, 1991; Robbins, 1983).

Complexity refers to the degree of differentiation and integration that exist within a work system's structure. *Differentiation* refers to the degree to which the work system is segmented into parts; *integration* refers to the number and types of mechanisms that are used to integrate the segmented parts for the purposes of communication, coordination, and control.

Complexity: Differentiation

Three common types of differentiation are employed in structuring work systems: vertical, horizontal, and spatial. Increasing any one of the three increases a work system's complexity.

Vertical Differentiation.   Vertical differentiation refers to the number of hierarchical levels separating the chief executive position from the jobs directly involved with the system's output. In general, as the size of an organization increases, the need for greater vertical differentiation also increases (Mileti, Gillespie & Haas, 1977); for example, in one study, size alone accounted for 50 to 59% of the variance (Montanari, 1976). A primary reason for this strong relationship is the practical limitations of *span of control*. Any manager is limited in the number of subordinates that he or she can direct effectively (Robbins, 1983). As a result, as the number of first-level employees increases, the number of first-line supervisors also must increase. This, in turn, requires more managers at each successively higher level, thus requiring more hierarchical levels in the work system's structure.

Although span of control limitations underlie the size–vertical differentiation relationship, these limitations can vary considerably, depending on a number of other factors; for example, depending on its sociotechnical characteristics, relatively larger or smaller spans of control may be appropriate. A major factor affecting span of control is the *degree of professionalism*, or education and skill requirements, designed into employee jobs. In general, as the level of professionalism increases, employees are able to function more autonomously, and thus need less supervision. As a result, the manager can supervise a larger number of employees. Other factors affecting span of control are the type of technology employed, degree of formalization, psychosocial variables, and environmental characteristics. These factors and their effects on work system structure will be discussed in Chapter 3.

Horizontal Differentiation. Horizontal differentiation refers to the degree of departmentalization and specialization within a given work system. The primary reason horizontal differentiation increases complexity is because it requires both a greater number and more sophisticated and expensive methods of control. Nevertheless, specialization is common to most work systems because of the inherent efficiencies in the division of labor. This point was demonstrated by Adam Smith over 200 years ago (1970/1876). Smith observed that 10 workers, each doing particular tasks (job specialization), could produce about 48,000 pins per day. If, instead, the 10 each worked separately and independently, performing all of the production tasks, they would be lucky to make 200.

One outcome of division of labor is that it creates groups of specialists, or *departmentalization*. The most common ways of designing departments into work systems are on the basis of (1) simple numbers, (2) function, (3) product or services, (4) client or client class served, (5) process, and (6) geography. Most large corporations will use all six (Robbins, 1983).

There are two commonly used ways to determine whether or not a work group should be divided into one or more departments. These are the degree of commonality of (1) *goals*, and (2) *time orientation*. The more that subgroups differ either in goals or time orientations, the greater is the likelihood that they should be structured as separate departments; for example, research and development (R&D) employees differ from salespersons on both of these dimensions. First, the two have very different goals: One is concerned with developing products, and the other with selling them. Secondly, the time orientation of sales personnel usually is short (1 year or less), whereas it usually is long (3 or more years) for R&D personnel. Thus, they clearly should be in separate departments, and usually are (Robbins, 1983).

Spatial Dispersion. Spatial dispersion refers to the degree that an organization's activities are performed in more than one location. Three different measures commonly are used to quantify the degree of spatial dispersion: (1) the number of geographic locations comprising the total work system, (2) the average distance of the separated locations from the organization's headquarters, and (3) the proportion of employees in these separated units in relation to the number in the headquarters (Hall, Haas & Johnson, 1967). In general, increasing any of these three measures increases complexity.

Complexity: Integration

Integration refers to the number of mechanisms designed into a work system for ensuring communication, coordination, and control among the differentiated elements. In general, as a work system increases in differentiation, the need for integrating mechanisms also increases. This happens because greater differentiation increases the number of units, levels, and departments that must communicate with one another, coordinate their separate activities, and be controlled for efficient operation. The most common types of integrating mechanisms that can be designed into a work system are formal rules and procedures, committees, task teams, liaison positions, and system integration offices. Vertical differentiation, in itself, is a primary integrating mechanism (i.e., a manager at one level typically serves to coordinate and control the activities of several lower-level groups). It also should be noted that computerized information and decision support systems can be designed to serve as integrating mechanisms.

Once the differentiation aspects of a work system's structure have been determined, a major task for the macroergonomic professional is to work with the organization's personnel to determine the kinds and number of integrating mechanisms to design into the work system. Having too few integrating mechanisms will result in inadequate coordination and control among the differentiated elements; too many will stifle the work system's functioning and increase costs. As described in Chapter 3, a systematic analysis of technological and personnel subsystem factors, and of characteristics of the external environment, all can be used to help determine the optimal number and types of integrating mechanisms.

Formalization

From a macroergonomics perspective, *formalization* can be defined as the degree to which jobs within the work system are standardized. Highly formalized work systems allow for little employee discretion over what is to be done, how it is to be accomplished, or when it is to be done (Robbins, 1983). Highly formalized designs are characterized by explicit job descriptions, extensive rules, and clearly defined, standardized procedures covering work processes. Ergonomists serve to increase formalization by designing jobs, machines, and software so as to standardize procedures and allow little opportunity for employee decision discretion.

It also is possible to design human–job, human–machine, and human–software interfaces ergonomically to permit greater flexibility and scope to employee decision making, thus creating work systems having low formalization. Employee behavior thus is relatively unprogrammed and the work system allows for greater use of employees' mental abilities. Under these conditions, greater reliance is placed on the employees' professionalism.

As a general rule, the simpler and more repetitive the jobs to be designed into the work system, the higher should be the level of formalization; however, consideration should be given to not making the work system so highly formalized that jobs lack any intrinsic motivation, fail to utilize employee skills effectively, or degrade human dignity. In contrast, the more nonroutine or unpredictable the work tasks and related decision making, the less amenable the work system is to high formalization. Instead, reliance must be placed on designing jobs requiring a relatively high level of professionalism. In general, because they do allow for more autonomy and greater use of one's mental abilities, more highly professionalized jobs tend to be more intrinsically motivating for growth-oriented employees.

Centralization

Centralization has to do with where formal decision making occurs within the work system. In highly centralized work systems, formal decision making is concentrated in a relatively few individuals, group, or level, usually high in the organization; and lower-level supervisors and employees have only minimal input into the decisions affecting their jobs (Robbins, 1983). In highly decentralized work systems, decisions are delegated downward to the lowest level having the necessary expertise. Highly decentralized work systems thus require lower-level employees to have a relatively higher level of professionalism.

Work systems carry out two basic forms of decision making, strategic and tactical, and the degree of centralization often is quite different for each: *Tactical* decision making has to do with the day-to-day operation of the organization's business; *strategic* decision making concerns broader policy and long-range planning for the organization. When the sociotechnical characteristics of the organization call for low formalization and high professionalism, tactical decision making may be highly decentralized, whereas strategic decision making may remain highly centralized. Under these conditions, there still can be less centralization of strategic decision making than at first appears to be the case. This is because the information required for strategic decision making often is controlled and filtered by middle management or even lower-level personnel. Thus, to the extent that these persons can reduce, selectively omit, summarize, or embellish the information that gets fed to top management, the less is the actual degree of centralization of strategic decision making (Hendrick, 1997).

As a general rule, centralization is desirable (1) when operating in a highly stable and predictable external environment, (2) for financial, legal, or other decisions which clearly can be done more efficiently when centralized, (3) when a comprehensive perspective is needed, (4) when the decisions have little effect on employees' jobs or are of little employee interest, and (5) when it provides significant economies. Decentralized decision making is desirable (1) when an organization needs to respond rapidly to changing or unpredictable conditions at the point where change is occurring, (2) when "grass roots" input to decisions is desirable, (3) to gain greater employee commitment to, and support for, decisions by involving employees in the process, (4) to more fully utilize the mental capabilities and job knowledge of employees, (5) to provide employees with greater intrinsic motivation, job satisfaction, and sense of self-worth, (6) when it can reduce stress and related health problems by giving employees greater control over their work, (7) when it can avoid overtaxing a given manager's capacity for human information processing and decision making, and (8) to provide greater training opportunity for lower-level managers (Hendrick, 1997).

MACROERGONOMIC CONSTRUCTS

The Sociotechnical Systems Model

Macroergonomics has its roots in sociotechnical systems theory and the empirical research that led to its development and refinement. This research was developed initially in the late 1940s and 1950s by Trist and Bamforth (1951) and their colleagues at the Tavistock Institute of Human Relations in the United Kingdom. In the United States, extensive follow-on sociotechnical systems research was carried out by Daniel Katz and Robert Kahn of the Survey Research Center at the University of Michigan, as well as by many others. This body of research served to confirm and refine the sociotechnical systems model of work systems. The classic Tavistock research that led to the initial development of sociotechnical systems theory, including the coining of the term "sociotechnical systems" by Trist and Bamforth, was on Welsh deep seam coal mining (DeGreene, 1973).

The Tavistock Studies

Prior to the 1950s the traditional way of mining coal was labor intensive. It consisted of teams of small, relatively autonomous groups of miners using hand tools. Control over the work was largely internal to the group. Each miner performed a variety of tasks, which were interchangeable within the group. The group derived considerable satisfaction from being able to complete the whole task. In addition, the close group interaction enabled ample opportunity for satisfying social needs on the job. As a result, the psychosocial and cultural characteristics of the workforce, the task requirements, and the work system's design all were *congruent*.

Following World War II, a major technological change occurred: The manual, or *shortwall*, method of coal mining was replaced by mechanical coal cutters. As a result, mining no longer was restricted to working a short face of coal. Instead, using the cutters, miners now could extract coal from a long wall. This new *longwall* method was expected to be far more efficient and less costly than the old manual method of extracting coal. Instead, it proved less efficient. There were frequent breakdowns in the process, and absenteeism and grievances greatly increased. The Tavistock studies revealed that the new longwall system resulted in a work system design that was not congruent with the psychosocial and cultural characteristics of the workforce. Rather than working in small work groups, the miners now were required to work in shifts of 10 to 20 men. Jobs were designed to include a very narrowly defined set of tasks, and job rotation was not possible. Opportunities for social interaction were severely limited. Instead of being autonomous, there was a high degree of interdependence among the tasks of the three shifts; and problems carried over from one shift to the next, thus holding up the labor stages in the extraction process. This highly rigid work system design proved very sensitive to both productivity and social disruptions. Instead of the expected large increase in productivity, low production, absenteeism, and intergroup rivalry became common (DeGreene, 1973).

Follow-on studies of other coal mines were carried out by the Tavistock Institute (Trist, Higgin, Murray & Pollock, 1963) in which the conventional longwall method was compared with a *composite* longwall method. The work system of the composite method was designed to utilize a combination of the new technology and features of the old psychosocial work structure of the manual system. As compared with the conventional longwall work system, the composite system reduced the interdependence of the shifts, increased the variety of skills utilized by each worker, created opportunities for satisfying social needs on the job, and permitted self-selection by workers of their team members. Productivity for the composite system was found to be significantly higher than for either the conventional longwall or the old manual system. Grievances, absenteeism, and other measures of poor morale and dissatisfaction dropped dramatically (DeGreene, 1973).

Based on these Tavistock Institute studies, Emery and Trist (1960) concluded that *different organizational designs can utilize the same technology*. The key is to select a work system design that is compatible with (1) the psychosocial and skill characteristics of the people who will constitute the personnel portion of the system and (2) the characteristics of the relevant external environment; and then employ the available technology in a manner that achieves congruence with these characteristics.

Joint Causation and Subsystem Optimization

Sociotechnical systems theory views organizations as open systems engaged in transforming inputs into desired outcomes (DeGreene, 1973). *Open* means that the boundries of work systems are permeable and, thus, are exposed to the environments in which they exist. These environments enter or permeate the organization along with the inputs to be transformed. Environmental changes enter the organization primarily in three ways: through its marketing or sales function, through the people who work in it, and through its materials or other input functions (Davis, 1982).

Organizations continually are interacting with their external environment. As transformation agencies, they receive inputs from their environment, transform these into desired outputs, and export these outputs to their environment. Organizations bring two critical factors to bear on this transformation process: technology in the form of a *technological subsystem*, and people in the form of a *personnel subsystem*. The design of the technological subsystem primarily defines the *tasks* to be performed, whereas the design of the personnel subsystem prescribes the *ways* in which tasks are performed. These two subsystems interact with one another at every human–job, human–machine, and human–software interface. Thus, the technological and personnel subsystems are mutually interdependent. Both subsystems are affected by causal events in the external environment. In other words, they operate under *joint causation*.

Joint causation underlies a related key sociotechnical systems concept, *joint optimization*. Because the technological and personnel subsystems respond jointly to causal events, optimizing one subsystem and then fitting the second to it results in suboptimization of the joint work system. Consequently, joint optimization requires the *joint design* of the technical and personnel subsystems in order to develop the best possible fit between the two, given the objectives and requirements of each, and of the overall work system (Davis, 1982). Inherent in this joint design is developing an optimal structure and related processes for the overall work system, given the work system's key sociotechnical characteristics. This will be discussed in detail in Chapters 3 and 4.

Joint Optimization and Human-Centered Interface Design

The concept of joint optimization initially may appear to conflict directly with that of human-centered interface design, which is central to the human factors/ ergonomic design philosophy. In human factors/ergonomics, we traditionally speak of changing the person through training, or changing the system through design of human–system interfaces, but in both paradigms we are designing to support the human. The issue, then, is whether jointly optimizing the technological and personnel subsystems compromises the human's dominant position in the human–machine relationship. As envisioned by the original (nonergonomist) sociotechnical systems researchers and practitioners in the post–World War II era, the social system, including the human, had equal billing with the technical system—at least in theory. In practice, the sociotechnical systems movement sometimes has been criticized for being technology-driven. In the macroergonomics subdiscipline, we engage the sociotechnical systems literature for a theoretical foundation. But we emphatically hold true to the notion that we design to support human capabilities, limitations, and other characteristics, and studiously avoid a technology-driven design mindset. For ergonomists, the principle of joint optimization is the avoidance of maximizing any single sociotechnical system element. Maximizing the technological subsystem, such as by automating and giving the human the leftover functions to perform, suboptimizes the overall system. Maximizing the personnel subsystem, such as by designing in jobs requiring a high level of professionalism without consideration of the organization's technology or other sociotechnical characteristics, suboptimizes the overall system. Attempting to maximize the organizational design by constantly restructuring in the absence of a macroergonomically sound reason (e.g., for purely political reasons) also suboptimizes the overall system. And finally, maximizing the external environment by allotting too much time with external stakeholders at the expense of internal operations also will suboptimize the total system. To achieve the appropriate balance, joint optimization is operationalized through (a) joint design, (b) a human-centered approach to function and task allocation and task design, and (c) attending to the organization's sociotechnical characteristics. The next four chapters on macroergonomic methods offer some pragmatic ways to achieve jointly optimized sociotechnical work systems.

The Sociotechnical System Elements

The design of a work system's structure and related processes involves consideration of three major sociotechnical system elements that interact with one another and the work system's design: (1) the *technological subsystem*, (2) *personnel subsystem*, and (3) *relevant external environment*. Each of these elements has been studied in relation to its effects on the various organizational design dimensions, described later in this chapter, and empirical models have emerged. These models can be used as macroergonomic tools in analyzing organizations and developing or modifying their work system designs. Some of these models that have proven particularly useful to the author will be described in Chapter 3.

Mutual Interdependence of the Elements

One of the most consistent findings in sociotechnical systems research is that the four basic sociotechnical system elements are mutually interdependent. A change in some characteristic of one of the four elements will affect the other three. Thus, if some aspect of the personnel subsystem is changed, it will impact on the technological subsystem, the work system's interaction with the external environment, and the structure and/or processes of the work system. Of critical importance from a macroergonomic perspective, if one does not anticipate and plan for these impacts on the other sociotechnical system elements, the impacts are likely to affect the work system in unanticipated and suboptimal ways.

Based on assessments of over 200 organizational units in which I have been involved, a common observation has been the failure of managers to recognize this interdependence of the sociotechnical system elements. Often, I have observed managers perceiving a problem with some aspect of one of the sociotechnical system elements, then developing tunnel vision and "fixing" that specific work system problem. The next thing the managers experience is a series of "ripple effects" on the other sociotechnical system elements which, in turn, create other problems. Not infrequently, these other problems turn out to be greater than the one the manager attempted to "fix" in the first place. I recall one such case in a hazardous industry in which, as a direct result of the senior manager's "fix," the serious accident rate went from an average of one per year to over one per month!

Relation of Macro- to Micro-Ergonomic Design

As a result of defining the design characteristics of the overall work system through a systematic macroergonomic analysis, many of the characteristics of the jobs to be designed into the system, and of the related human–job, human–machine, and human–software interfaces, already have been prescribed. Some examples are as follows (Hendrick, 1991a).

1. Horizontal differentiation decisions determine the degree of specialization and departmentalization to be designed into the work system. These decisions thus prescribe how narrowly or broadly jobs are to be designed and, often, how they should be grouped or departmentalized.

2. Decisions concerning the level of formalization and centralization will dictate the degree of routinization and employee discretion to ergonomically be designed into the jobs and related human–machine and human–software interfaces and, as a result, the level of professionalism to be designed into each job. These decisions also help prescribe many of the design requirements for the communications and decision support systems, including what kinds of information are required by each position and networking requirements.

3. Vertical differentiation decisions largely determine the number of managerial personnel required. When these decisions are coupled with those concerning horizontal differentiation, spatial dispersion, formalization, and centralization,

they prescribe many of the design characteristics of the managerial positions. These include span of control, decision authority, nature and scope of decision responsibility, information and decision support requirements, and qualitative and quantitative educational and experience requirements.

In summary, although it is not a sequential linear process, effective macro-ergonomic design ultimately drives much of the micro-ergonomic design of the work system. As a result, it insures *optimal ergonomic compatibility* of the system components with the work system's overall structure. In sociotechnical system terms, this approach enables joint optimization of the technological and personnel subsystems from top to bottom throughout the organization. The result is *harmonization* of the work system's elements with its overall design and relevant external environment. This harmonization enables the likelihood of optimal system functioning and effectiveness, including productivity, quality, system safety, and employee psychosocial comfort, health, intrinsic motivation, commitment, and perceived high quality of work life.

On the other hand, when a purely micro-ergonomic approach is taken—or more commonly, a technology-centered approach—there is a high probability of creating work systems in which the personnel subsystem is forced to adapt to the system's technology and structure in a "pounding square pegs into round holes" fashion.

Starting with the previously cited classic longwall coal mining studies by the Tavistock Institute (Trist & Bamforth, 1951), the organizational literature is full of examples that show this lack of compatibility to have an adverse effect on system productivity and attendant employee job satisfaction and commitment (e.g., see Argyis, 1971 for a classic summary of the findings of a number of prominent U.S. researchers).

It is important to note that, in actual practice, an organization, jobs, and attendant human–machine and human–software interfaces often already exist. As a result, once the design modifications to the overall work system are determined, the next step is to review existing jobs and related human–system interfaces to determine if they are congruent with it. Where they are not, decisions must be made on how to modify them to make them congruent. As is described in Chapter 2, this aspect of the macroergonomic process usually should involve active employee participation.

Organizational Synergism

Systems theorists and researchers generally agree that all complex systems are *synergistic*; that is, that the whole is more than the simple sum of its parts. Because organizations are complex systems, they, too, are synergistic. Because of this synergism, the following should tend to occur in our complex work systems (Hendrick, 1991a, 1997).

When Work Systems Have Incompatible Designs.   When either the structure or processes of a work system are not compatible with its sociotechnical system characteristics, and/or the human–job, human–machine, or human–software interfaces are incompatible with the work system's design, the whole is less than the sum of its parts. Accordingly, we can expect some combination of the following to be relatively deficient: (1) productivity, (2) quality, (3) lost time accidents and injuries and adherence to safety standards and procedures, and (4) motivation and related aspects of job satisfaction and perceived quality of work life (e.g., perceived stress, psychosocial comfort, etc.). Further, these detriments may be greater than a simple sum of their parts would indicate.

When Work Systems Have Compatible Designs.   When a work system has been effectively designed macroergonomically, and that effort has been carried through to the micro-ergonomic design of human–job, human–machine, and human–software interfaces, the result should be a work system that is fully harmonized. When this happens, synergistic functioning can occur, and organizational effectiveness measures will be much greater than the simple sum of the parts would indicate.

Assuming the above two theoretical propositions are true, macroergonomics potentially can improve virtually all measures of organizational effectiveness exponentially, including productivity, scrap rates, quality, safety, health, employee motivation and commitment, and the perceived quality of work life. In 1991, I theorized that, instead of the 10 to 25% improvements in these system effectiveness measures that many of us have experienced from our successful micro-ergonomic interventions, we should see improvements of 60 to 90%, or more (Hendrick, 1991a). As documented with actual cases in Chapters 7 and 8, this prediction is proving to be accurate.

PROBLEMS WITH TRADITIONAL
WORK SYSTEM DESIGN PRACTICES

During the past 20 or so years, I have been involved in assessing more than 200 organizational units. Based on these assessments, I have been able to identify three highly interrelated work system design practices that frequently underlie dysfunctional work system development and modification efforts. These are (1) technology-centered design, (2) a "leftover" approach to function and task allocation, and (3) a failure to consider an organization's sociotechnical characteristics and integrate them into its work system design (Hendrick, 1995b).

Technology-Centered Design

Designers typically try and exploit technology by incorporating it into some form of hardware or software to achieve some desired purpose. If those who must operate or maintain the hardware or software are considered, it usually is in terms of what skills, knowledge, and training will be required. Even these kinds of skills and training considerations often are not well thought through ergonomically. Consequently, the intrinsic motivational aspects of jobs, psychosocial characteristics of the workforce, and other related work system design factors rarely are considered. Yet, paradoxically, these are the very factors that can significantly improve work system effectiveness.

When a technology-centered design approach is taken, ergonomic aspects of design typically are not considered until after the equipment or software already is designed. Then the ergonomist may be called in to modify some of the human–system interfaces to reduce the likelihood of human error, eliminate awkward postures, or improve comfort. Even this level of involvement frequently does not occur until testing of the newly designed system reveals serious human–system interface design problems. Because of cost and schedule considerations at this point in the design process, the ergonomist is severely limited in terms of making fundamental changes to improve the work system. Instead, the ergonomist is restricted to making a few "band aid" fixes of specific human–machine, human–environment, or human–software interfaces. Unfortunately, the ultimate outcome is a suboptimal work system.

A well-known relationship exists between when professional ergonomic input occurs in the design process and the value of that input in terms of system performance: The earlier the input occurs in the design process, the greater and more cost effective is the impact on system effectiveness.

There is another problem with a technology-centered approach when redesigning existing work systems: Employees are not actively involved in the planning and implementation process. The organizational change literature frequently has shown that this lack of employee involvement often leads to a poorly designed work system. Equally important, it also can lead to a lack of commitment and, not infrequently, either to overt or passive-aggressive resistance to the changes. From my personal observations, when a technology-centered approach is taken, if employees are brought into the process at all, it is only after the work system changes have been designed. The employee's role becomes that of doing usability testing of the designed system. As often happens, when employees find serious problems with the changes, cost and schedule considerations prevent any major redesign to eliminate or minimize the identified deficiencies.

Because most of the so-called *re-engineering* efforts of the early 1990s used a technology-centered approach, it is not surprising that a large majority of them were unsuccessful. Keidel notes that these efforts failed to address the "soft" (i.e., human) side of engineering and often ignored organizational effects (Keidel, 1994).

"Leftover" Approach to Function and Task Allocation

When a technology-centered approach is taken, it often leads to treating the persons who will operate and maintain the system as impersonal components. The focus is on assigning to the "machine" any functions or tasks that its technology will enable it to perform. Then, what is leftover is assigned to the people components to perform. Consequently, the function and task allocation process fails to consider the characteristics of the workforce and the nature of the relevant external environment. The consequence often is a poorly designed work system that fails to make effective use of its human resources. A good early example of this was the implementation of the longwall coal mining system, cited previously (Trist & Bamforth, 1951).

As noted earlier, effective work system design requires joint design of the technical and personnel subsystems (DeGreene, 1973). Put in ergonomic terms, joint optimization requires a human-centered approach. With respect to function and task allocation, Bailey (1989) refers to it as a *humanized task* approach. He notes that

> this concept essentially means that the ultimate concern is to design a job that *justifies* using a person, rather than a job that merely can be done by a human. With this approach, functions are allocated and the resulting tasks are designed to make full use of human skills and to compensate for human limitations. The nature of the work itself should lend itself to internal motivational influences. The leftover functions are allocated to computers. (p. 190)

Failure to Consider the System's Sociotechnical Characteristics

As was noted earlier, the sociotechnical systems literature identifies four major characteristics or elements of work systems: (1) the technological subsystem, (2) the personnel subsystem, (3) the external environment, and (4) organizational design. The sociotechnical systems literature further documents that these four elements interact with one another, so a change in any one affects the other three (and, if not planned for, often in dysfunctional or unanticipated ways). Consequently, the characteristics of each of the first three elements affect the fourth: the organizational design of the work system. As will be described in Chapter 3, empirical models of these relationships have been developed that can be used to determine an effective work system structure.

As was first documented by the Tavistock studies over four decades ago (Emory & Trist, 1960; Trist & Bamforth, 1951), a technology-centered approach to the organizational design of work systems does not adequately consider the key characteristics of the other three sociotechnical system elements. Not surprisingly, a suboptimal work system design often is the result.

Criteria for an Effective Work System Design Approach

Based on these observations, several criteria can be gleaned for selecting an effective work system design approach (Hendrick, 1995b).

1. *Joint design.* The design approach should be human centered. Instead of designing the technological subsystem and requiring the personnel subsystem to conform to it, the approach should require design of the personnel subsystem jointly with the technological subsystem. Further, the approach should allow for extensive employee participation throughout the design process.
2. *Humanized task approach.* The function and task allocation process first should consider whether there is a need for a human to perform a given function or task before making the allocation to humans or machines.
3. *Integrate the organization's sociotechnical characteristics into the design.* The approach should systematically evaluate the organization's key sociotechnical system characteristics, and then integrate them into the work system's design.

Macroergonomics is an approach that meets all three of these criteria. As was noted earlier, conceptually, macroergonomics is a top-down sociotechnical systems approach to work system design, and the carry-through of the over-all work system design characteristics to the design of human–job, human–machine, and human–software interfaces. As will be described later, it is a top-down sociotechnical systems approach in that it begins with an analysis of the relevant sociotechnical system variables and then systematically utilizes these data in designing the work system's structure (Chapter 3) and related processes (Chapter 4). Macroergonomics is human-centered in that it systematically considers the worker's professional and psychosocial characteristics in designing the work system; and then carries the work system design's characteristics through to the ergonomic design of specific jobs and related hardware and software interfaces. Integral to this human-centered design process is joint design of the technical and personnel subsystems, using a humanized task approach in allocating functions and tasks. As will be described in the next chapter, and illustrated by an actual case in Chapter 8, a primary methodology of macroergonomics is *participatory ergonomics.* It is a methodology that involves employees at all organizational levels in the design process. In spite of its young age as a formally identifiable subdiscipline, I believe it is because it does meet these criteria that macroergonomics has enjoyed its considerable successes.

2

Macroergonomic Methods: Participation

Ogden Brown, Jr.
University of Denver

INTRODUCTION

This chapter is concerned with participation and participatory practices as the principal methodology in the design and analysis of work systems. Several closely related terms and concepts, such as participation, participatory ergonomics, employee involvement, stakeholder commitment, participative management, and various other participatory approaches appear throughout wide-ranging literature in the fields of ergonomics, psychology, and management, and are often used interchangeably. Cotton (1993), for example, uses the term "employee involvement" in a broad sense, and defines it as "a participative process to use the entire capacity of workers, designed to encourage employee commitment to organizational success" (p. 3). He points out that it is not a true unitary scientific concept, but rather a useful catchall term for a variety of approaches, all of which employ participative techniques. The principal problem appears to be that each approach to employee involvement seems to be studied in isolation. It is possible to compare and contrast them, but not to test them comparatively (Cotton, 1993).

With respect to participation as a macroergonomic method, then, I propose that participatory ergonomics be regarded as an approach to employee involvement that is concerned with ergonomic design and analysis. From this point of

25

view, participatory ergonomics is an approach or scheme that belongs in the catchall category noted above and fits the definition given by Cotton. One may thus infer that participatory ergonomics is the involvement approach unique to the field of ergonomics (and the subdiscipline of macroergonomics).

This chapter begins with a comprehensive examination of the background, history, and context of the wide-ranging area of worker participation and involvement. A theory of participation is presented, and several typologies (or forms) of participative approaches to employee involvement are discussed. The emphasis then turns to the concept of participatory ergonomics as the involvement approach that is unique to ergonomics. Requirements for participatory ergonomics are identified which include the dimensions of participatory ergonomics, a general framework for structuring participatory initiatives, and a discussion of the process. Finally, some applications of participatory ergonomics and critical requirements in the implementation of participatory ergonomics are presented.

Background

Participation is certainly not a new concept. For many years the literature has reflected concern with topics such as better worker relations, more supportive supervision, and more meaningful and interesting work. In the United States, early proponents of increased participation came from the human relations school of psychology (Argyris, 1957; Likert, 1961; McGregor, 1960) which evolved in reaction to Taylor's Scientific Management. Human growth and fulfillment were viewed as a desirable outcome of participation. However, this normative view was not supported by empirical research except for a few field studies which seemed to support the premise that employee acceptance, understanding, and implementation of decisions were enhanced by participating in the decision. It thus appeared that the results of early research on participation were equivocal, that advocates of participation were biased, and that the research methodologies were seriously flawed (Locke & Schweiger, 1979).

These early concerns unfortunately did not have much of an effect on how managers performed their duties and how organizations were managed. However, many organizational practitioners took participation to the field in the belief that society not only ought to attend to both economic and productivity aspects, but also to the quality of work life that organizations offered to employees. At the same time, a growing body of literature began to suggest that job characteristics such as autonomy, task significance, and experienced meaningfulness of one's work were related to motivation and job performance (Hackman & Lawler, 1971; Lawler, 1986). There was also growing recognition that involvement in the process of change is critical to the acceptance and institutionalization of change. Today, change is everywhere: many managers and organizations are now practicing what has been advocated for years. New organizational philosophies have emerged that are designed, developed, and operated with the participation of the employees involved. Employees are recognized by the organization as valuable resources for problem solving and for the design, development, and implementation of technology to enhance organizational effectiveness, improve product quality, and the overall quality of work life (Brown, 1993b).

History

During the first half of the last century the United States was at the peak of its economic growth and power. Little change occurred because there was no real need to change. Historically, American organizations achieved profits through a strategy of growth. This served them well for many years, but the way business is done today has changed drastically and permanently. Because growth is no longer assured, profit must now be realized through both quality and productivity. Many of the largest organizations in America have lost markets and have suffered due to losses in productivity, poor product quality, and a lack of innovation (Brown, 1993b).

A principal reason for this poor performance has been the concentration of power at top levels of management (Lawler, 1986). Concentration of knowledge and information at top levels and absolute control of reward systems serve to exacerbate the problem. Add in the turbulent environments of today's market conditions, foreign competition, the new composition of the workforce, and rapidly changing societal values, and one finds very powerful forces that argue for change in organizational philosophies, strategies, and management practices. This, in turn, calls out for organizationwide programs of change to provide the opportunity for employees to participate actively in the activities and decisions that directly affect them.

Context

A great amount of empirical research has been compiled that shows when, where, and how to implement participative practices. Organizations must become more aware of the tools and techniques available, and they must understand the strengths and weaknesses of different applications and forms of participation. Informed choice is necessary in order to employ participatory techniques effectively. Critical to such informed choice are the professional skills and knowledge of the ergonomist as well as those of the worker.

In the real world, no single approach that is participatory in nature is universally effective: a contingency approach is preferred. Underlying a contingency model are the assumptions that no single approach will be effective under all circumstances, and that most participatory approaches in use today will be effective only under a particular set of conditions (Brown, 1993b). Cohen (1996) identifies several "shaping factors" that should be considered in deciding what type of participation is appropriate in a given context. And then, of course, there are situations in which a participatory approach is unfeasible or inappropriate. Neumann (1989), for example, has observed that some organizations may not be suited to participatory approaches and that some people may not wish to participate because of deeply held beliefs.

THE THEORY OF PARTICIPATION

Economic Perspective

Neoclassical economic theory holds that the decision-making rights of an organization are vested in its owners. The owners delegate some of their rights to appointed agents (or managers) to act on their behalf. A basic problem for the owner is how to develop an incentive contract for the managers so that they will use their superior access to information to act in the owners' best interests. This incentive problem leads to so-called agency costs—costs the owners incur in order to motivate managers to act on their behalf, and the reduction in value of the firm resulting from imperfections in the motivational arrangements adopted.

From an agency perspective, delegation of decision-making rights by managers to the worker may well have negative effects on organizational performance from the owners' point of view. Agency theory holds that as the number of decision-makers or agents increases, the costs of monitoring performance increase. Further, an organization's residual loss is likely to be greater if the delegation of decision-making rights is accompanied by some form of profit sharing to motivate the participating employees. Hence, an agency framework leads to the conclusion that participatory arrangements are inevitably inefficient (Levine & Tyson, 1990).

On the other hand, if employees have knowledge that the managers lack about the workplace or the technology, then participatory approaches that motivate workers to use that knowledge or communicate it to the managers can improve organizational performance. Quality circles and work team production techniques are the two types of participation most likely to have such effects. The link is that of information—such participation leads to an increase in productivity because it enhances the use and flow of information in the organization (Brown, 1993a; Levine & Tyson, 1990).

Behavioral Perspective

As previously stated, early arguments for increased participation came from the human relations school of psychology and have since been reinforced by researchers such as Hackman and Lawler (1971), Lawler (1986, 1992, 1996), and Mohrman (1982). Quality of work life research in the 1970s and 1980s also should be mentioned in this context (Brown, 1986). A growing body of literature indicated the value of participation, involvement, commitment, autonomy, the meaningfulness of work, and ownership in problem solving and decision making (Brown, 1991a, 1994a).

A complementary link between participatory practices and cooperative behavior has been advocated by some researchers, focusing on the effects of participation on factors such as employee commitment to organizational objectives, trust in the organization's managers, and a sense of good will towards one's coworkers. Yet another connection, between performance and participation, may be that increased trust, commitment, and good will resulting from employee participation may increase job satisfaction and thus reduce the disutility of effort: these effects, in turn, may lead to greater effort and improved performance (Deci, 1975).

EMPIRICAL RESEARCH ON EMPLOYEE PARTICIPATION

Two contextual factors in particular influence the impact of participation: form and methodology (Cotton, 1993). There seems to be a basic problem with participation research with respect to these contextual variables. Locke and Schweiger (1979), in what is perhaps the best known review of worker participation, and Leana, Locke, and Schweiger (1990) focused on the methodologies used in the studies they reviewed. Cotton, Vollrath, Froggatt, Lengnick-Hall, and Jennings (1988) and Cotton, Vollrath, Lengnick-Hall, and Froggatt (1990) focused on the type or form of participation in their reviews. The enigma arises from the fact that studies that examine the most powerful forms of employee participation are also the studies that tend to include factors in addition to participation and therefore have the weakest methodologies.

A debate has since evolved with respect to these contextual factors. The unfortunate fact is that form and methodology are correlated, and neither side may thus prove it is right. Leana, et al. (1990) argued that positive results of employee participation were due to looser research methodology, while Cotton, et al. (1988, 1990) argued that positive results were due to the form of participation. This debate serves to point up what may be the fundamental problem in research on participation: There is no single form of participation. Many studies use indices of participation that encompass a wide range of participatory approaches (Brown, 1993a). Finally, there are many diverse forms of participation, each with its own issues and results (Cotton, 1993).

A second basic problem noted in the literature is that very few empirical studies provide quantitative assessments of the effects of participation on productivity (Brown, 1993a). However, there is a large body of empirically derived literature that examines how participatory arrangements affect more easily defined and measured outcomes such as absenteeism, turnover, and job satisfaction. The results have been consistently more favorable for such measures (Brown, 1993a).

Another problem that arises in research on participation is that many studies use indices of participation encompassing a wide range of participatory approaches which may mask or confound the effects of any one particular form of participation. Further, participatory arrangements are often confounded with other changes in the workplace such as changes in technology, changes in reward structures, more training, and a myriad of other interventions. Most studies do not distinguish effects of the participatory arrangements from effects of the other changes (Brown, 1993a).

APPROACHES TO PARTICIPATION: TYPOLOGIES

An examination of suggested approaches to participation reveals at least three different approaches to participatory arrangements, all of which are designed to encourage employee participation even though they result in very different types of involvement. An organization that is interested in adoption of some form of worker participation and involvement should be aware of the differences between these approaches and select the approach that offers the best fit with the organization (Brown, 1994b; Lawler, 1991). It should be noted that the concept of *fit* is an extremely important concept in the literature on organizational design. Major elements that need to "fit" are organizational structure, reward systems, information processes, the technology, and the people (Lawler, 1992). Organizational structure is critical in determining how involvement oriented the organization can and should be. Lawler believes some organizational designs make it virtually impossible to create an involvement-oriented organization, while other designs almost demand an organization be involvement oriented.

The three major approaches to participation are parallel suggestion involvement (consultative participation): job involvement (substantive participation): and high involvement (Brown, 1994b; Lawler, 1991). A fourth approach, representative participation, may also be employed. Basically, these approaches differ in the degree to which they propose that information about (1) organizational performance, (2) worker knowledge, (3) the reward system, and (4) the power to act and reach decisions that influence organizational policies and practices should be moved to the lowest possible level of the organization. These key features are also a useful way to consider fit among various parts of the organization in terms of how the parts affect them (Lawler, 1992).

When these four features are moved downward in the organization, employee participation and involvement is being practiced: the high involvement approach does the most to move them downward; the parallel suggestion involvement approach does the least. Simply put, there is no more fundamental change in an organization than to move information, knowledge, rewards, and power to lower levels: this is the very essence of participation and involvement (Brown, 1995b). It serves to alter the basic nature of the work itself and directly affects the job of every employee through enablement and empowerment (Brown, 1993b). And, in turn, it impacts directly on effectiveness of the whole organization.

Parallel Suggestion Involvement

In this approach, employees are asked to solve problems and produce ideas that will influence the operation of the organization. Such programs are a parallel structure to normal work relationships because they place people in a separate new structure or situation that operates differently from traditional organizational structure. Basically, suggestion involvement programs serve to alter the relationship between organization and worker. Research suggests that this approach can lead to improvement in organizational performance and job satisfaction, and that it may give workers the opportunity to influence things they would not normally influence. It is, however, expensive and it is difficult to maintain momentum using this approach due to resistance by middle management (parallel structures may be perceived as a threat to their power) and the lack of expertise to solve complex problems (Brown, 1994b, 1995b; Lawler, 1991).

Quality Circles. Quality circles are a popular approach to suggestion involvement and may even be part of a larger quality program. Circles are voluntary groups whose basic purpose is to improve product quality through suggestions about product quality or work methods. They have no formal authority and no financial rewards are given. They are a parallel structure to the traditional organization. Day-to-day activities thus remain intact and pose no real threat to the basic management structure. They are a formal program of direct face-to-face participation with a moderate level of influence (Brown, 1993b). Research findings are, at best, mixed. Anecdotal evidence is relatively positive, but careful research is less optimistic (Cotton, 1993). A large number of circles fail early, and the greatest period of success is between 6 and 18 months (Griffin, 1988).

Quality of Work Life Programs. A quality of work life program is another suggestion involvement scheme. Of all the many approaches to employee participation, quality of work life (QWL) is the most difficult to define because it means so many different things to different people (Brown, 1986). Because of the problems of multiple definition, many researchers focus on QWL as a joint labor–management program aimed at increasing worker participation. It differs from employee problem solving in that it recognizes the need to bring two sometimes adversarial groups together to identify areas of mutual concern (Brown, 1993b). It usually employs a parallel structure that exists at multiple organizational levels from top to bottom. On the whole, case studies indicate that QWL programs tend to improve labor–management relations and product quality, but there is little evidence of increased productivity as a result of such programs (Cotton, 1993).

Job Involvement

Job involvement approaches focus on the design of work in ways that motivate better performance on the job. The job involvement approach has significant implications for how the organization is structured and managed. Essentially, individuals are given new skills and knowledge, new feedback, additional decisions to make, and may even be rewarded differently (Lawler, 1991).

The job involvement approach represents a major change in the basic operations of the organization. People at the lowest levels get new information, skills, power, and may be rewarded differently. The new information, knowledge, power, and rewards relate to particular work tasks. Typically, the job involvement approach does not have to do with the structure and operation of the entire organization. Job involvement differs from parallel suggestion approaches in that the daily work activities of all people are affected: it is the standard way in which the organization conducts its business (Lawler, 1991).

Job Enrichment. Job enrichment, a micro approach to job design, is one job involvement strategy. Job enrichment attempts to make individual jobs more interesting and challenging, and focuses on creating individual tasks which provide feedback to the worker: it also increases workers' influence over how the work is performed, requires them to employ a variety of skills, and gives them a complete piece of work to perform by increasing critical job dimensions such as autonomy and task identity (Hackman & Oldham, 1980). Job enrichment does not eliminate any levels of management (desirable though that may be). It does not create a parallel structure, but rather it makes changes in the way work is performed. For this very reason it may make a more powerful and dramatic change (Brown, 1993b). Further, because of its micro focus, several of the other participatory approaches typically include job enrichment in their programs (Cotton, 1993).

Research on job enrichment is more extensive and more rigorous than that for any other major participatory approach. Job enrichment enjoys major theoretical models which have stimulated research. Empirical studies and reviews have indicated that enriched jobs serve to increase the incumbent's job satisfaction (Cotton, 1993). In particular, quality improves because when people are responsible for the work they do, they are more motivated to produce a high-quality product (Lawler, 1992). Improvements in performance may also happen with job enrichment, but they are much less positive, and empirical research results are mixed (Brown, 1993b). This finding should not be too surprising. Should performance improve just because a job is made more interesting or has more responsibility? Any relationship between performance and enrichment is probably complex and nonlinear (Schwab & Cummings, 1976).

Further, individual differences such as a desire for enrichment or *growth need strength* have proven to be moderators of the relationship between job satisfaction and job enrichment (Hackman & Oldham, 1980). It would seem that the ability or skills to perform the enriched job would also appear to be a necessary requirement!

Self-Directed Work Teams. An increasingly popular job involvement approach to participation is that of the self-directed work team (also called autonomous or semiautonomous work groups). The early history of work teams is decidedly European, and the concept is deeply imbedded in the sociotechnical systems approach to the design of work (Emery & Trist, 1960; Trist, 1981). Self-directed work teams involve a formal system of direct employee participation and a high degree of control (Cotton, 1993). The work team is empowered to make decisions concerning daily work operations that would normally be made by a manager. They perform much like individual job enrichment programs in the effects they have on the organization. Depending on their degree of autonomy, work teams may move considerable power, knowledge, and information to the lowest organizational levels. They make an important difference in the participative structure of the organization. Work teams may gain the power, knowledge, and information that workers at lower levels do not enjoy in traditional organizations (Lawler, 1986, 1992).

There are certain distinctive characteristics inherent in work teams. They are designed so that the task for which they are responsible comprises a whole, meaningful piece of work. Typically, team members are cross-trained to perform most (if not all) tasks within the team's area of responsibility. To feel responsible for their work, teams must make the important decisions concerning how the work is to be performed: they must feel that they are in control of the work process (Hackman, 1989). Two kinds of training are critical to work teams: extensive technical training, as well as training in team and interpersonal skills. Similar to job enrichment, work teams are most effective when their members are individuals who desire challenging and complex work (Hackman & Oldham, 1980).

Implementation issues are an important consideration with the use of self-directed work teams because the organization must make major changes which require the support of both worker and management to be successful. Problems encountered in implementing work teams include inadequate training for all team members, resistance of lower- and middle-level managers, and lack of support from top management. Self-directed work teams are an extreme example of moving direct daily decision making to the lowest organizational levels (Brown, 1995a). Thus they are probably one of the most difficult changes from a traditional organizational structure (Cotton, 1993). On the other hand, work teams would seem to be one of the more effective approaches in the development and implementation of technology through participatory ergonomics (Brown, 1993b).

Research results for self-directed work teams are somewhat similar to those for job enrichment. Team members have a high concern for quality and usually have the information and knowledge to make meaningful improvements in the quality of their services or products (Lawler, 1992). In a summary of 156 studies that examined self-directed work teams, Cotton (1993) found that the vast majority achieved positive results: just 6 results were negative and 17 null with respect to outcome variables. This review indicates that work team do have positive effects, but it does not suggest what is required for success. Organizations that have implemented self-directed work teams have generally found them to be an effective participatory approach to improve quality, job satisfaction, commitment, and organizational effectiveness (Brown, 1993b).

High Involvement

The high involvement or "commitment" approach builds on what we have learned from the parallel suggestion and job involvement approaches. This scheme structures an organization such that those at the lowest levels will have a sense of involvement, not just in how well they perform their jobs or how effectively their work team performs, but in terms of the total performance of the entire organization (Brown, 1994b). High involvement goes much further than either of the other approaches in moving knowledge, information, rewards, and power to the lowest levels of the organization. The high involvement approach can create an organization in which employees genuinely care about the performance of their organization because they know about it, are able to influence it, are rewarded for it, and enjoy the knowledge and skills to contribute to it (Lawler, 1991).

A high involvement approach argues for consistent and continuing change in just about every part of an organization. Workers must be involved in decisions about their jobs and work activities. The implications for job design and redesign and methodologies for work station design are relatively clear. However, workers should also be empowered to play a role in decisions concerning organizational structure, strategy, and other such major decisions. This would appear to argue for structures with fewer vertical levels and wider spans of control. Rewards, in turn, should be based on the performance of the entire organization, but individuals also need to be rewarded for their contributions. It would thus seem that skill-based pay for every individual is warranted. In turn, this requires expanded training programs for both technical training as well as for interpersonal and team skills (Brown, 1994b).

All members of the organization must acquire expertise in problem analysis, decision making, group process, and self-management. It is most important in a high involvement approach that information systems operate effectively both horizontally and vertically so that every member of the organization has access to operating data which will inform them of how the work is being performed and allow them to make decisions that directly affect or impact their work. A new technology, for example, might require a task analysis, job or work station redesign, or work restructuring which the worker might be empowered to do. Without the necessary information, the worker may not be enabled to make the required changes (Brown, 1994b).

Creating a high involvement organization is clearly a more complicated and quite different task than is implementing a parallel suggestion or job involvement program. Almost every feature of a mechanistic, bureaucratic, control-oriented organization must be redesigned. Design innovations may even be required simply because the right approaches are not developed and available (Lawler, 1991).

One should always remember that no single involvement approach that employs participatory techniques will be universally effective. As stated earlier, a contingency model is deemed appropriate when one is attempting to implement any program of worker participation. An organization that espouses and fosters high involvement practices calls for the enablement and empowerment of every individual from the highest to the lowest organizational levels. And, if the organization is involved with complex technologies that call for complex knowledge and that require many interdependent work relationships, high involvement ergonomics would seem to be the proper contingency approach to employee participation (Brown, 1994b; Lawler, 1992).

Representative Participation

A fourth participatory approach, representative participation, is a form of indirect employee involvement. This approach includes worker councils, joint labor–management committees, and employee representation on boards of directors (Brown, 1993a). It is much more common in Europe than in the United States and has typically been achieved through national legislation. By the 1980s almost every nation in Western Europe had some type of legislation requiring works councils, board representatives, or both. In fact, representative participation is the most widely legislated form of employee participation and involvement around the world. In the United States, however, having workers or worker representatives on a board of directors (except for union leaders) is considered a strange concept at best (Cotton, 1993).

Dachler and Wilpert (1978) have categorized this approach as "indirect participation" in their theoretical model of dimensions to describe types of employee involvement. In this form of participation, most employees do not directly participate. Instead, they elect or are represented by a small group of employees who actually do participate in decisions that can range from daily issues to organizational policy issues. However, research on various forms of representative participation has not demonstrated many positive effects. Although it would seem to be a powerful type of employee involvement, there are few indications that representative participation has an impact on the organization or its employees. The evidence suggests that management still dominates: equal participation does not mean equal influence in decision making. And even if employee influence is increased, there is no evidence that any positive effects filter down to individual employees. The greatest value of representative participation would therefore seem to be symbolic (Cotton, 1993). It likely will never achieve the positive results associated with the other approaches to participation.

THE CONCEPT OF PARTICIPATORY ERGONOMICS

Participatory ergonomics as a concept has been defined in several differing but complementary ways. As Wilson and Haines (1997) point out, participatory ergonomics (and participation) may be regarded as a philosophy, an approach or strategy, a program, or a set of tools and techniques. Imada (1991a) defines it as a macroergonomic approach to the implementation of technology in organizational systems that requires end users to be highly involved in developing and in implementing the technology. Brown (1993b) views participatory ergonomics as an emerging new organizational philosophy designed, developed, and operated with the participation of the employees involved. Nagamachi (1995) sees participatory ergonomics as active involvement by workers in complementary ergonomic knowledge and procedures in the workplace.

Clearly, participatory ergonomics is a complex concept which involves a number of different dimensions (Haines & Wilson, 1998). There is no consensus or general agreement about what the term actually means. Wilson (1995) has derived a working definition of participatory ergonomics as: "The involvement of people in planning and controlling a significant amount of their own work activities, with sufficient knowledge and power to influence both processes and outcomes in order to achieve desirable goals" (p. 37). And as Hendrick and Kleiner (2001) observe, when the participation or involvement involves ergonomic design and analysis, employee involvement may be said to constitute participatory ergonomics. This position is in agreement with that proposed in the introduction to this chapter: that is, participatory ergonomics is an approach to employee involvement that is unique to the field of ergonomics.

Whatever the definition, it is quite evident that participatory ergonomics is the principal and most often used methodology in the optimization of organizational and work system design. Wilson and Haines (1997) observe that since 1985 each successive Congress of the International Ergonomics Association has had an increased number of papers and sessions presented on the topic of participatory ergonomics, and that *Ergonomics Abstracts* also shows a marked increase.

Hendrick (1997) states that the most widely used methodology in macroergonomic analysis and organizational design interventions is participatory ergonomics. Finally, an analysis of the content of the proceedings volumes for a series of six international symposia entitled, "Human Factors in Organizational Design and Management" (Bradley & Hendrick, 1994; Brown & Hendrick, 1986; Brown & Hendrick, 1996; Hendrick & Brown, 1984; Noro & Brown, 1990; Vink, Koningsveld & Dhondt, 1998) reveals that participatory ergonomics was the most widely used methodology employed by the authors of the studies and applications therein.

A CONCEPTUAL FRAMEWORK

Dimensions of Employee Involvement

Following the general approach taken to participation in this chapter, from the broader concept of worker involvement to the more specific concept of participatory ergonomics, it is appropriate here to consider the model proposed by Dachler and Wilpert (1978) which identifies the dimensions that may be used to describe the properties of employee involvement. They proposed five general dimensions: formal-informal, direct-indirect, level of access, content of involvement issues, and social range of the involvement (Dachler & Wilpert, 1978).

Formal-Informal. Formal involvement refers to a "system of rules . . . imposed on or granted to the organization" (Dachler & Wilpert, 1978, p. 10). Informal involvement is a consensus that arises in a casual way. A quality circle program or a work team would be formal forms of involvement, whereas a supervisor who casually allows workers to make decisions about how work is done would be informal involvement.

Direct-Indirect. Direct involvement refers to "immediate personal involvement of organizational members" (Dachler & Wilpert, 1978, p. 12), a one-on-one involvement with immediate and personal impact. Indirect involvement involves some kind of worker representation in which the worker's representative is involved. Quality circle programs exemplify direct involvement, while a worker council would be indirect involvement.

Level of Access. This refers to the amount of influence that organizational members can exert while making a decision. It is a continuum of access from no advance information given to employees to a decision completely in the hands of employees.

Content of the Issues. Even though most employee involvement programs focus on issues and decisions directly related to one's work, this is not always true; for example, workers may make suggestions about aspects of work or policy issues not necessarily related to one's job responsibilities.

Social Range of Involvement. This dimension refers to who is involved. It may also refer to whether involvement is on an individual or group level.

These dimensions may be used to describe and categorize various forms of employee involvement, and there are many possible combinations.

Dimensions of Participatory Ergonomics

As stated earlier, participatory ergonomics is a complex concept, and there is no consensus about what the term actually means. Participatory ergonomics initiatives can take a wide variety of forms. Haines and Wilson (1998) have identified some of the dimensions across which participatory ergonomic initiatives might vary (see Table 2.1). These dimensions follow without further citation. Haines and Wilson caution that the perspective of Western industrialized society is taken, and that in other parts of the world some of these dimensions might not be relevant or might not apply (Haines & Wilson, 1998). The dimensions appear to be closely related to those of Dachler and Wilpert (1978) but more specific to participatory ergonomics (another example of the differences in approaches to involvement noted by Cotton (1993).

Extent/Level. This dimension incorporates the earlier dimensions of level and focus (Wilson & Haines, 1997) and is concerned with where ergonomics is applied: across an organization, a work system, a work station, or a product (macro- to microparticipation).

Purpose. This dimension considers whether participatory ergonomics is being used to implement a particular change or whether it is the method of work organization itself.

TABLE 2.1
Dimensions of Participatory Ergonomics

Extent/Level	Organization......Worksystem.......Workplace........Product
Purpose	Work organization............Design............Implementation
Continuity	Continuous ....Discrete
Involvement	Direct (full/partial)....................................Representative
Formality	Formal ....Informal
Requirement	Voluntary ....Compulsory
Decision-Making	Workers decide..............Consensus..............Consultation
Coupling	Direct...Remote

(Adapted from Haines & Wilson, 1998.)

Continuity of Use.   Continuity of use depends on whether the process has a continuous (repetitive) or discrete (periodic) timeline.

Involvement.   This dimension is concerned with who actually takes part in the process. It may range from full direct participation (all affected stakeholders become participants), through partial direct participation (representation by a subgroup of stakeholders), to representative participation (nominated or elected representatives of stakeholders who participate in the process).

Formality.   Participation may range from formal (work teams, committees) to informal, where managers casually allow workers to make decisions about their work.

Requirement for Participation.   Involvement may be either voluntary or compulsory. Voluntary participation is the usual form wherein workers volunteer their contributions and are involved in initiating the process. Compulsory participation is found where involvement in quality circles and continuous improvement programs is obligatory.

Decision-Making Structures.   This dimension is dependent on the degree of centralization-decentralization of decision-making authority in the organization and the degree of empowerment accorded the worker.

Coupling.   This final dimension refers to how directly participative methods are applied. Direct coupling involves the application of participants' views and suggestions, while remote coupling involves some filtering of participants' views.

Structure: A General Framework

Haines and Wilson (1998) have developed a "first general framework" for initiating and structuring participatory ergonomic initiatives (Fig. 2.1). It begins with an organizational decision to implement and employ participatory ergonomics. They point out certain motivational factors that may contribute to this decision: legislation, availability of expert advice, external recommendations, awareness of the importance of ergonomic problems, management philosophy, workforce/union negotiations, worker complaints and claims, and product/market advantage (Haines & Wilson, 1998). After deciding to implement participatory ergonomics, some type of initiative will then be implemented, the structure of which may be defined across the dimensions of participatory ergonomics described earlier. Criteria that influence the structure of the initiative include organizational size and culture, nature of workplace problems, the time frame and resources available, stakeholders, and workforce education and training.

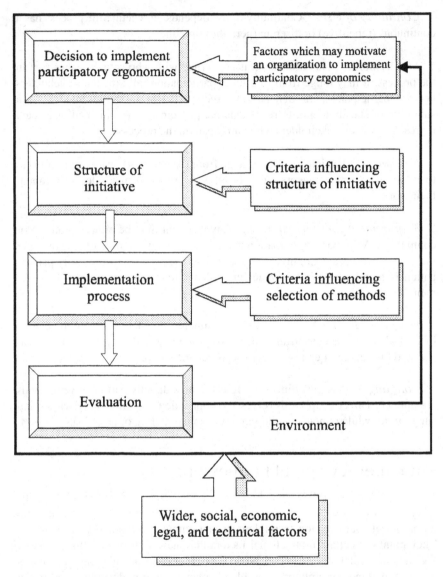

FIG. 2.1. General framework for developing and implementing a participatory ergonomic initiative (adapted from Haines & Wilson, 1998).

Once the decision to implement participatory ergonomics has been made, implementation methods may be selected. Factors influencing the selection of methods include the type of problem, knowledge of and experience with methods chosen, resources available, and the number of participants. Finally, a feedback loop will be required for ongoing continuous improvement, the sociotechnical principle of *incompletion* (Hendrick & Kleiner, 2001). From a macroergonomic point of view, this framework is an *open system* and is subject to factors from the external environment (such as economic, technical, social, and legal factors). During macroergonomic design these factors are analyzed and taken into consideration along with other relevant variables (Hendrick & Kleiner, 2001).

APPLICATIONS OF PARTICIPATORY ERGONOMICS

As noted earlier, participatory ergonomics may be regarded as the involvement approach unique to the field of ergonomics and its subdiscipline of macro-ergonomics. The ergonomic literature is replete with case studies that employ a wide variety of participative tools and techniques. There is a myriad of specific applications which may vary according to the goals or objectives of the partici-patory initiative, the structure of the initiative or intervention, the organizational or workplace setting, participatory methods and approaches employed, and the expertise of the stakeholders involved.

Participation is the most widely used macroergonomic methodology: many studies have been successfully conducted and reported across a broad spectrum of different applications and for many different purposes. Hendrick and Kleiner (2001) discuss applications of participation in individual and team decision mak-ing and problem solving, product and system design, training system design, and work system analysis and design. Many initiatives have used participatory tech-niques aimed at reducing or preventing work-related musculoskeletal disorders (Kuorinka & Patry, 1995; Liker, Joseph & Ulin, 1991). Other studies have been concerned with design for manufacturing (Nagamachi & Yamada, 1992; Salvendy & Karwowski, 1994) and with large-scale change efforts (Kleiner, 1996). Still other approaches have been used in workplace redesign interventions (Nagamachi, 1995; Wilson, 1995).

Hendrick (1996) cites studies of representative participation in materials han-dling systems in the steel industry, an ergonomic training program for imple-menting a participatory ergonomic process in seven manufacturing companies, and an extensive worker participation program at an agricultural equipment man-ufacturer. Haims and Carayon (1996) conducted a study in an office environment to develop a theoretical model and related design principles as guides for the de-sign and implementation of participatory programs. Zink (1996) reported on the implementation of new technology in an electronics organization which required the support and cooperation of every department and which employed a mixture of bottom-up and top-down approaches. Imada and Stawowy (1996) performed a workplace redesign intervention in two food service stands at Dodger Stadium that resulted in increased productivity and reduced customer transaction time. Rooney, Morency and Herrick (1993) used participatory ergonomics in applying macroergonomics as an approach and methodology for introducing total quality management (TQM) at the L.L. Bean Corporation. Among other improvements, this intervention resulted in a large reduction of lost-time accidents and injuries. Finally, see Chapter 8 for an interesting case study of a successful macro-ergonomic analysis and intervention program using a participatory approach in the petroleum industry.

REQUIREMENTS FOR PARTICIPATORY ERGONOMICS

It is essential to adoption of participatory practices that the highest levels of management be unequivocally committed and supportive (Brown, 1990, 1993b). It must be clear to management that there is a need for cooperation, that both workers and managers are able to deal effectively with participatory practices, that participation is relevant to worker interests as well as management objectives, and that the benefits of participation will outweigh any costs or problems associated with implementation (Brown, 1990). An organizational philosophy must be adopted that advocates participatory practices not only for business reasons, but also for development of human resources, and establishment of mutual trust, equity in the exchange relationship, and involvement of every stakeholder (Brown, 1991b, 1993b). The earlier discussion on high involvement ergonomics is particularly relevant here.

The structuring of a participatory initiative is generally defined across the dimensions of participatory ergonomics, described earlier, and should consider organizational factors such as culture, size, structure, available resources, training needs, commitment of stakeholders, and the like. Haines and Wilson (1998) recommend creating an initiative sufficiently structured and yet flexible enough to respond as the participants and the process develop, and also to allow the program to adapt to change.

A successful participatory ergonomic program will include appropriate methods and processes and foster a climate of continuous improvement (Wilson & Haines, 1997). Requirements may vary in degree depending on the extent of participation (Hendrick & Kleiner, 2001). In the high involvement approach, for example, workers need to be involved in decisions about their jobs and work activities and be empowered to play a role in organizational-level decisions concerning strategy and structure (Brown, 1994b). In turn, this leads to a requirement for expertise by all members of the organization in problem analysis, decision making, group processes, and self-management (which argues for expanded training programs in both technical and interpersonal and team skills).

Three interdependent requirements are identified by Wilson and Haines (1997): motivation, knowledge, and confidence. As participants gain in knowledge and ability, confidence will increase in both self and the participative

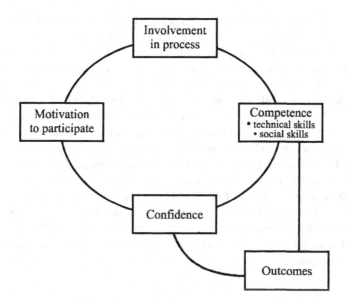

FIG. 2.2. Participatory ergonomics cycle (from Haines & Wilson, 1998).

process. This then should motivate participants to become more involved in the initiative (Haines & Wilson, 1998). Over a period of time, adjusting the participatory ergonomic cycle of involvement may be needed (Fig. 2.2), based on evaluative feedback, by changing the technical and/or social skill mixture, group process, and the like (Hendrick & Kleiner, 2001).

A final requirement is to evaluate the effectiveness (or lack thereof) of participatory programs. Outcome measures, a reduction in risk factors, process measures, and cost-benefit analyses are several methods used in evaluating participatory programs (Haines & Wilson, 1998). Hendrick (1996) presents several ergonomic applications that employed participatory approaches and in which costs and benefits were documented.

CONCLUSION

It should be reemphasized that there is no one best approach to the use of worker involvement and participation, and that participation and involvement are not universally good for all organizations (Brown, 1994b, 2000). However, Hendrick and Kleiner (2001) observe that there seems little doubt that the participative nature of macroergonomic interventions accounts for many of the impressive results realized (e.g., see Chapter 8). As Haines and Wilson (1998) point out, participatory ergonomics is a diverse and complex concept. There is a plethora of different definitions, a range of models, and a multiplicity of tools and methods employed in participatory ergonomic initiatives.

Successful implementation of participatory ergonomics requires the empowerment of people to make decisions and to implement and evaluate them (Imada, 1991a). Organizationwide change is necessary, and it is imperative that top management be unequivocally committed (Brown, 1991b, 1994b). The implications and potential of participatory ergonomics for the analysis, design, integration, and implementation of new technology into organizational systems are boundless. Beliefs and core values of participatory practices advocate people being involved. If one assumes that stakeholders should participate in decisions that affect them, and they have the knowledge and skills to contribute, then there appears to be ample reason and sufficient evidence to believe that participatory ergonomics offers great promise for enhanced quality and productivity and a far better realization of human potential (Brown, 1994b, 1996). Chapter 3 discusses additional macroergonomic methods, many of which assume participation by individuals or groups.

3

Macroergonomic Methods: Assessing Work System Structure

Hal W. Hendrick
University of Southern California

Effective design of a work system's structure involves consideration of three major sociotechnical system elements that interact and affect optimal work system functioning: (1) the *technological subsystem*, (2) *personnel subsystem*, and (3) *relevant external environment*, or that portion of the external environment that permeates the organization on which it depends for its survival and success. Each of these elements has been studied in relation to its effects on the three organizational design dimensions described in Chapter 1, *complexity, formalization*, and *centralization*, and empirical models have emerged. These models can be used as macroergonomic tools in assessing and developing or modifying the design of a given work system. The models included in this chapter have proven particularly useful to the author.

TECHNOLOGICAL SUBSYSTEM ANALYSIS

As a determinant of work system structure, technology has been operationally defined in several distinctly different ways that are useful to macroergonomics: (1) by the mode of production, or *production technology;* (2) by the action individuals perform on an object to change it, or *knowledge-based technology;* (3) by the way it reduces uncertainty, or *strategy for reducing uncertainty;* and (4) by the degrees of automation, workflow rigidity, and quantitative specificity of evaluation of work activities, or *workflow integration.* A major generalizable model of the technology–work system design relationship has been empirically derived from each of these empirically derived classification schemes.

Woodward: Production Technology

The first persons to study technology as a determinant of organizational structure empirically were Joan Woodward and her associates in the United Kingdom (1965). Woodward and her colleagues were looking for differences between successful and less successful organizations within the same industry, across a variety of industries. They studied 100 manufacturing firms in South Essex, England, having at least 100 employees. The companies varied in size, managerial levels (2 to 12), span of control (2 to 12 at the top; 20 to 90 at the first-line supervisory level), and ratio of line employees to staff personnel (less than 1:1 to more than 10:1). Using interviews, systematic observations, and review of company records, the following factors were among those noted for each firm: (1) the organization's mission and significant historical events; (2) the manufacturing processes and methods utilized; and (3) the organization's success, as measured by changes in market share, relative growth or stagnation within its industry, and fluctuation of its stock prices.

Woodward found that the various industries studied could be classified in terms of their modes of technology: (1) *unit,* (2) *mass,* and (3) *process* production. Further, these modes could be conceptualized as representing categories on a scale of increasing *technological complexity.* At the least complex end were the unit and small batch producers that manufacture custom-made products. Next were the large batch or mass production firms. These companies produced automobiles, refrigerators, and other more or less standardized products using predictable, repetitive production steps. Highest in technological complexity were heavily automated process production firms, such as oil and chemical refineries. Within each type of production mode, the only thing that discriminated between the successful and less successful firms was the characteristics of their organizational structure. Specifically, in terms of optimal design, three organizational structure variables were found to increase as technological complexity increased. First, as *technological complexity increased, the degree of vertical differentiation also increased.* The successful firms within each technology mode tended to cluster around the median number of hierarchical levels for that mode. In Woodward's sample, the optimum number of levels for unit producers was three, for mass production it was four, and for process plants, six. The less successful companies within each production category had a significantly greater or lesser number of levels. Second, *as technological complexity increased, the optimal ratio of administrative support staff to industrial line employees increased.* Third, *as technological complexity increased, the span of control of the top-line managers increased.* More specifically, Woodward's findings for the successful firms in each production technology mode were as follows.

Unit production firms had low complexity with little line and staff differentiation and widely defined role responsibilities. Both formalization and centralization were low. High formalization and centralization apparently were not feasible because of the custom-made, nonroutine nature of the work.

Mass production units had high complexity with clear line and staff differentiation and narrowly defined role responsibilities. Formalization and centralization both were high. In comparison with the other two production modes, the proportion of skilled workers was relatively small.

Process production units had high vertical differentiation with little line and staff differentiation and widely defined role responsibilities. Formalization and centralization both were low. High formalization and centralized control were not needed because of the heavily automated, inherently tightly controlled nature of process technology.

Several follow-up studies have lent support to Woodward's conclusions (Harvey, 1968; Zwerman, 1970). However, there are a couple of cautions with respect to generalizing Woodward's findings. First, Woodward implies causation when, in fact, her methodology really only establishes correlation. Second, Woodward's data were collected from within a single culture and at a particular point in time. In a different culture, or at some other time, the psychosocial and other environmental factors might conceivably result in somewhat different interactions with production mode in terms of their influence on work systems.

Perrow: Knowledge-Based Technology

Although it can be a useful analytical tool for macroergonomics, a shortcoming of Woodward's model is that it applies only to manufacturing firms, which constitute less than half of all organizations. Perrow (1967) developed a more generalizable model of the technology–work system structure relationship that uses a *knowledge-based* rather than a production classification scheme. Perrow began by defining technology as the action one performs on an object in order to change the object. This action requires some form of technical knowledge. Using this approach, Perrow identified two underlying dimensions of knowledge-based technology. The first dimension is *task variability,* or the number of exceptions encountered in one's work. The second one concerns the type of search procedures one has available for responding to task exceptions, or *task analyzability.* These search procedures can range from "well defined" to "ill defined." At the well-defined end of the continuum, problems are solved using rational-logical,

		Task	Variability
		Routine with few exceptions	High variety with many exceptions
Problem	Well defined and Analyzable	Routine	Engineering
Analyzability	Ill-defined and Unanalyzable	Craft	Non-routine

FIG. 3.1.   Perrow's classification scheme.

quantitative, and analytical reasoning. At the ill-defined end there are no readily available formal search procedures, and one must rely on experience, judgment, and intuition to solve problems. Dichotomizing these two dimensions yields a matrix with four cells, as shown in Fig. 3.1. Each cell represents a different knowledge-based technology.

1. *Routine* technologies have well-defined problems with few exceptions. Mass production units typify this category, as do some types of service organizations in which the nature of the servicing is largely repetitive. Routine technologies lend themselves to standardized coordination and control procedures and, thus, are associated with high formalization and centralization.
2. *Nonroutine* technologies have many exceptions and difficult to analyze problems. Combat aerospace operations is an example. Critical to these technologies is flexibility. Therefore, they need to be highly decentralized and have low formalization.
3. *Engineering* technologies have many exceptions but can be handled using well-defined rational-logical processes. Consequently, they lend themselves to moderate centralization but need the flexibility that is achievable through low formalization.
4. *Craft* technologies involve fairly routine tasks, but problem solving relies heavily on the experience, judgment, and intuition of the individual craftsperson. Thus, decisions must be made by those with the particular expertise. This requires decentralization and low formalization.

Perrow's model has been supported by empirical research in both the private and public sectors (e.g., Hage & Aiken, 1969; Magnusen, 1970; Van deVen & Delbecq, 1979). I have found this model to be particularly useful for analyzing an organization's technology to determine its implications for the work system's structure.

Thompson: Technological Uncertainty

Thompson (1967) has found that the type of technology determines a strategy for reducing uncertainty in work systems, and that specific structural arrangements facilitate uncertainty reduction. Based on the tasks that a work system performs,

Thompson identified three types of technology: long-linked, mediating, and intensive.

Long-Linked. Long-linked technology accomplishes its tasks through a sequential interdependence of its units, such as an automobile assembly line. Because it consists of a fixed sequence of repetitive steps, the only major uncertainties are at the input and output sides. Consequently, management responds to uncertainty by controlling the inputs and outputs. The primary means of effecting this control is through planning and scheduling which, in turn, suggests a moderately complex and formalized structure.

Mediating. A mediating technology is one that links clients on both the input and output sides. Accordingly, it performs a mediating or interchange function. Examples are banks, post offices, and utility companies, which link units that otherwise are independent. These otherwise independent organizations are bound together by rules, regulations, and standard operating procedures. As a result, they perform best with low complexity and high formalization.

Intensive. An intensive technology is one that provides a customized response to a diverse set of contingencies. A variety of techniques are drawn on to transform an object from one state to another. The particular techniques selected are, in part, based on feedback from the object itself. A classic example is a hospital, where the object being transformed is the patient. The particular techniques employed are dependent on the condition of the patient and responses to previously used techniques. Thus, the major uncertainty is the object itself. Flexibility of response, such as having many alternatives to employ, is essential to effective functioning. Consequently, an intensive technology performs best when the structure has high complexity but low formalization.

Unfortunately, Thompson's model has not been well tested empirically. Therefore, no definitive conclusions can be drawn regarding the model's validity (Robbins, 1983). The one study of note analyzed 297 subunits for 17 business and industrial firms (Mahoney & Frost, 1974). The results provided partial support for the model by demonstrating that long-linked and mediating technologies were associated closely with formalization and advanced planning, whereas well-functioning intensive technologies were characterized by mutual adjustments to other units.

Aston Studies: Workflow Integration

Based on their studies of a wide range of manufacturing and service organizations, a team of researchers at the University of Aston in the United Kingdom concluded that technology can be defined in terms of three basic characteristics:

automation of equipment, or the extent to which work activities are performed by machines; *workflow rigidity,* or the extent to which the sequence of work activities is inflexible; and *specificity of evaluation,* or the degree to which work activities can be analyzed by specific, quantitative means. The Aston researchers further found the three characteristics to be highly related, so they combined them into a single scale labeled *workflow integration* (Hickson, Pugh, & Pheysey, 1969). For smaller organizations, workflow integration was found weakly related to organizational structure: As workflow integration increases, specialization, formalization, and decentralization of tactical authority also should increase for optimal functioning. For larger organizations, this relationship was not as apparent.

The most important finding of the Aston studies was that, although technology affects organizational structure, it appears to have significantly less impact than the other two sociotechnical system elements (i.e., the personnel subsystem and relevant external environment). Of particular importance for macroergonomics, and consistent with the Tavistock findings cited in Chapter 1, the Aston studies further demonstrated that the so-called *technological imperative*—the view that technology has a compelling influence on structure and, thus, should determine work system design—greatly overstates the case (Baron & Greenberg, 1990). In spite of these rather conclusive findings, the myth of technological determinism continues to persist.

Other Technological Considerations

Over the past several decades, we have seen major advances in computer and communications technology. Two forms of this technology have major implications for work system design: advanced information technologies (AIT) and computer integrated manufacturing (CIM). AIT tends to facilitate decentralizing operational or *tactical* decision making while enhancing the efficiency of centralized *strategic* decision making (Bedeian & Zammuto, 1991). Computer-based AIT links employees electronically, which better enables them to participate in the tactical decision-making process. Consequently, AIT enhances the efficiency of decentralization and greater professionalism. Because lower-level employees often select and filter the information and structure the databases, AIT also enables them to have a greater indirect influence on strategic decision making.

CIM, by its very nature, results in a very high level of integration of workflow processes and, thus, in a high level of interdependence among differentiated units. This interdependence increases the need for effective integrating mechanisms across functional units. Often, CIM also increases the need for market-based unit grouping, such as product task teams. This is particularly the case during the product design phase (Bedeian & Zammuto, 1991; Drucker, 1988).

PERSONNEL SUBSYSTEM ANALYSIS

There are at least three major characteristics of the personnel subsystem that are sensitive to the design of an organization's work system structure: (1) the degree of professionalism, (2) demographic characteristics, and (3) psychosocial aspects of the workforce (Hendrick, 1997).

Degree of Professionalism

Robbins (1983) notes that formalization can take place either on the job or through the process of professionalization. On the job, formalization is external to the employee, and is what is meant by the term "formalization." Rules, procedures, and human–system interfaces are designed into the work system to limit employee discretion. As a result, the skill requirements of jobs tend to be low. Professionalism, on the other hand, creates internal formalization of behavior through a socialization process that is an integral part of the education and training process: Persons learn the values, norms, and expected behavior patterns of the job before entering the organization.

From a macroergonomic design standpoint, there is a trade-off between formalization of the work system and professionalization of the jobs in the work system. Where the work system is designed to allow for low formalization and, thus, considerable employee discretion, jobs should be designed to require persons with relatively greater professional training or education. The rationale for this is rather straightforward: In the absence of formal decision rules and procedures, employees need to have the necessary professional knowledge and skills to make the decisions. Most often, it is the need to have employees that can deal with unique, nonroutine, or unanticipated situations that creates the need for low formalization and more highly professionalized jobs.

Demographic Factors

Demographic characteristics of the workforce that comprise the organization's personnel subsystem can also potentially interact with the work system's design. Those characteristics that are most striking within the United States and most industrialized countries are: (1) the "graying" of the workforce, (2) demographic shifts in psychosocial characteristics, (3) the broadening of the cultural diversity of the workforce, and (4) the recent large increase in the number of women in the workforce.

Graying of the Workforce. As the post–World War II baby boom bulge continues to move through their working careers, the average age of the workforce has been increasing at the rate of about 6 months per year. This trend began in the late 1970s and continued through the 1990s. As a result, we now have an older, more experienced, more mature, and better-trained workforce than was the case 20 years ago. For example, according to the U.S. Bureau of Labor Statistics, in 1970, 35% of the U.S. workforce was without a high school diploma. In 1997, less than 11% did not have a diploma. During this same period, the number in the workforce with education beyond high school has doubled. Put simply, the workforce has become more professionalized. Consequently, if employees are to feel fully utilized and remain motivated toward their work, work system structures need to accommodate this change by becoming less formalized and decentralizing more of the decision making. Macroergonomic practitioners need to pay particular attention to these factors in designing those work systems and jobs in which high formalization and centralization traditionally have characterized the work system's structure.

Value System Shifts. Yankelovich (1979), based on extensive longitudinal studies of workforce attitudes and values dating back to the 1940s, noted that those workers born after World War II have views and feelings about work that are very different from those of their predecessors. He further noted that these different conceptions and values would have a profound effect on work systems in the 1980s and beyond. Because their attitudes and values are so different, Yankelovich refers to this group of workers as the "new breed." This new breed of workers has three principal values that distinguish them from those of the mainstream of older workers: (1) the increasing importance of leisure, (2) the symbolic significance of the paid job, and (3) the insistence that jobs become less depersonalized and more meaningful. According to Yankelovich, the two aspects of work that are particularly important to new breed employees are being recognized as an individual and the opportunity to be with pleasant people with whom they like to work. From a work system design standpoint, these values translate into (1) a need for less hierarchical, less formalized, and more decentralized organizational structures; (2) for an attendant greater level of professionalism to be designed into individual jobs and human–system interfaces than found in traditional, highly formalized bureaucracies; and (3) for work systems and job designs that encourage group interaction and participatory decision making. Operationally, these design characteristics allow for greater individual recognition and respect for an employee's worth and provide the opportunity to satisfy social needs on the job.

Cultural Diversity. In a number of urban areas in the United States, such as the greater Los Angeles, San Francisco, Miami, and New York City areas, immigration has resulted in the development of workforces that are far more culturally diverse than existed a few decades ago. Unless organizations accommodate to this diversity, it will adversely affect employee motivation and commitment (Jackson, 1992; Thomas, 1991). Much of the needed accommodation has to do with changing organizational cultures to be more inclusive. One structural change to the work system that can facilitate this accommodation process is decentralizing decision making to allow greater employee control over their work environment and related policies and procedures. This includes the use of participatory ergonomics in designing or modifying work systems.

Women. During the 1980s and 1990s, women not only have been entering the workforce in progressively increasing numbers, but entering occupations and jobs that traditionally were staffed almost entirely by males. As yet, there is no clear indication as to how these demographic changes will, or should, affect work system design. I have noted in my consulting that, as women have moved into traditionally male positions, they have tended to emphasize the importance of modifying work systems and jobs so as to allow for greater social interaction—particularly in situations in which opportunity for social interaction previously had been limited. In addition, I have observed that women often have been the leaders in their organizations in pushing for work system changes that enable more flexible work hours (e.g., flextime, shared jobs, etc.).

Psychosocial Factors

Harvey, Hunt, and Schroder (1961) have identified the higher-order structural personality dimension of concreteness-abstractness of thinking, or *cognitive complexity,* as underlying different conceptual systems for understanding reality. Like organizational complexity, described in Chapter 1, cognitive complexity has the same two major dimensions: differentiation and integration. Here, *differentiation* refers to the number of differentiated conceptual categories one has developed for storing experiential information; and to the number of differentiated subcategories, or shades of gray, one has developed within each conceptual category. In a sense, it refers to the number of conceptual file drawers (conceptual categories) and files (subcategories) one has developed. *Integration* refers to the number of conceptual rules and combinations of rules one has developed for integrating data from the various conceptual categories to gain insight into complex situations, problems, and issues, and to derive creative, insightful solutions.

In general, the degree to which a given culture or subculture (1) encourages by its childrearing and educational practices an inquiring or active exposure to new experiences or diversity, and (2) through affluence, education, communications media, transportation systems, etc., provides opportunities for exposure to diversity, the more cognitively complex the persons of that culture will become.

Research consistently has shown that persons having a relatively low level of cognitive complexity tend to be *concrete* in their conceptual functioning. They tend to see the world as relatively more static and unchanging than do their more cognitively complex colleagues, and will resist change. They tend to interpret things more literally, and have a high need for structure and order, and for simplicity and consistency. They are more authoritarian and absolutist in their thinking, have a low tolerance for ambiguity, tend to be paternalistic, and tend to hold ethnocentric views. In contrast, cognitively complex persons tend to be *abstract* in their conceptual functioning. In comparison with concrete functioning persons, they see the world as dynamic and constantly changing. They have a lower need for structure and order and for consistency and stability. They tend not to be authoritarian, are relativistic rather than absolutist in their thinking, and have a high tolerance for ambiguity. They also possess a greater capacity for empathy and are more people-oriented, less paternalistic, and less ethnocentric than concrete functioning persons.

Research has shown that the majority of the U.S. workforce born after World War II is functioning at a more cognitively complex level than the mainstream of their older colleagues (Harvey, et al., 1961; Harvey, 1963; Hendrick, 1996). It is not that their older colleagues have regressed in cognitive complexity. In fact, they probably have increased somewhat in complexity over time. Rather, the following factors appear to account for much of the change: (1) the shift since World War II to more permissive childrearing practices that tend to encourage relativistic thinking and an openness to new experiences; (2) a concomitant shift in primary school teaching practices toward active, experiential learning and encouraging children to explore their environment rather than simply accepting a set of absolutes provided by the teacher; (3) a higher level of general education; (4) greater affluence; (5) the development of far superior communications, library, and transportation systems since World War II; and (6) new technology that enables greater exposure to diversity (e.g., television, the personal computer, and the Internet). (See Hendrick [1996, 1997] for a more detailed discussion of the nature of cognitive complexity and its relation to organizational design and management.)

As the World War II workforce has moved out of our organizations and more of the "new breed" have moved in, the workforce has, and should continue, to become progressively more cognitively complex in its conceptual functioning. One result has been a progressively increasing demand by workers for better-designed jobs, greater participation in decision making, and less formalization. Another result is a greater ability of the workforce to cope with ambiguity and change.

I have found evidence to suggest that relatively concrete work groups and managers function best in mechanistic organizations, characterized by moderately high vertical differentiation, centralization, and formalization, and in which the work system structure and processes are unambiguous and relatively slow to change. In contrast, although they can perform well in mechanistic organizations, cognitively complex or abstract persons prefer more organic organizational designs, characterized by relatively low levels of vertical differentiation, formalization, and centralization (Hendrick, 1979, 1981, 1990).

Personnel Subsystem Implications for Work System Design

Much of the data on personnel subsystem determinants of work system design are in the form of attitude survey results or projections from psychosocial and demographic studies. Although their nature is somewhat tenuous, there is a strong convergence of these data. At least within the industrialized countries, these data indicate a need for work systems to be as vertically undifferentiated, decentralized, and lacking in formalization as their technology and external environments will permit. Given the current trend toward highly dynamic virtual organizations and attendant work systems, these personnel subsystem data are very supportive.

EXTERNAL ENVIRONMENT ANALYSIS

Critical to the survival of organizations is their ability to adapt to their external environment. As open systems, organizations require monitoring and feedback mechanisms to follow and sense changes in their relevant task environments and a capacity to make timely, responsive adjustments. *Relevant task environments* refers to those aspects of the external world that can positively or negatively influence organizational effectiveness.

Types of External Environments

Negandhi (1977), based on field studies of 92 industrial firms in five different countries, has identified five types of external environments that affect organizational functioning.

1. *Socioeconomic.* Particularly the degree of stability, nature of the competition, and availability of materials and qualified workers.
2. *Educational.* The availability of facilities and programs, and the educational level and aspirations of workers.
3. *Political.* Governmental attitudes toward (a) business (friendliness versus hostility), (b) control of prices, and (c) "pampering" of industrial workers.
4. *Cultural.* Social status and caste system, values and attitudes toward work, management, etc., and the nature of trade unions and union–management relationships.
5. *Legal.* The degree of legal controls, restrictions, and compliance requirements.

For each organization, the relevant task environments will be different in type, qualitative nature, and importance. For any given organization, the particular combination of relevant task environments, each weighted in terms of its importance or degree of influence, constitutes the organization's *specific task environment*. A major determinant of any organization's specific task environment is its *domain*, or the range of products or services offered and market share (Robbins, 1983). Domain determines the points at which the organization depends on its specific task environment (Thompson, 1967).

Another determinant of an organization's specific task environment is its *stakeholders*. These include the company's stockholders, lenders, customers, employees, governmental agencies, and the local community, among others. Each of these entities has an interest in the organization.

For a given organization, the broader the domain and the more stakeholders it has, the more complex is its specific task environment.

Environmental Uncertainty

Of critical importance to work system design is the fact that all specific task environments vary along two highly critical dimensions: change and complexity (Duncan, 1972). *Degree of change* refers to the extent to which a given specific task environment is dynamic or remains stable over time, and the predictability of change. The *degree of complexity* is operationally defined by the number of components that constitute an organization's specific task environment (i.e., does the company interact with few or many government agencies, customers, suppliers, competitors, etc.?). Environmental change and complexity, in combination, determine the *environmental uncertainty* of an organization. Figure 3.2 illustrates this relationship for four different levels of uncertainty.

Two major models have been derived empirically for assessing environmental uncertainty as a determinant of work system structure. The first model focuses directly on environmental uncertainty. The second treats environmental uncertainty as one of several key environmental dimensions affecting work system structure, albeit the most important.

Burns and Stalker: Environmental Uncertainty.   Based on studies of 20 English and Scottish industrial companies, Burns and Stalker (1961) found that the type of work system structure that worked best in a relatively stable and simple organizational environment differed from that required for a more

Degree of Complexity	Degree of Change	
	Stable	Dynamic
Simple	Low uncertainty	Mod. high uncertainty
Complex	Mod. Low uncertainty	High uncertainty

FIG. 3.2   Environmental uncertainty dimensions.

dynamic and complex environment. For stable, simple environments, *mechanistic structures,* characterized by relatively high to moderately high vertical and horizontal differentiation, formalization, and centralization, worked best. Mechanistic structures typically have routine tasks, programmed behaviors, and can respond to change only slowly. A strong emphasis is placed on stability and control. Conversely, for dynamic, complex environments, *organic structures,* characterized by flexibility and quick adaptability, worked best. Organic work systems stress lateral rather than vertical communication, influence based on knowledge and expertise rather than position and authority, information exchange rather than directives from above, conflict resolution by interaction rather than by superiors, and loosely, rather than tightly, defined job descriptions and responsibilities. Organic work systems thus require low vertical differentiation and formalization, decentralized tactical decision making, and a relatively high level of professionalism. These findings are similar to those that were implicit in Emory and Trist's (1965) analysis of the effects of environmental instability on sociotechnical systems.

Lawrence and Lorsh: Subunit Environment and Design Complexity. A common way organizations deal with complex specific task environments is to develop specialized units to deal with particular parts of the environment. Lawrence and Lorsh (1969) conducted field studies to determine what type of work system structure was best for coping with different economic and market environments. They studied companies in a number of industries that varied considerably in their degree of environmental uncertainty, including food, plastics, and containers. From their studies, Lawrence and Lorsh identified five major environmental variables that can be assessed to determine the optimal degree of horizontal differentiation: (1) uncertainty of information (low, moderate, high), (2) time span of feedback (short, medium, or long), (3) pattern of goal orientation (focus of tasks), (4) pattern of time orientation (short, medium, or long), and (5) pattern of interpersonal relationships (task or social). In general, the more dissimilar the functions are on one or more of these dimensions, the greater is the likelihood that they should be differentiated into separate subunits (departmentalized) for effective functioning.

Lawrence and Lorsh also found that (1) the greater the differentiation, the greater the need for designing integrating mechanisms into the work system, and (2) the level of environmental uncertainty was of foremost importance in determining the most effective work system structure. In particular, subunits with more stable environments (e.g., production) tended to have high formalization, whereas those operating in less predictable environments (e.g., research and development) had low formalization.

Lawrence and Lorsh's empirical findings are especially valuable to macroergonomics because they demonstrate that whenever a work system's structure does not fit its mission, external environment, or resources, its functioning is likely to suffer.

INTEGRATING THE RESULTS
OF THE SEPARATE ASSESSMENTS

The separate analyses of the key characteristics of a given organization's technological subsystem, personnel subsystem, and specific task environment each
should have provided guidance about the structural design for the work system.
Frequently, these results will show a natural convergence. At times, however, the
outcome of the analysis of one sociotechnical system element may conflict with
the outcomes of the other two. When this occurs, the macroergonomic specialist
is faced with the issue of how to reconcile the differences. Based both on the
suggestions from the literature and my personal experience in evaluating over
200 organizational units, the outcomes from the analyses can be integrated by
weighting them approximately as follows: If the technological subsystem analysis is assigned a weight of "1," give the personnel subsystem analysis a weight of
"2," and the specific task environment analysis a weight of "3." For example,
let's assume that the technological subsystem falls into Perrow's "routine" category, the personnel subsystem jobs call for a high level of professionalism, and
the external environment has moderately low complexity. Weighting these three
as suggested above would indicate that a moderately formalized and somewhat
centralized work system would work best. Accordingly, the results would
indicate that most jobs should be redesigned to require a somewhat lower level
of professionalism, and attendant hardware and software interfaces should be
designed/redesigned to be compatible.

As noted in the aforementioned research of Lawrence and Lorsch, the specific
functional units of an organization may differ in the characteristics of their
technology, personnel, and specific task environments—particularly within
larger organizations. Therefore, the separate functional units may, themselves,
need to be analyzed as though they were separate organizations, and the resultant
work systems designed accordingly.

CONSIDERING JOB DESIGN
CHARACTERISTICS
IN MACROERGONOMICS

In designing or modifying a work system, it is important to be aware continually
of what impact decisions about work system structure (and processes) are likely
to have on the design of individual jobs. Hackman and Oldham (1975) have
empirically identified five specific job characteristics critical to intrinsic job
motivation, employee self-worth, stress reduction, and satisfaction for growth-
oriented employees: (1) *task variety,* or having different (meaningful) things to
do in one's work; (2) *identity,* or sense of job wholeness; (3) *significance,* or

perceived job meaningfulness; (4) *autonomy,* or control over one's work; and (5) *feedback,* or knowledge of results. The absence of these characteristics often is seen as resulting in dehumanized jobs that reduce psychological meaningfulness and felt responsibility. Such jobs can to lead to high stress, demotivated employees, job dissatisfaction, absenteeism, and reduced productivity (Organ & Bateman, 1991).

The importance of these five job characteristics repeatedly has been noted across many types of organizations and work situations. For example, Bammer (1990, 1993) conducted a meta-analysis of the field studies of upper extremity work-related musculoskeletal disorders (WMSDs) among computer operators that were reported in the literature during the 1980s. Bammer's results showed no consistent relationship of nonwork factors to employee musculoskeletal disorders. She did find that biomechanical factors are important, and noted that (microergonomic) efforts to improve them should be encouraged, but that, by themselves, biomechanical improvements are an insufficient means to reduce work-related musculoskeletal disorders adequately. The factors that consistently related to musculoskeletal disorders across studies were work organization variables. Bammer concluded that "improvements in work organization to reduce pressure, and to increase task variety, control, and the ability for employees to work together must be the main focus of prevention and intervention." She further noted that "ironically, such improvements in work organization generally also lead to increased productivity" (Bammer, 1993, p. 35). In summary, what Bammer identified as key correlates of work-related musculoskeletal disorders were Hackman and Oldham's job characteristics plus the opportunity to satisfy social needs on the job.

Based both on the literature and my personal consulting experience, in designing or modifying work systems one must be careful not to preclude designing individual jobs having Hackman and Oldham's five job characteristics. Further, it usually is important to ensure that the work system design will enable persons to satisfy social needs on the job.

CHOOSING THE RIGHT STRUCTURAL FORM

As described earlier, structural analysis involves considering how sociotechnical system variables should shape the basic dimensions of the work system. However, from a macroergonomic perspective, it also involves integrating these work system dimensions into an overall structural form. Fortunately, there are a variety of types of proven structural forms that are available to work system designers. As previously noted for the design of the individual dimensions of work system structure, the different types of structural form also can enhance or inhibit organizational functioning, depending on the organization's specific sociotechnical characteristics. Thus, the key is to select the specific type of structural form for the work system that best fits the organization's sociotechnical characteristics and related work system dimensions.

The four general types of overall organizational structure most commonly found are: (1) classical machine bureaucracy, (2) professional bureaucracy, (3) matrix organization, and (4) free-form design (Robbins, 1983). Most large, complex organizations have relatively autonomous units with different structural forms. In general, the larger the organization, the more likely it is to use more than one type of work system structure. In this section, the four general types of organizational structure will be described, along with the advantages and disadvantages of each, plus two new types. Finally, guidelines are provided for determining when each type is, or is not, likely to be appropriate.

Classical or Machine Bureaucracy

The classical bureaucratic form of work system has its roots in two streams of thought: *scientific Management and the ideal bureaucracy.*

Scientific Management. At the beginning of the twentieth century, industrial technology was developing rapidly in America and Europe, and labor was becoming highly specialized. As a result, engineers were being called on to help design work systems and optimize efficiency. One of these engineers, Frederick W. Taylor (1911), developed the concept of *Scientific Management,* which had a major impact on the shaping of classical organizational theory. Taylor's concepts of work system structure are implicit in his four basic principles of management (Szilagyi & Wallace, 1990, p. 662).

> *First.* Develop a science for each element of man's work that replaces the old rule-of-thumb method.
> *Second.* Scientifically select and train, teach, and develop the workman. In the past he chose his own work and trained himself as best he could.
> *Third.* Hardily cooperate with the men in order to ensure all of the work is being done in accordance with the principles of the science that has been developed.
> *Fourth.* Provide equal division of work and responsibility between the management and the workmen. The management takes over all work for which they are more qualified than the workmen. In the past, almost all the work and the greater part of the responsibility were thrown upon the men.

Ideal Bureaucracy. The classical bureaucracy type of structure was conceptualized by Max Weber. He recommended that organizations adhere to the following work system design principles (1946, p. 214).

1. All tasks necessary to accomplish organizational goals must be divided into highly specialized jobs. A worker must master his trade, and this expertise can be more readily achieved by concentrating on a limited number of tasks.
2. Each task must be performed according to a "consistent system of abstract rules." This practice allows the manager to eliminate uncertainty due to individual differences in task performance.
3. Offices or roles must be organized into hierarchical structure in which the scope of authority of superordinates over subordinates is defined. This system offers the subordinates the possibility of appealing a decision to a higher level of authority.
4. Superiors must assume an impersonal attitude in dealing with each other and subordinates. This psychological and social distance enables the superior to make decisions without being influenced by prejudices and preferences.
5. Employment in a bureaucracy must be based on qualifications, and promotion is to be decided on the basis of merit. Because of this careful and firm system of employment and promotion, it is assumed that employment will involve a lifelong career and loyalty from employees.

Weber believed that adhering to these principles was the "one best way" to achieve efficiency and meet organizational objectives. He assumed that implementing a structure that emphasized administrative efficiency, stability, and control was the key to obtaining optimal effectiveness (Szilagyi & Wallace, 1990).

Collectively, the two sets of theoretical principles of Taylor and Weber resulted in what today we know as the *machine bureaucracy* type of work system structure. Its basic characteristics are as follows (Robbins, 1983):

1. *Division of labor.* The work system is comprised of narrowly defined jobs having well-defined, relatively routine tasks.
2. *A well-defined hierarchy.* A clearly defined, formal hierarchical structure in which each lower office is under supervision and control of a higher one. Tasks primarily are grouped by function. Line and staff functions are well defined and kept separate.
3. *High formalization.* Extensive use is made of formal rules and procedures to ensure uniformity and regulate employee behavior.
4. *High centralization.* Decision making is reserved for management. Employees have relatively little decision discretion.
5. *Career tracks for employees.* Because members are expected to pursue their careers within the organization, career tracks form part of the work system's design for all but the most unskilled positions.

Advantages. The three primary advantages of the machine bureaucracy are administrative efficiency, stability, and control over the work system's functioning. Having narrowly defined jobs with formalized tasks minimizes the likelihood of error, as well as prerequisite skills and related training time and costs. These design features also better enable employees to know their own function and those of others. Formalization ensures stability, control, and a well-integrated and consistent pattern of functioning. Centralization also ensures control and enhances stability.

Disadvantages. The machine bureaucracy design has two inherent major disadvantages. First, by their inherent nature, machine bureaucracies are slow and inefficient in responding both to environmental change and nonroutine situations. Second, although careful macroergonomic design can overcome or greatly reduce it, machine bureaucracies result in jobs that often violate Hackman and Odham's job characteristics, described earlier. In particular, such jobs often fail to utilize adequately the mental and psychological capacities of the workers. As a result, jobs tend to lack intrinsic motivation.

When to Use. When (1) the specific task environment is comparatively simple, stable, and/or predictable, (2) employee education and skill levels are relatively low, and (3) system operations are repetitive or otherwise can be routinized. To the extent that these conditions do not exist, one of the other forms of work system structure is preferred.

Professional Bureaucracy

The professional bureaucracy overcomes some of the disadvantages of the machine form by relying on a relatively high degree of professionalism. Jobs are more broadly defined, less routine, and allow for greater employee decision discretion (Robbins, 1983). Accordingly, there is less need for formalization, *tactical decision making* can be decentralized, and fewer levels of hierarchy are needed. They are similar to machine bureaucracies in that they are hierarchical, positions are grouped by function, and *strategic decision making* typically remains centralized.

Advantages. Compared with machine bureaucracies, the professional bureaucracy has at least three major advantages. First, it can cope with complex environments and nonroutine tasks more efficiently. Secondly, jobs are designed to better utilize the mental and psychological capabilities of employees, thus making them more intrinsically motivating. Thirdly, professional bureaucracies require less managerial tactical decision making and control, thereby enabling management to give greater attention to long-range planning, strategic decision making, dealing with employee concerns, and reinforcing good employee performance.

Disadvantages. Professional bureaucracies are not as efficient as machine bureaucracies for coping with simple, stable environments. Second, they require a more highly skilled work force and related additional training time and expense. Third, control is less tight, and the distinction between line and staff functions is more ambiguous. Fourth, the management skills required by professional bureaucracies tend to be more sophisticated; for example, a greater reliance is placed on the use of persuasion and facilitation skills rather than a simple and direct authoritarian style.

When to Use. When the external environment is fairly complex, somewhat unstable or unpredictable, and there is an available applicant pool of professionalized workers. A professional bureaucratic form is less optimal than the machine form for highly repetitive operations with a simple, stable environment. It also is less desirable if the available management pool is highly authoritarian and concrete in its functioning.

Adhocracies

Although it is worse for machine bureaucracies, both bureaucratic forms tend to be inefficient in responding to highly complex or dynamic specific task environments. Because of this shortcoming, other types of work system structures have evolved. These newer types are known collectively as adhocracy designs. An *adhocracy* is a "rapidly changing adaptive, temporary system organized around problems to be solved by groups of relative strangers with diverse professional skills" (Bennis, 1969, p. 45). Adhocracies are characterized by moderate complexity, low formalization, and decentralization (Robbins, 1983). Although horizontal differentiation can be high, vertical differentiation invariably is moderate to low. Rather than the tight administrative control afforded by centralized decision making and formal rules and procedures, flexibility and rapidity of response are emphasized. Thus, a high level of professionalism and fewer administrative layers are essential.

When to Use. When the ability to be innovative or respond rapidly to changing situations is essential, and when these responses require collaboration of persons possessing different specialties, the adhocracy forms of organization are considerably more effective than the bureaucratic forms (Robbins, 1983). On the other hand, adhocracies have at least three major disadvantages. First, because of less clear boss–subordinate relationships, the lines of authority and responsibility are ambiguous. Consequently, *conflict* is an integral part of adhocracy. The second disadvantage is *psychosocial stress*. The composition of teams or units in adhocracies is dynamic, while the establishing and dismantling of human relationships is a slower psychosocial process. Consequently, this process becomes stressed any time there is significant work system change.

Concrete functioning employees are especially likely to have difficulty coping. Thirdly, adhocracies lack the tight administrative control that comes with routine functions and structural stability. It is only when these losses in administrative efficiency are more than offset by the competitive gains from greater efficiency of responsiveness or innovation that some form of adhocracy is to be preferred. The two most common forms of adhocracies are the *matrix* and *free-form* designs.

Matrix Design

The matrix form of work system combines departmentalization by function with departmentalization by project or product line. Matrix organizations are like bureaucracies in that they have functional departments that tend to be lasting. They are unlike bureaucracies in that members of the functional departments are farmed out to project or product teams as new projects or product lines develop. This is done to meet the project's need for the combined technical expertise of the various functional departments. When the need for a given functional department's professional input no longer is required, or the level of effort reduces, individuals either return to their "home" department or transfer to another project or product team. Although the project or product manager supervises the team's interdisciplinary effort, each team member also has a functional department supervisor. The matrix design thus violates a fundamental bureaucratic concept: *unity of command.*

Advantages.   The primary advantage of the matrix type of work systems is that it enables the best of two worlds: the stability and professional depth afforded by functional departmentalization and the rapid response capability of interdisciplinary teams.

Disadvantages.   In addition to the disadvantages that characterize all adhocracy designs, cited earlier, employees in matrix organizations must serve two bosses: their functional department head, who tends to be relatively long-term oriented and remote from the team member's immediate tasks, and the project team director, who tends to be short-term oriented and immediate to the employees' tasks. Serving two masters with differing goals, responsibilities, and time orientations frequently creates conflict and can disrupt organizational functioning. For employees, there is another potential problem. When assigned too long to a project, they may have difficulty keeping technically current. They also may lose contact with their respective functional departments. Both of these consequences can have an adverse impact on their careers.

When to Use.   The matrix organization is especially well suited for responding to dynamic and complex specific task environments. Under these conditions, both rapid interdisciplinary responsiveness and providing for functional depth in individual disciplines can be essential to the organization's ability to compete.

Free-Form Design

The newest of the four common forms of work system structure is the free-form design. In its pure form, the free-form organization is like an amoeba in that its shape continually changes in order to survive (Szilagyi & Wallace, 1990). Of paramount importance to free-form work systems is the ability to respond rapidly to change in highly dynamic, complex, and competitive environments.

In free-form work systems, a *profit center* arrangement replaces functional departmentalization. Profit centers are managed by teams and are highly results oriented. Accordingly, free-form work systems have low hierarchical differentiation, very low formalization, and highly decentralized decision making. A very heavy reliance thus is placed on professionalism. Like matrix organizations, project teams are created, changed, and disbanded as required to meet organizational goals and problems. Unlike matrix organizations, free-form work systems have no underlying functional departmentalization or "home" structure. Instead, employees "float" from one project team or cost center to another as their services are needed.

In order to function effectively in a free-form work system, managers and employees alike need to possess a great deal of personal flexibility, a low need for structure and stability, and a high tolerance for ambiguity. The ability to handle change without becoming unduly stressed is essential. Based on my personal consulting experience, too often, prospective employees are not screened on these characteristics and the organization eventually goes through a crisis because of it (which usually is when I get called in to help).

Advantages.   Free-form work systems have one major advantage over other designs: the ability to respond to highly dynamic, complex, and competitive specific task environments with great speed and innovation.

Disadvantages.   The one major advantage of free-form designs, cited previously, comes at a cost. Free-form work systems have all of the disadvantages of matrix adhocracies, only to a greater extent. Conflict, psychosocial stress, and inherent administrative inefficiency are an integral part of a continuously changing and amorphous work system structure.

When to Use.   A free-form design should be considered whenever the organization's survival depends on innovation and speed of response, and a highly professionalized and flexible workforce is available. Most often, these features characterize (1) small to medium-sized high-technology organizations and (2) semi-autonomous "outlaw" subunits of large bureaucratic or matrix organizations (e.g., the Lockheed "skunkworks," which designed many of the U.S. military aircraft during the last half of the twentieth century—and did so faster, cheaper, and with many fewer engineers than conventional aircraft design groups) operating in highly dynamic and complex competitive environments.

New Adhocracy Forms

Because of recent technological advances, new variations of the matrix and free-form designs have begun to emerge. Of particular note are the *modular* and *virtual* types (Dess, Rasheed, McLaughlin, & Priem, 1995).

Modular Form.   The modular type of adhocracy structure outsources nonvital functions while retaining full strategic control. Outside units are used to manufacture parts, handle logistics, maintenance, or accounting activities, and/or to perform housekeeping services. The "organization" thus consists of a central hub surrounded by networks of outside suppliers and specialists. What constitutes the organization is amorphous, in that modular parts readily can be added or deleted as the immediate needs require. This flexibility enables modular adhocracies to reduce uncertainty by adapting rapidly to changing market conditions with little cost.

Virtual Organizations.   Rather than the central hub of the modular organization, the virtual type consists of a continually evolving network of independent companies. In its purest form, a virtual organization does not have a central office or hierarchical structure. In order to pursue common strategic objectives, suppliers, customers, and even competitors will link together to share skills, costs, and access to one another's markets (Dess, et al., 1995). The major advantage of the virtual organization is that each firm brings a particular set of competencies to the alliance, thereby enabling them collectively to be a more competitive yet highly flexible (virtual) entity. Essentially, the virtual organization copes with uncertainty by using a collectivist strategy.

4

Macroergonomic Methods: Assessing Work System Processes

Michelle M. Robertson

*Liberty Mutual Research Center
for Safety and Health*

Brian M. Kleiner

*Virginia Polytechnic Institute
and State University*

Michael J. O'Neill

Herman Miller Inc.

Broadly speaking, work systems can be categorized as skill-based versus knowledge-based. Skill-based work systems are typically observed in industrial environments. Knowledge-based work systems are typically office environments which can exist within service or manufacturing industries. Traditional approaches to work system design in both environments have exhibited technology-centered design, a "leftover" approach to function and task allocation, which fails to consider an organization's sociotechnical characteristics and integrate them into its work system design. In this chapter, we present two general methodologies for macroergonomic assessment of work system process: The first, macroergonomic analysis and design (MEAD) technology, has been used more typically in industrial environments, and the second, identified simply as a "systems analysis tool," more typically in office environments. However, both can be applied to a variety of work environments. Both employ macroergonomic principles and values for improved performance and human well-being.

MACROERGONOMIC ANALYSIS
AND DESIGN (MEAD) METHODOLOGY

Background

The macroergonomic analysis and design (MEAD) methodology has been developed based in part on the contributions of Emery and Trist (1978), Taylor and Felton (1993), Clegg, Ravden, Corbertt, and Johnson (1989), and experience with large-scale change in academia, industry, and government (Kleiner, 1996). The approach integrates sociotechnical systems theory and microergonomics because sociotechnical system approaches historically did not directly address micro-ergonomic issues, and micro-ergonomics historically failed to address the larger system's environmental and organizational issues.

The MEAD 10-Step Process

There are 10 steps in this particular methodology:

1. Scanning the environmental and organizational design subsystem
2. Defining production system type and setting performance expectations
3. Defining unit operations and work process
4. Identifying variances
5. Creating the variance matrix
6. Creating the key variance control table and role network
7. Performing function allocation and joint design
8. Understanding roles and responsibilities perceptions
9. Designing/redesigning support subsystems and interfaces
10. Implementing, iterating, and improving

Step 1: Scanning the Environmental
and Organizational Subsystems

The first phase of sociotechnical analysis of work system process is to scan the system, then the environment and organizational subsystems. Because the external environment, operating under the principle of joint causation, may be the most influential subsystem in determining whether the sociotechnical system will be successful, achieving a valid organization/environment fit and joint optimization is essential.

Within the system scan, there is often a gap between what the organization professes as its defining characteristics and its actual identity as observed from organizational behavior. It is thus useful to assess the nature and extent of this variance. To accomplish this, the formal company statements about mission (i.e., purpose), vision, and principles are identified and evaluated with respect to their components. With regard to performance objectives, it is instructive to see whether and to what extent the organization places emphasis on targeted criteria; for example, is the welfare of employees emphasized in the professed values or guiding principles? Does the mission statement speak of "quality" products or processes, and if so, do processes support achievement of these goals?

System scanning involves defining the workplace in systems terms, including defining relevant boundaries. Several tools are available to assist with scanning. The organization's mission is detailed in systems terms (i.e., inputs, outputs, processes, suppliers, customers, internal controls, and feedback mechanisms). The system scan also establishes the initial boundaries of the work system. As described by Emery and Trist (1978), there are throughput, territorial, social, and time boundaries to consider.

Entities outside the boundaries identified during the system scan are part of the external environment. In the environmental scan, the organization's subenvironments and the principle stakeholders within these subenvironments are identified. Their expectations for the organization are identified and evaluated. Conflicts and ambiguities are seen as opportunities for process or interface improvement. Variances are evaluated to determine design constraints and opportunities for change. The work system itself can be redesigned to align itself with external expectations, or conversely, the work system can attempt to change the expectations of the environment to be consistent with its internal plans and desires. According to sociotechnical systems theory, the response in part will be a function of whether the environment is viewed by the organization as a source of provocation or inspiration (Pasmore, 1988). Much of the time the gaps between work system and environmental expectations are gaps of perception, and communication interfaces need to be developed between subenvironment personnel and the organization. Design focuses on design or redesign of interfaces among the organizational system and relevant subenvironments to improve communication and decision making. These interfaces are referred to as organization–or work system–environment interfaces. The type of information obtained appears in Table 4.1. As can be seen, consistent with a sociotechnical systems approach, variances are the focus. Several variances are noted between the current state and the desired (i.e., future) state.

It is useful to develop organizational design hypotheses based on the environmental and system scans. By referring to the empirical models of the external environment (Hendrick & Kleiner, 2001), optimal levels of complexity (both differentiation and integration), centralization, and formalization can be hypothesized.

Step 2: Defining Production System Type and Setting Performance Expectations

The work system's production type can help determine optimal levels of complexity, centralization, and formalization. The system scan performed in the

TABLE 4.1
System and Environmental Scan

Item	Current	Desired
Purpose	Produce products for national security and make a profit for company.	Be the nation's model for restoration of land systems and application of advanced restoration technologies.
Philosophy	• Management controls • Do what it takes to get throughput	• Emphasize quality and safety • Respect the environment
Objectives		
Technical	Make a profit, maintain contract with customer, reduce costs.	• Restore the environment for profitable use (education, agriculture, businesses). • Comply w/environmental regulations. • Develop marketable environmental technologies.
Social	Avoid strikes with union or stockpile products to avoid supply interruption.	• Develop cooperation among employees and with external stakeholders. • High levels of quality and safety
Outputs	Weapons-grade materials, products	• Environmental restoration expertise • Reservation environment safe for profitable or recreational use
Inputs	Unprocessed materials, product components, other raw materials	• Contaminated environment • People willing to work and learn
Boundaries	Suppliers, products ready for	Input: government regulations
Throughput	shipping are at the boundaries.	Output: knowledge base in libraries
Territorial	Nuclear complex	Nuclear complex and nearby schools doing training
Social	Company production employees are in the system.	Researchers supplying knowledge; contractors, customer, and state become part of the work system.
Time	Fiscal year budget/contract period	20 years/ contract period
Expectations		
Union-to-system	Abide by contract. Don't interrupt work, give	Work together to increase capabilities, safety and contributions of
System	concessions.	employees and their families.
News media	Give information for big story.	Give us a shot at good stories.
System	Don't make us look bad.	Help change our image, publicity.
State	Abide by regulations, keep honest records.	Joint: showcase what can be done through cooperation, publicity.
System	Cut us some slack to make products.	
Customer	Products whenever needed	No political embarrassment

TABLE 4.1

(Continued)

Item	Current	Desired
System	Profit, jobs, contract	Contract, profit, showcase
Local community	Jobs, pride in community	Jobs, pride in reclamation, safety
System	Labor force, no hassles	Labor force, cooperation
Presenting problems	• Lack of cooperation and relationships with internal and external environment • Management needs to develop consistency between espoused vision and values and what it does. • Workforce may lack skills and desire to pursue clean-up • Lack of knowledge about what is in the ground or tanks • Lack of knowledge about how to clean up environmental hazards	
Future scenario	Realistic: site closes, someone else comes to do cleanup, we get lawsuits for contamination, stock value depressed.	Idealistic: get new contract, become showcase on how to clean up, resulting in new contracts elsewhere, favorable impact on stock value.

(Adapted from Groesbeck, Sienknecht & Merida, 1998. Used with permission.)

previous phase should help in this regard, and the analyst should consult available production models. In this context, key performance criteria related to the organization's purpose and technical processes are identified. This requires a determination of success factors for products and services, but may also include performance measures at other points in the organization's system, especially if decision making is important to work process improvement. As described in Kleiner's (1997) framework adapted from Sink and Tuttle (1989), specific standardized performance criteria guide the selection of specific measures which relate to different parts of the work process. Measures can be subjective, as in the case of self-reports, or measures can be objective, measured from performance.

As illustrated in Fig. 4.1, Sink and Tuttle (1989) suggested organizational performance can be measured or assessed using seven performance criteria or clusters of measures: efficiency, effectiveness, productivity, quality, quality of work life (QWL), innovation, and profitability or budgetability. The seven performance criteria relate to specific parts of the organization as represented by an input–output model similar to that proposed by Deming (1986). Within a given performance criterion, specific measures can be derived. Data sources for each measure can be subjective, as in the case of self-reports, or can be based on objective data. Kleiner (1997) contributed a flexibility criterion which related to each of these checkpoints as well, due to the increasing need to manage and measure flexibility in systems.

According to Sink and Tuttle (1989), QWL includes safety as a criterion; however, the need for a healthy and safe working environment is differentiated

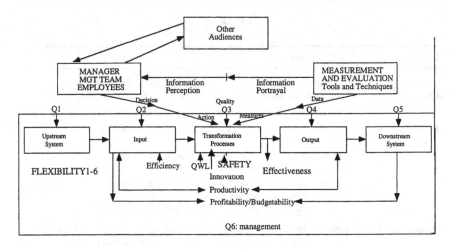

FIG. 4.1. Performance criteria in a work system (adapted from Sink & Tuttle, 1989).

from QWL, the affective perception of the total work environment. The efficiency criterion focuses on input or resource utilization. Effectiveness focuses on whether objectives are realized. Productivity is operationalized as outputs/inputs. Innovation refers to creative changes to process or product that result in performance gains. Profitability is a standard business management criterion. For not-for-profit organizations, Sink and Tuttle (1989) introduced "budgetability" or expenditures relative to budget to replace the profitability criterion. Quality Checkpoints 2 and 4 correspond to traditional measures of quality control, traditionally assured through inspection of inputs and outputs, respectively. Quality Checkpoints 1, 3, and 5 are quality criteria popularized by Deming and the total quality management (TQM) movement. In essence, a TQM approach to quality moves resources from Checkpoints 2 and 4 exclusively and share resources with the other system checkpoints. Checkpoint 1 emphasizes the quality of suppliers, which has been operationalized within the quality movement in the form of supplier certification programs and processes. Checkpoint 3 in process control pertains to the use of statistical quality control charts to monitor and control processes. Checkpoint 5 refers to customer satisfaction, operationalized as the customer getting what is wanted and needed. Checkpoint 6 corresponds to TQM or the method by which the other criteria are managed.

Once the type of production system has been identified and the empirical production models consulted, the organizational design hypotheses generated in the previous phase should be supported or modified until the personnel subsystem can be thoroughly analyzed as well. In terms of function allocation, requirements specifications can be developed at this juncture, including micro-ergonomic requirements. Also included are system design preferences for complexity, centralization, and formalization. Clegg, et al. (1989) also suggested the use of scenarios that present alternative allocations and associated costs and benefits.

Step 3: Defining Unit Operations and Work Process

Unit operations are groupings of conversion steps that together form a complete piece of work and are bounded from other steps by territorial, technological, or temporal boundaries. Unit operations often can be identified by their own distinctive subproduct and typically employ 3 to 15 workers. They can also be identified by natural breaks in the process, i.e., boundaries determined by state changes (transformation), or actual changes in the raw material's form or location (input) or storage of material. For each unit operation or department, the purpose/objectives, inputs, transformations, and outputs are defined. If the technology is complex, additional departmentalization (horizontal differentiation) may be necessary. If collocation is not possible or desirable, spatial differentiation and the use of digital integrating mechanisms may be needed. If the task exceeds the allotted schedule, then work groups or shifts may be needed. Ideally, resources for task performance should be contained within the unit, but interdependencies with other units may complicate matters. In these cases, job rotation, cross-training, or relocation may be required. Figure 4.2 illustrates the

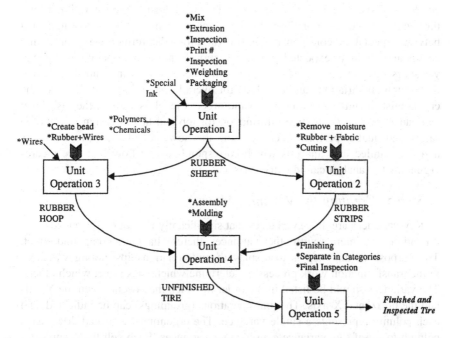

FIG. 4.2. Unit operations in tire manufacturing (adapted from Blanco & Duggar, 1998; used with permission).

identification of unit operations or groupings of sociotechnical systems and associated technical processes.

The current workflow of the transformation process (i.e., conversion of inputs to outputs) should be flowcharted, including material flows, workstations, and physical as well as informal or imagined boundaries. In linear systems such as most production systems, the output of one step is the input of the next. In nonlinear systems such as many service or knowledge work environments, steps may occur in parallel or may be recursive. Unit operations are identified. Also identified at this stage are the functions and subfunctions (i.e., tasks) of the system (Clegg, et al., 1989). The purpose of this step is to assess improvement opportunities and coordination problems posed by technical design or the facility.

Identifying the work flow before proceeding with detailed task analysis can provide meaningful context in which to analyze tasks. Once the current flow is charted, the macroergonomist or analyst can proceed with a task analysis for the work process functions and tasks. Figure 4.3 illustrates some of the technical production processes seen in tire manufacturing.

Step 4: Identifying Variances

A variance is an unexpected or unwanted deviation from standard operating conditions, specifications, or norms. STS distinguishes between input and throughput variances. For throughput variances, Deming (1986) distinguished between special or common causes of variation, the former being abnormal causes and the latter expected system variation from normal operations. Special variances need to be tackled first to get the work process in control, at which time common variation can be tackled for overall system improvement. For the ergonomist, identifying variances at the process level as well as the task level can add important contextual information for job and task redesign to improve safety and quality performance. By using the flowchart of the current process and the detailed task analysis which corresponds to the flowchart, the macroergonomist or analyst can identify variances.

Step 5: Creating the Variance Matrix

Key variances are those variances that significantly impact performance criteria and/or may interact with other variances, thereby having a compound effect. The purpose of this step is to display the interrelationships among variances in the transformation work process to determine which ones affect which others. The variances should be listed in the order in which they occur down the Y-axis and the horizontal X-axis. The unit operations (groupings) can be indicated, and each column represents a single variance. The ergonomist can read down each column to see if this variance causes other variances. Each cell then represents the relationship between two variances. An empty cell implies two variances are

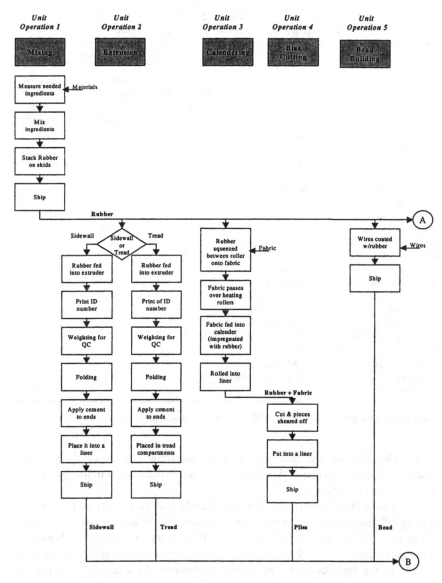

FIG. 4.3. Technical processes in tire manufacturing (adapted from Blanco & Duggar, 1998; used with permission).

unrelated. The analyst or team can also estimate the severity of variances by using a Likert-type rating scale. Severity would be determined on the basis of whether a variance or combination of variances significantly affect performance. This should help identify *key variances*.

TABLE 4.2
Identification of Key Variances in Tire Manufacturing

Key Variance	Unit Operation	significant downstream impact	numerous relationships with other variances	significant impact as a single variance
Raw material consistency	Mixing	√	√	√
Raw material composition	Mixing	√	√	√
Mixing duration	Mixing	√	√	√
Rubber quality	Tread extrusion	√	√	√
Tread thickness	Tread folding	√	√	
Adhesive strength	Tread cementing	√	√	
Quantity adhesive used	Tread cementing	√	√	
Rubber to fabric adhesion	Rubber impregnation	√	√	√
Bond consistency	Rubber impregnation	√	√	
Wire to rubber adhesion	Bead building	√	√	√
Bond consistency	Bead building	√	√	
Wire surface quality	Bead building	√	√	
Wire to rubber adhesion	Wire enveloping	√	√	√
Component dimensions	First-stage building	√	√	√
Press force	First-stage building	√	√	√
Press duration	First-stage building	√	√	√

(Adapted from Groesbeck, Sienknecht & Merida, 1998. Used with permission.)

We identified key variances for the tire manufacturer, based on the variance matrix on the previous page. The types of variances discovered in this procedure are illustrated in Table 4.2.

A variance is considered *key* if it significantly affects quantity of production, quality of production, operating costs (utilities, raw material, overtime, etc.), social costs (dissatisfaction, safety, etc.), or if it has numerous relationships with other variances (matrix). Typically, consistent with the Pareto principle, only 10 to 20% of the variances are significant determinants of the quality, quantity, or cost of product.

Step 6: Creating the Key Variance Control Table and Role Network

The purpose of this step is to discover how existing variances are currently controlled and whether personnel responsible for variance control require additional support. The key variance control table includes: the unit operation in which variance is controlled or corrected; who is responsible; what control activities are currently undertaken; what interfaces, tools, or technologies are needed to support control; and what communication, information, special skills, or knowledge are needed to support control. A key variance control table is illustrated in Table 4.3 to demonstrate the types of data collected for key variances in order to better support the focal roles.

A job is defined by the formal job description which is a contract or agreement between the individual and the organization. This is not the same as a work role, which is comprised of actual behaviors of a person occupying a position or job in relation to other people. These role behaviors result from actions and expectations of a number of people in a role set. A role set is comprised of people who are sending expectations and reinforcement to the role occupant. Role analysis addresses who interacts with whom, about what, and how effective these relationships are. This relates to technical production and is important because it determines the level of work system flexibility.

In a role network, the role responsible for controlling key variances is identified first. Although multiple roles may exist that satisfy this criterion, there is often a single role without which the system could not function.

With the focal role identified within a circle, other roles can be identified and placed on the diagram in relation to the focal role. Based on the frequency and importance of a given relationship or interaction, line length can be varied, where a shorter line represents more or closer interactions. Finally, arrows can be added to indicate the nature of the communication in the interaction. A one-way arrow indicates one-way communication, and a two-way arrow suggests two-way interaction. Two one-way arrows in opposite directions indicate asynchronous (different time) communication patterns. To show the content of the interactions between the focal role and other roles and an evaluation of the presence or absence of a set of functional relationships for functional requirements, labels are used to indicate the Goal of controlling variances; Adaptation to short-term fluctuations; Integration of activities to manage internal conflicts and promote smooth interactions among people and tasks; and Long-term development of knowledge, skills, and motivation in workers. Also, the presence or absence of particular relationships is identified as Vertical hierarchy; Equal or peer; Cross-boundary; Outside; or Nonsocial.

The relationships in the role network are then evaluated. Internal and external customers of roles can be interviewed or surveyed for their perceptions of role effectiveness as well. Also, the organizational design hypotheses can be tested against the detailed analysis of variance and variance control. The role analysis and variance control table may suggest, for example, a need to increase or decrease formalization or centralization. If procedures are recommended to help control variances, this increase in formalization must be evaluated against the more general organizational design preferences suggested by the environmental and production system analyses.

TABLE 4.3

A Variance Control Table for Tire Manufacturing

Key Variance	Unit Operation			Controlling Role	Controlling Action	Information Needed
	Where Occurs	Where Observed	Where Controlled			
4: Width of rubber strip	Unit Operation 2: Extrusion	Unit Operation 2 or 7: Extrusion or tire building	Unit Operation 2: Extrusion	*Machine maintenance *Machine operator *Quality inspectors	*Machine adjustments *Machine adjustments or call maintenance personnel *Stop production or reject failed products	*Reports from machine operators and quality inspectors *Visual inspection acceptance standards, product observation *Weight of product, weight standard
6: Weight of rubber strip	Unit Operation 2: Extrusion	Unit Operation 2 or 7: Extrusion or tire building	Unit Operation 2: Extrusion	*Machine maintenance *Machine operator *Quality inspectors	*Machine adjustments *Machine adjustments or call maintenance personnel *Stop production or reject failed products	*Reports from machine operators and quality inspectors *Visual inspection acceptance standards, product observation *Weight of product, weight standard
7: Amount of cement applied	Unit Operation 2: Extrusion	Unit Operation 2 or 7: Extrusion or tire building	Unit Operation 2 or 7: Extrusion or tire building	*Machine maintenance *Machine operator	*Machine adjustments *Machine adjustments or call maintenance personnel	*Reports from machine operators and quality inspectors *Visual inspection acceptance standards, product observation

8: Quality of fabric	Unit Operation 3: Calendering	Unit Operation 3 or 7: Calendering or tire building	Unit Operation 3: Calendering	*Quality inspectors	*Stop production or reject failed products	*Visual inspection to determine if cement is holding
				*Incoming-Quality Inspector	*Reject batches of inferior quality and/or call supplier	*Product observation, measurements, acceptance standards, and incoming-product/supplier certification
26: Shape after molding	Unit Operation 8: Curing	Unit Operation 8 or 9: Curing or finishing	Unit Operation 8 or 9: Curing or finishing	*Machine operator	*Call incoming-quality inspector	*Visual observation and acceptance standards
				*Machine maintenance	*Machine adjustments	*Reports from machine operators and quality inspectors
				*Machine operator	*Machine adjustments or call maintenance personnel	*Visual inspection acceptance standards, product observation
				*Quality inspectors	*Stop production or reject failed products	*Product observation, results from roundness and balance inspection, and acceptance standards

(Adapted from Blanco & Duggar, 1998. Used with permission.)

Step 7: Performing Function Allocation and Joint Design

Having previously specified system objectives, requirements, and functions, it is now time to allocate systematically functions and tasks to human and machine or computer. It is helpful to review the environmental scan data to check for any subenvironment constraints (e.g., political, financial, etc.) before making any mandatory allocations (Clegg, et al., 1989). Next, provisional allocations can be made to the human(s), machine(s), both, or neither. In the latter case, a return to developing requirements may be appropriate using four groups of criteria: technical feasibility; health and safety; operational requirements (i.e., physical, informational, performance); and function characteristics (i.e., criticality, unpredictability, psychological). (See Kleiner (1998) for a review of macroergonomic directions and issues in function allocation.)

Technical changes are made to at best prevent, or at worst, control key variances. Human-centered design of the following may be needed to support operators as they attempt to prevent or control key variances: interfaces, information systems to provide feedback; job aids; process control tools; more flexible technology; work station or handling system redesign; or integrating mechanisms.

After considering human-centered system changes in the previous step, it is time to turn our attention to supporting the person directly by addressing knowledge and/or skill requirements of key variances and any selection issues that may be apparent. In the variance control table, we identified who controls variances and the tasks performed to control these variances. At this stage, we suggest personnel system changes to prevent or control key variances. This may entail specific skill or knowledge sets that can be acquired through technical training, formal courses, workshops, or distance learning.

At this point in the process, organizational design hypotheses have been generated and iteratively adjusted as new analyses are performed. It is now time to take the specifications for organizational design levels of complexity, centralization, and formalization and produce specific structures. Depending on the level of work system process analysis, this may require design/redesign at the organizational level or at the group/team level or at both levels.

Step 8: Understanding Roles and Responsibilities Perceptions

It is important to identify how workers perceive their roles documented in the variance control table, especially if the table was initially constructed by those who do not occupy the roles identified. Through interviews, role occupants can participate in an analysis of their perceptions of their roles. The analyst can compare expected roles with perceived roles and can identify whether there are any gaps. If there are any discrepancies, these variances can be managed through training and selection as well as technological support. Essentially, two role networks are operating, the one needed and the one perceived. Any variation between the two can be reduced through participatory ergonomics, training, communication, interface design, or tool design.

Step 9: Designing/Redesigning Support Subsystems and Interfaces

Consistent with the STS design principle, "support congruence" (Taylor & Felton, 1993), now that the work process has been analyzed and jointly designed, other internal organizational support subsystems may require redesign (e.g., management system, reward system, maintenance). The goal is to determine the extent to which a given subsystem impacts the sociotechnical production system; to determine the nature of the variance; to determine the extent to which the variance is controlled; and to determine the extent to which tasks should be taken into account in redesign of operating roles in the supporting subsystem units.

According to the Clegg, et al. (1989) method of function allocation, individual and cumulative allocations made on a provisional basis earlier can be further evaluated: requirements specifications (including the scenarios developed earlier); resources available at the time of implementation (including human and financial); and the sum total outcome. In addition to a check of function allocation, interfaces among subsystems should be checked and redesigned at this juncture.

Especially at the team and individual levels of work, the internal physical environment should be ergonomically adjusted if necessary to promote human well-being, safety, and/or effectiveness. Evaluating the technical and personnel variance analyses, we can assess whether there are physical environmental changes that will promote improvement. These changes might include changes to temperature, lighting, humidity, noise control/hearing protection, etc.

Step 10: Implementing, Iterating, and Improving

At this point, it is desirable to execute or implement the work process changes prescribed, design interfaces, and allocate functions. Because in most cases, the macroergonomic team will not have the authority to implement the changes suggested by the analysis, proposals with recommendations for change may be required for presentation within the formal organizational structure. Such proposals should be consistent with the macroergonomic principles and should include, for example, both technical and social objectives, participatory ergonomics, and multidimensional performance improvement predictions. Based on the proposal feedback, modifications to the proposal may be necessary, which will require a return to the earlier step that represents a challenged assumption or design.

This process is iterative. For continuous improvement (i.e., the STS principle of "incompletion"), evaluations may suggest a return to an earlier step in the process for renewed partial or full redesign. Once the proposal for change is accepted and implementation begins, regular reviews of progress are required. To complement the weekly formative evaluations performed by the implementation team, semiannual formative evaluations should be performed by an objective outside party. This evaluation should be presented to the implementation team, and a constructive dialogue about expectations and progress-to-date should be conducted.

OFFICE WORK SYSTEMS

Background

Intensive use of computer and information technology for long periods of time in the office workplace calls for an examination of employee performance and possible work-related health and stress problems. The nature of these technologies and the work environment influence employee musculoskeletal fatigue, discomfort, and pain (e.g., Cakir, Hart & Stewart, 1978; Smith & Carayon, 1989; Sauter, Dainoff & Smith, 1990) and may have an adverse effect on work effectiveness, stress, and work-related musculoskeletal disorders (WMSDs) (e.g., Bongers, De Winter, Kompier & Hildebrandt, 1993; Kuorinka & Forcier, 1995).

The method we describe in this section for work system process assessment has most often been applied to office work systems (e.g., Kleiner, 1998; Robertson & O'Neill, 1987; Robertson & Rahimi, 1990) but potentially could be used for industrial work systems.

Office systems, or "knowledge work systems," by their nature are complex to study because of their multivariate characteristics. Understanding the potential causal factors of problems arising from poorly designed office work systems requires a macroergonomic approach using systems analysis and processes. The traditional micro-ergonomic approach to office work environments emphasizes the microscale aspect of the work environment at the individual workstation (e.g., keyboard, screen image, manuscript placement, etc.) (Robertson & Rahimi, 1990; Sauter, et al., 1990). If we broaden this approach to a macroergonomic level, other factors such as: physical environmental variables (e.g., layout, storage, and adjustability); work tasks; work organization; organizational structure; technology characteristics; and psychosocial variables are recognized to impact individual, group, and organizational performance (O'Neill, 1998; Robertson & O'Neill, 1997, 1999; Robertson & Rahimi, 1990; Smith & Sainfort, 1989; Swanson & Sauter, 1999). Understanding the interrelationships of these elements of the work system, and their effects on health, stress, work, and organizational goals, is necessary to provide a comprehensive perspective of office environments.

With a macroergonomic approach, systems tools and processes are used to develop strategic criteria for the design of an office work system in which the social and technical subsystems are aligned with the organizational mission. The strategic criteria are akin to a "performance specification" for the entire work system. Using these strategic criteria, problem-solving tools, such as ergonomic analysis for workstation or job design, can be employed to enhance organizational effectiveness. Prior research suggests that there is the potential for translating the findings from such a macroergonomic approach into relevant office system design and organizational planning interventions (e.g., see Chapter 3; also, Hendrick, 1997; O'Neill, 1998; Robertson & Courtney, 2001; Robertson & O'Neill, 1997; Robertson & Robinson, 1995).

This section provides an overview of a systems analysis tool (SAT) that employs a macroergonomic approach to identify problems and probable causal factors related to office work environments. SAT also provides a process for developing strategic and systematic solutions for solving problems arising in a computer-intensive office environment. The interventions and solutions presented here are based on 20 years of research and case studies drawn from a global sample of office environments. In addition, SAT guides the development of alternative solutions, a cost/benefits assessment of each alternative solution, information to support the selection and implementation of solutions, and feedback and measurement of changes in worker performance due to the process. By applying SAT, the interaction and fit of the office worker, job tasks, organizational structure, and physical environment are aligned around organizational goals, resulting in increased worker health and effectiveness.

Applying the Systems Analysis Tool— Understanding Office Work Systems

In applying systems analysis, the level of analysis is targeted at the business unit or departmental level, where the business mission and objectives are identified, as well as individual and group goals in support of that mission.

The systems analysis tool we describe here is frequently used in business and industry and based on the work of Mosard (1982) and the early work of Hall (1969). There are six analytic steps:

1. Defining the problem: the problem factor tree
2. Setting objectives, developing an evaluation criteria table, and developing alternatives: the objective/activity tree
3. Modeling the alternatives: the input-output flow diagram
4. Evaluating alternatives: the criteria scorecard: cost/benefit analysis
5. Selecting an alternative: the decision table: selecting an alternative based on future conditions
6. Planning for implementation, evaluation, and modification: scheduling and project management flow

Step 1: Defining the Problem—The Problem Factor Tree

Complexity is inherent in performance problems, therefore, a systematic analytic approach to issue definition is necessary. From data previously collected or recently observed, a problem factor tree (PFT) is constructed which identifies the problem, subproblems, and causal factors, including their interrelationships (Hall, 1969; Mosard, 1982).

A completed problem factor tree depicts a hierarchical, logical structure identifying the elements of the problem (Fig. 4.4). To develop a PFT, work issues and problems are precisely stated and associations between the issues are linked using causal and logical inference (Mosard, 1982). The lower-level causal factors in the PFT are selected as being potential contributors to the major, higher-level problem. Feedback loops may also be incorporated. In a recent project, we created a high-level problem statement for an office work system, stated as: "An increase in turnover, lost work days, and claims disabilities, and a decrease in performance and effectiveness of office workers associated with occupational stress from office technologies and office system design" (see Fig. 4.4). Implicit in this statement, we hypothesized that this high-level problem was caused by occupational stress from office technologies and workplace design. Further research revealed two underlying subproblems: (1) psychosomatic stress, and (2) physiological stress. In addition, two subproblems relating to psychosomatic stress were found: (1) psychosocial disturbances, and (2) perceived lack of environmental control (see Fig. 4.4).

Figure 4.4 also shows other potential causal factors that contribute to the defined problem. These include lack of job content, poor job design, and lack of flexible workstation design. These are depicted at the base of the hierarchical structure. For the potential causal factors in job content and job design, several individual and group level factors are identified and shown in the middle of the tree. Elements of poor job design may include teamwork or collaborative problems at the departmental level, such as cross-functional teams, or at the individual level of a small, informal group gathering. Job content and job design are viewed as the main element of the social subsystem. Potential casual factors in the technical subsystem are primarily located in the workstation design, layout configuration of the workspace, and VDT design. Of the many subproblems identified in regard to physiological stress, two are of paramount importance: visual and musculoskeletal discomforts. Other potential factors are shown, contributing to both of these subproblems, including lack of flexible workstation design, improper VDT screen design, and improper layout and design of the workstation (see Fig. 4.4).

The problem factor tree depicts the associated factors and subproblems that represent the integration of micro-ergonomic and macroergnomic office system components and subsystems, including organizational and job design issues. This initial step of the system analysis lends itself to developing an understanding of the technical and social subsystems as well as the work environment subsystem. Other robust and valid systems tools based on sociotechnical processes can be applied together with this particular system approach and model (e.g., Hendrick, 1997; Smith & Sainfort, 1989).

Step 2: Setting the Objectives and Developing Alternatives—The Objective/Activity Tree

Objectives and evaluation criteria are developed for use in selecting the best alternative to address the causal factors. An objective tree is a hierarchical, graphical structure of objectives that address previously identified problems (Mosard, 1983). The tree is created by identifying major needs, goals, objectives, and subobjectives. In Fig. 4.5, the objective tree is depicted in the upper half of the figure and shows the major goals as: (1) decrease lost work days, claims disabilities, and turnover, and (2) increase performance and effectiveness of office workers by alleviating occupational stress from office system design and technologies.

The objective tree shown in Fig. 4.5 addresses the inherent psychological and physiological health problems identified in the system. Alternatives are defined as a specified set of activities, tasks, or programs designed to accomplish an objective (Mosard, 1982). The objective/activity tree in Fig. 4.5 illustrates four alternatives, listed as A through D, and an associated set of activities for each. Four subobjectives are defined: (a) redesign the job and job content, (b) ergonomically redesign the workstation and environment, (c) to train managers, and (d) ergonomically redesign the workstation and environment, train managers in job and workstation redesign and awareness, and write an office ergonomics manual. The degree of interaction among objectives, constraints, and the persons/groups involved in the process should also be analyzed (Mosard, 1982). Hybrid alternatives may be created that incorporate the best features of any of the initially identified alternatives. Alternative C is a hybrid alternative representing one of the many possible combinations of common activities. The four alternatives depicted were derived from case studies, field research, and longitudinal studies representing typical approaches implemented by companies to achieve the objective listed at the top of Fig. 4.5 (e.g., O'Neill, 1998; Robertson & Robinson, 1995; Robertson, Robinson, O'Neill & Sless, 1998; Smith & Sainfort, 1989).

After the objectives and alternatives are selected, a preliminary decision criteria table is developed. This table is used to evaluate the "usefulness" of each of the alternatives as methods for accomplishing the objectives. Decision criteria typically include risks, costs, expected benefits, and measure of effectiveness, based on short-term and long-term perspectives (Mosard, 1982). Table 4.4 presents an example of a preliminary decision criteria table for evaluating office ergonomic alternatives.

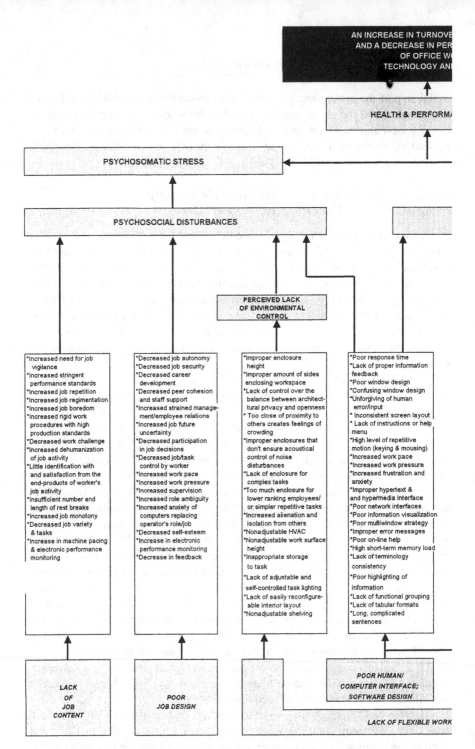

FIG. 4.4. A problem factor tree depicting micro- and macroergonomic subproblems and associated factors with office work systems. The major problem is defined at the top with the

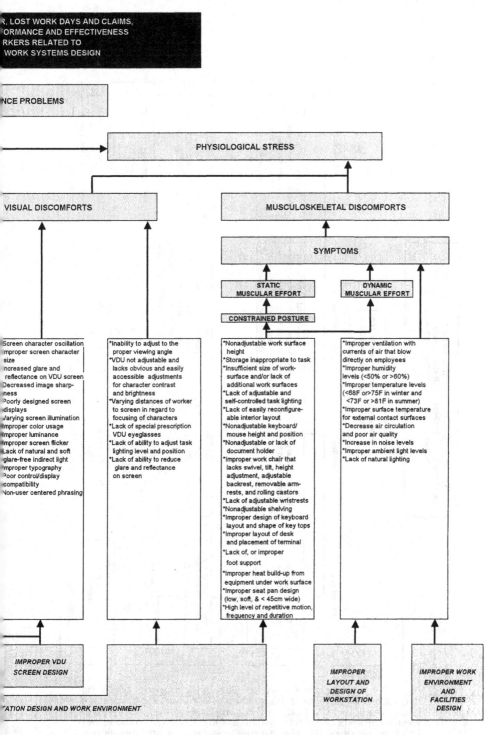

NCE PROBLEMS

PHYSIOLOGICAL STRESS

VISUAL DISCOMFORTS

MUSCULOSKELETAL DISCOMFORTS

SYMPTOMS

STATIC
MUSCULAR EFFORT

DYNAMIC
MUSCULAR EFFORT

CONSTRAINED POSTURE

Screen character oscillation
mproper screen character
size
ncreased glare and
 reflectance on VDU screen
Decreased image sharp-
ness
Poorly designed screen
displays
Varying screen illumination
Improper color usage
Improper luminance
Improper screen flicker
Lack of natural and soft
glare-free indirect light
Improper typography
Poor control/display
compatibility
Non-user centered phrasing

*Inability to adjust to the
 proper viewing angle
*VDU not adjustable and
 lacks obvious and easily
 accessible adjustments
 for character contrast
 and brightness
*Varying distances of worker
 to screen in regard to
 focusing of characters
*Lack of special prescription
 VDU eyeglasses
*Lack of ability to adjust task
 lighting level and position
*Lack of ability to reduce
 glare and reflectance
 on screen

*Nonadjustable work surface
 height
*Storage inappropriate to task
*Insufficient size of work-
 surface and/or lack of
 additional work surfaces
*Lack of adjustable and
 self-controlled task lighting
*Lack of easily reconfigure-
 able interior layout
*Nonadjustable keyboard/
 mouse height and position
*Nonadjustable or lack of
 document holder
*Improper work chair that
 lacks swivel, tilt, height
 adjustment, adjustable
 backrest, removable arm-
 rests, and rolling castors
*Lack of adjustable wristrests
*Nonadjustable shelving
*Improper design of keyboard
 layout and shape of key tops
*Improper layout of desk
 and placement of terminal
*Lack of, or improper
 foot support
*Improper heat build-up from
 equipment under work surface
*Improper seat pan design
 (low, soft, & < 45cm wide)
*High level of repetitive motion,
 frequency and duration

*Improper ventilation with
 currents of air that blow
 directly on employees
*Improper humidity
 levels (<50% or >60%)
*Improper temperature levels
 (<68F or>75F in winter and
  <73F or >81F in summer)
*Improper surface temperature
 for external contact surfaces
*Decrease air circulation
 and poor air quality
*Increase in noise levels
*Improper ambient light levels
*Lack of natural lighting

IMPROPER VDU
SCREEN DESIGN

IMPROPER
LAYOUT AND
DESIGN OF
WORKSTATION

IMPROPER WORK
ENVIRONMENT
AND
FACILITIES
DESIGN

ATION DESIGN AND WORK ENVIRONMENT

subproblems and associated factors shown below, indicating
the hierarchical, logical structure of the encompassing problem
elements.

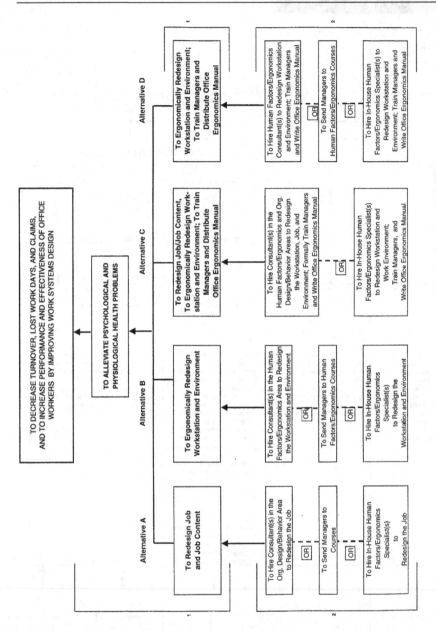

Fig. 4.5   An objective/activity tree illustrating four sets of alternatives/activities or organizational interventions for office systems. The lower-level objectives contribute to the attainment of the middle-level objectives, which in turn contribute to the upper-level needs or goals. The alternatives/activities are specified to identify the necessary expertise to design the office work systems intervention.

TABLE 4.4

Preliminary Decision Criteria (Short and Long Term)

Scope	Risk of Failure (Obstacles to Success)	Costs: Short & Long Term	Benefits/Effectiveness
1. Will help the entire organization internally and externally	1. Employees resistant to change	1. Materials & equipment	1. Increase in morale; quality of work life
2. Long-term program effectiveness and efficiency	2. Employees inability to learn new skills	2. Resources (physical and human)	2. Decrease in absenteeism (e.g., lost work days)
	3. Lack of management acceptance and support	3. Decrease in performance	3. Decrease in worker's compensation cases (medical-related costs)
	4. Training programs lag behind current knowledge	4. Ergonomic job design and training programs	4. Increase in product quality; customer satisfaction
	5. Lack of active employee participation	5. Decrease in product quality	5. Increase in productivity
	6. Current job design inappropriate for new workplace designs	6. Production downtime	6. Decrease in turnover
	7. Stress from changes in the environment/technology/ work organization		7. Decrease in job stress
	8. Fail to utilize training manuals and materials		8. Increase in health and well-being
	9. Lack of management reinforcement and feedback		9. Decrease in number of claims
			10. Decrease in insurance costs
			11. Increase group collaboration

This table presents a preliminary decision criteria which will be used in step 4 to evaluate each alternative. Creating this table begins the process of identifying critical decisions criteria.

Step 3: Modeling the Alternatives—
The Input–Output Flow Diagram

A predictive model representing either each alternative set of activities or representing the entire system is developed next. The purpose of this approach is to allow alternative configurations to the systems to be analyzed. The system element interrelationship and/or gross resource requirements are depicted in order to determine the effectiveness of each alternative set of activities (Mosard, 1983).

Modeling techniques such as flowcharts, simulation, and systems dynamics modeling may also be used in this step. The model used in this analysis is an input-output flow diagram (Mosard, 1982). In an input–output flow diagram, the inputs consist of people, resources, and information, and the outputs are the results and products of the system. These outputs can, in themselves, become the sources of inputs to other subsystems, and thus extend the diagram to represent fully the entire system being analyzed. Figure 4.6 illustrates the two phases of this model: the redesign phase, and the operation phase. Inputs for the redesign phase include contributions from two general areas, human resources and finance. In the human resources component, individuals such as industrial psychologists, managers, employees, human resource managers, ergonomists, facility operations, trainers, and health and safety managers are included. The activities for each redesign project or program are listed for phase 1, shown in the left input box in Fig. 4.6. At the end of the redesign phase, the outputs become the inputs for the second phase, the operation phase; for example, in the job redesign program, managers and employees have acquired new skills and the jobs have been analyzed and redesigned. The managers and the employees will now interact within their own work systems, and the results of these interactions are presented in the outputs (e.g., increase in performance, decrease in job stress, decrease in injury claims and worker's compensation cases). Overall, these outputs fall into three categories: changes in employee and group behaviors, organizational factors, and reduction in business costs.

Step 4: Evaluating the Alternatives—
The Criteria Scorecard: Cost/Benefit Analysis

The alternatives are evaluated by measuring each alternative set of activities and then comparing them against each other. For this process we refer to the major decision criteria and the preliminary decision table. These criteria generally include: project cost, risk of failure, effectiveness, and benefits for all appropriate future conditions. An evaluation criteria "scorecard" is developed for use in evaluating and comparing alternatives (Table 4.5). The scorecard incorporates the preliminary decision criteria defined in Step 2, and is biased to take a long-term perspective.

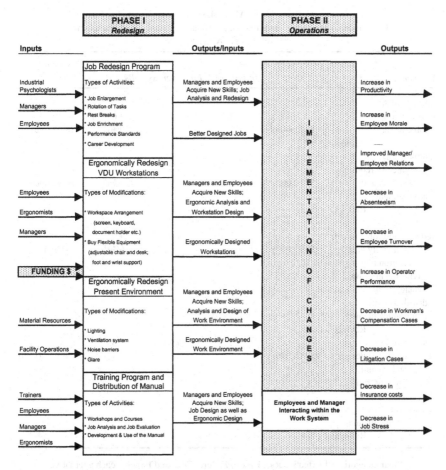

FIG. 4.6. A model presenting an input-output flow analysis. The two phases of this model for alternative C are the redesign phase and the operation phase. Alternative C is: "To redesign job/job content; to ergonomically redesign workstation and environment; to train managers; and to distribute an office ergonomics manual."

The model in Fig. 4.5 shows that four alternatives were created based on projects conducted with numerous business organizations (Robertson & O'Neill, 1995; Robertson, et al., 1998; Smith, Sauter & Dainoff, 1990). To complete a comprehensive cost-effective analysis for these alternatives, an economic advantage analysis was conducted. The economic advantage analysis identifies costs and benefits to provide an economic analysis for each alternative (Robertson & O'Neill, 1995; Robertson & Rahimi, 1991). This analysis provides additional detail, which links the costs for each proposed program or change to the physical

TABLE 4.5
Evaluation Criteria Scorecard: Economic Advantage Analysis of Each Office
Work System Alternative[1]

Alternatives	Project Cost	Risk of Failure	Effectiveness	Benefits	Overall Rating
Alternative A:					
Redesign job/	−3	−2	+5	+6	6
job content	($175,000)[2]			(10%)[3]	(Moderate)
Ratings:					
Alternative B:					
Ergonomically redesign	−4	−3	+5	+6	4
workstation and	($370,000)[2]			(10%)[3]	(Moderate)
environment					
Ratings:					
Alternative C:					
Redesign job/job content,	−8	−1	+9	+9	9
ergonomically redesign	($590,000)[2]			(26%)[3]	(High)
workstation and					
environment, train					
managers and					
distribute office					
ergonomics manual					
Ratings:					
Alternative D:	−6	−2	+7	+8	7
Ergonomically redesign	($440,000)[2]			(17%)[3]	(Moderate–High)
workstation and					
environment, train					
managers and					
distribute office					
ergonomics manual					
Ratings:					

[1]In the four categories of Effectiveness, Risk of Failure, Cost, and Benefits each alternative (A thru D) was subjectively rated on a scale ranging from 0 to 10 or 0 to −10. Because Risk of Failure and Project Cost are potentially negative characteristics, a rating scale ranging from 0 to −10 was used. A 0 rating indicates no cost; a −10 rating indicates a high cost. Effectiveness and Benefits are positive outcomes and therefore were rated on a 0 to 10 scale. A 0 rating indicates a low level of benefit/effectiveness; a 10 rating indicates a high level of benefit/effectiveness. The scores from each of the four categories were summed to determine the "Overall Rating" from each alternative. Along with the numerical ratings. The table shows subjective descriptors of the ratings in parentheses below each number in the "Overall Rating" column. Along with the numerical ratings, the table shows the potential percentage increase in employee performance for each alternative in parentheses below the ratings. Along with the numerical ratings, the table shows the approximate "Project Cost" in dollar amounts for each alternative in parentheses below the ratings.

[2]Costs are calculated per 100 employees.

[3]Benefits are evaluated in terms of percent increase in worker and group performance.

work environment to the potential gain in productivity or reduction in health costs resulting from the program investment.

The economic advantage analysis also identifies costs and effectiveness metrics, including: (1) human resource costs (e.g., compensation, salary, turnover, and absenteeism, workers' compensation costs, injury costs); (2) facilities costs (work environment) (e.g., rentable space, operating costs, annual facility costs, furniture investments, technology and information investments, work environment strategy costs, construction costs); and (3) effectiveness measures—organizational (e.g., process efficiency, work environment changes, customer satisfaction, space utilization), unit/department (e.g., product development time, successful projects, number of customers) and group and individual (e.g., error rates, amount of completed work, quality). These cost/benefit metrics are used for each proposed alternative to determine the economic advantage of each alternative or program/ activities and may be expressed as a percentage of annual compensation demonstrated over "x" years.

Step 5: Selecting an Alternative—
The Decision Table

The value of each alternative program or intervention is established using weighted values. These weighted values are based on objective and subjective measures. A decision table similar to Table 4.6 is used to structure this information, with alternatives listed on the "Y" axis and the future condition statements on the X axis. A probability is determined for each of the stated future conditions. An example of a future condition could be the probability and level of funding for the program. Each alternative would be evaluated in terms of level of funding and the probability of being funded.

Step 6: Implementation, Evaluation,
and Modification

A schedule and sequence of tasks, responsibilities, and requirements is developed for the implementation activities. This schedule might include a contingency plan with scheduled decision points and decision responsibilities. There are several scheduling techniques that are available that can be used as well as software applications for creating databases containing effectiveness measures, costs, and other pertinent metrics collected during the systems analysis.

One such implementation scheduling technique is the DELTA chart. The DELTA chart is a multipurpose flowchart that can be used to show aspects of major decisions (timing, responsibilities, and choices), events (timing and duration), and activities (timing, duration, and responsibilities) for the implementation time frame. Figure 4.7 presents a DELTA chart for the implementation of alternative C. This chart provides an overview of plans for initiating and implementing the selected

TABLE 4.6

Decision Table: Selecting an Office Work Systems Alternative Based on
Future Conditions

Future Conditions (e.g., funding) Probability of Funding	High Level of Funding 0.5	Moderate Level of Funding 0.25	Low Level of Funding 0.25
Alternative A: Redesign job/job content	4	2	2
Ratings*			
Alternative B: Ergonomically redesign workstation and environment	5	1	3
Ratings*			
Alternative C: Redesign job/job content, ergonomically redesign workstation and Environment; train managers and distribute office ergonomics manual	1	6	7
Ratings*			
Alternative D: Ergonomically redesign workstation and Environment; train managers and distribute office ergonomics manual	2	4	5
Ratings*			

*The numbers indicate the selection preference rankings based on the weighted criteria and overall rating score from Table 1. Each alternative was subjectively rated on a 0 to 10 scale, where a 0 rating indicates a low preference and a 10 rating indicates a high preference.

activities. This figure shows specific requirements for the project leaders (PL), including ergonomists, human resource managers, industrial psychologists, and facilities and real estate managers. Senior managers in the finance department, including other top managers, are shown as the major decision-makers and leaders.

Several other activities also occur in this step to define the process of providing feedback to the decision-maker in the company regarding the results of the program. Using information gathered from the evaluation and feedback processes, selected modifications and changes to the program also occur. This process is a continuous feedback loop of applying the systems analysis approach to solve problems and to measure the effectiveness of the program continuously.

CONCLUSION

A macroergonomic approach to office work system analysis identifies the salient variables that influence employee health, stress, and performance and creates a systematic method for addressing these issues in a cost-effective manner.

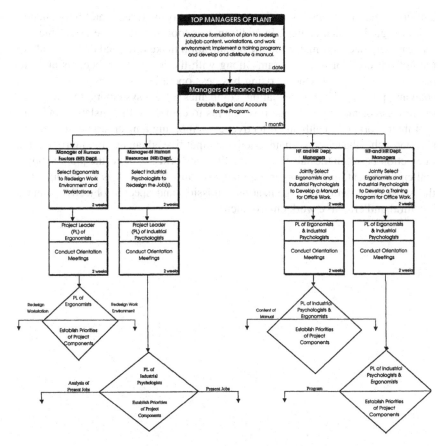

FIG. 4.7. This figure depicts a DELTA chart for the implementation of alternative C. This chart provides an overview of plans for initiating and implementing the selected activities.

Typically, companies focus on microscale solutions, but these tend to result in technology-centered designs, a "leftover" approach to function and task allocation, and a failure to consider an organization's sociotechnical characteristics and integrate them into its work system design (Hendrick, 1986a, 1986b). Such work system design deficiencies, in turn, lead to poor performance, job dissatisfaction, and increased risk to worker health and safety.

The important benefits of the macroergonomic approach are the understanding of work systems and the effective integration of micro-ergonomic and macroergonomic approaches for solving organizational problems related to work system environments. The systematic approach represented by the tool we outlined in this section can assist ergonomists, health and safety professionals, and organizational planners in the process of analysis, design, and implementation of organizational change. In addition, this tool integrates the use of metrics that can help decision-makers in the organization make informed decisions about the upfront costs of new programs along with the longer term benefits and impacts to the company. Due to rising business operating costs, as well as recent state and pending federal legislation, companies need a systematic tool to aid in the process of making strategic investments to comply with legislation and support their workers' health and effectiveness. Applying a macroergonomic analysis approach enhances an organization's competitive advantage by improving the interaction between the subsystems and fit of the office worker, the job, the physical environment, and the organizational structure. The result is an alignment of the subsystems with the organizational mission and specific objectives, and a healthful and effective work environment.

5

Other Methods for Assessing Work Systems

Hal W. Hendrick
University of Southern California

CLASSICAL METHODS

A number of the methods that historically have been used in organizational studies have been adopted for use in macroergonomic research and application. The most widely used of these classical methods are the *laboratory experiment, field study, field experiment, organizational questionnaire survey, interview survey,* and *focus group* (Hendrick, 1997; Hendrick & Kleiner, 2001). A brief description of these six methods, as modified for macroergonomic use, follows. Actual applications of a number of these methods are demonstrated in the studies summarized in Chapter 7.

Laboratory Experiment

In most scientific disciplines, the laboratory experiment is the researcher's primary approach for determining causal relationships. This method is called the *laboratory experiment* because the researcher typically conducts the study in some form of laboratory where the conditions can be controlled and changes precisely observed and measured. The researcher manipulates some independent variable of interest while controlling or otherwise discounting the effects of other variables that might affect the dependent variable. If a change occurs to a dependent variable of interest, the cause of the change can more clearly and efficiently be attributed to the independent variable than with other research approaches, such as those described below.

Advantages. The laboratory experiment has several major advantages. First, the researcher can be *fully prepared* to make accurate observations. Second, the researcher, can *repeat* the observations under the same conditions to see if the results are the same. If the results replicate, then the researcher can have more confidence that there is, in fact, a cause and effect relationship. Third, it enables other experimenters to *duplicate* the study and make an independent check of the results. Finally, the researcher can *vary the conditions* systematically and note the variation in results.

Disadvantages. The laboratory approach has one major disadvantage. That is its *lack of realism.* In real life, things do not occur in isolation. There often are *interactions* among the causal variables that determine the effects on the dependent or outcome variables. The sterile environment of the laboratory thus may change the nature not only of the situation, but also of the *motivation* of the participants, which may be quite different from that in the field environment. Also, laboratory subjects may not be representative of the actual worker populations they are intended to represent (e.g., using college sophomores as laboratory subjects).

Macroergonomic Applications. In actual practice, the magnitude of the realism shortcoming depends on the nature of the problem and one's ability to simulate actual field conditions in the lab. In macroergonomic applications, we typically attempt to simulate an actual work system environment in the laboratory that enables us to systematically manipulate either work system or sociotechnical variables of interest (e.g., complexity, centralization, formalization) and systematically observe and record the impact on various performance variables of interest (see Chapter 7).

Field Study Method

The field study method also is variously referred to as systematic or naturalistic observation and as real-life research. These terms, taken together, provide a good description of this approach. It involves going out into the *field* to *systematically observe* events as they occur *naturally* in *real life*.

Advantages. The field study method's primary advantage is *realism*. By observing things as they occur naturally, the researcher avoids the sterility and artificiality of the laboratory. However, a caution is in order here: It is important for the researcher to recognize that his or her very presence changes the situation, and thus can affect what happens. The researcher has to take great care to be as unobtrusive as possible. But because of this method's realism, when the researcher is able to establish cause and effect relationships we can have high confidence in the practical usefulness of the results.

Disadvantages. There are two major disadvantages to the field study method. First, the researcher has to *wait* for things to occur naturally. Consequently, the observation process may take a long time and incur considerable expense before any cause and effect relationships can be established. Second, the researcher may have to observe things occurring naturally many times under various conditions before extraneous variables can be eliminated as causal factors and the true causal variables teased out and their interactions identified.

Macroergonomic Applications. A common use of the field study approach in macroergonomics is to examine existing performance records for a given work system, and the specific conditions or nature of the work system under which those performances occurred. Through this approach, it often is possible to identify problem areas with the work system's design that are amenable to macroergonomic intervention—and to gain insight as to the nature of the macroergonomic intervention that is needed.

Field Experiment Method

Perhaps the most widely used of the classical methods in macroergonomic interventions is the field experiment. The field experiment differs from the field study in that, instead of passively observing events as they occur naturally, the macroergonomist acts as a *change agent*: Selected variables are deliberately and systematically manipulated and the effect on the outcome or performance variables of interest are observed.

Advantages. To a large extent, the field experiment combines the advantages of the field study and the laboratory experiment while overcoming their major disadvantages. Like the laboratory study, the researcher deliberately manipulates the dependent variable(s) of interest, thus overcoming the field study's problem of having to wait for things to occur naturally. In addition, the field experiment gains the field study's advantage of *realism*. In comparison with the field study, the field experiment is more efficient in terms of time and related costs.

Disadvantages. By "artificially" causing a desired change to occur, the researcher or change agent may be introducing extraneous variables that influence the effects of the change. In macroergonomic field experiments, one such problem can be the employees' perceptions of the purpose of the change. How employees perceive the intervention can alter how they respond to it and their related motivation. Also, how changes are implemented can sometimes determine the success or failure of the intervention.

Sometimes, the cost of using workers in field experiments or quasi-experiments may be seen by organizations as too prohibitive. For this reason, in presenting such a proposal to management, it is important to include a sound cost-benefit analysis.

Macroergonomic Applications. One effective way of using a field experiment is as a follow-on to an initial field study. An initial field study can suggest possible dependent variables that then can be manipulated in a field experiment; for example, a review of performance records and a sociotechnical analysis of the work system's organizational structure, such as described in Chapter 3, may indicate where a structural change is needed. The change then can be made and the impact on the work system's performance observed. Sometimes this work system change will be done in a particular part of the organization to test it. If the change proves effective, it then will be implemented on a larger scale. Because of the potential problems of employee acceptance and support of the change, the field experiment often is combined with a *participatory ergonomic* approach to effect the intervention (as noted in Chapter 2, employees are far more likely to accept and support changes where they are actively involved in the change process).

Organizational Questionnaire Surveys

The most widely used method for evaluating organizational functioning and identifying deficiencies is the written survey questionnaire. In fact, most large corporations and many smaller ones use some form of periodic questionnaire survey. Generic survey questionnaires have been developed and validated for assessing such things as organizational climate, supervisory or managerial behavior and practices, meetings, job satisfaction, and attitudes about such things as safety and fringe benefits. In addition, organizations develop their own surveys, or supplemental questions, to assess specific issues of interest, such as attitudes about recent or proposed changes in policies, practices, or procedures.

A widely used variation of the questionnaire survey is the *survey feedback* method. It differs from the pure survey questionnaire method in that data gained from the survey is summarized and subgrouped statistically by organizational level, department, project, etc., and then is "fed back" to the individual organizational units. Typically, the unit manager, who has been trained on how to conduct survey feedback sessions, then participates with the unit's employees in interpreting the data and deciding on what changes, if any, should be made to improve unit effectiveness.

Constructing an organizational survey that can yield valid, useful data is a very tricky process and requires the help of a trained expert, such as an industrial/organizational psychologist. Even with expert help in constructing the questionnaire, it is essential to pilot test the survey with a small, representative group to make sure that both the instructions and questions are understood as intended. Invariably, refinements will be required as a result of the pilot test.

Advantages. A major advantage of the questionnaire survey method of gathering information is *anonymity* and *confidentiality.* Questionnaires can be administered in a manner that assures the anonymity and confidentiality of the individual employee's responses. Therefore, employees are likely to feel freer to express their true opinions without fear of reprisal. As a result, important information that is not part of the existing management information system can be obtained.

The second major advantage is that the opinions and perspectives of a large number of employees can be obtained relatively quickly and inexpensively. This enables the researcher to ensure a large representative sampling of the employees, if not the entire workforce.

A third advantage is *replication*: The same questions can be asked over time and changes in employees' responses measured. Finally, questionnaire data can be analyzed *quantitatively.* Quantitative analysis enables the data for various work system units, levels, or subgroups to be compared, and areas of both consensus and variation to be noted. Such comparisons often yield valuable insights into problems and potential solutions.

Disadvantages. A major potential disadvantage is *ambiguity of purpose.* Upper management, staff groups, and/or consultants developing the survey may have difficulty in reaching consensus about the purposes of the survey, its content, or the procedures for analyzing, interpreting, and using the results. In addition, management and/or unions may be so committed to existing policies and practices that the survey may be seen as a meaningless exercise. Still another potential disadvantage is *organizational disturbance.* The survey may call attention to quiescent issues, thus needlessly reawakening them. Alternately, the survey may induce unrealistic expectations about subsequent actions. Either of these consequences could disrupt normal operations.

Finally, it is important to note that survey data can establish *correlations* but cannot establish *causation.* Thus, the survey may help in recognizing relationships between things and suggest ways of improving functions, but it will take actually making a change and seeing what happens to establish causation (i.e., a laboratory or, more likely, field experiment).

Macroergonomic Applications. Organizational questionnaire survey data can be an excellent tool for quickly and inexpensively identifying symptoms of work system design problems and where within the work system the problems exist (i.e., in specific units or throughout the entire work system). In some cases, a problem may be identified in a particular unit or level of the organization. A questionnaire survey then can be developed and administered to see how widespread the problem is throughout the organization. A questionnaire survey also can be used in its survey feedback form as an integral part of the participatory ergonomic process. The results can provide management, employees, and the ergonomist with data to help in identifying work system design problems and/or suggest what needs to be done to improve the work system's functioning. Finally, questionnaire surveys can be used as a pre- and postmacroergonomic intervention measure to assist in evaluating the effect of work system changes. These data also can suggest where further redesign may be needed.

Interview Surveys

Interviews constitute yet another frequently used method for gathering information in organizations. In organizational research, the most commonly used interview approach is the *stratified semistructured interview survey*. The interviews are stratified in that a representative sample of employees from all levels and units of interest are interviewed. Because of both time and expense, it usually is impractical to interview all employees, or even a large sample of them. Accordingly, a comparatively small, stratified sample offers a practical means of ensuring that the data gathered will be representative of the full group of interest. The approach is semistructured in that the interviewer has prepared key questions that he or she uses to tap the topics of interest systematically. The interviewer then will ask follow-up questions ad-hoc, depending on the response the interviewee gives to each of the key questions.

Advantages. In comparison with the questionnaire survey method, interviews have several major advantages. First, they enable the interviewer to observe both verbal and nonverbal responses to questions. Sometimes, the nonverbal responses provide even more important data than do the verbal. A second major advantage is that the interviewer can follow up on initial responses. This enables the interviewer to explore unanticipated responses or unanticipated issues, or go into a given issue in greater depth. Finally, if the interviewer is successful in establishing a good rapport with the interviewee, the interviewee may become more ego involved in the process and put more thought and effort into his or her responses than would be the case with an impersonal questionnaire.

Disadvantages. Two major disadvantages of interviews are *time* and *cost*. These not only include the time and costs associated with conducting the interviews and analyzing the results, but also the time and costs of training or hiring skilled interviewers. Whereas a written questionnaire survey can be administered by a single person to a large group at once, interviews can only be done with one or a very few persons at a time by a given interviewer. Further, it typically takes an hour or more to conduct an interview to gather organizational data.

A third disadvantage is that anonymity for the interviewee is lost. Thus, the interviewee may be inclined to play it "safe" and only give "sociably acceptable" answers to questions. I would note that this problem can be minimized by (1) using skilled, trained interviewers, (2) interviewing employees on their own "turf" rather than in the manager's or interviewer's office, (3) clearly explaining the purpose of the interview, what will be done with the results, and giving assurances of confidentiality of the person's individual responses.

Finally, the appearance, demeanor, and interpersonal skills of the interviewee can bias the interviewer's interpretations of the interviewee's responses—particularly if the interviewer is untrained.

Macroergonomic Applications. Semistructured interview surveys can be a highly effective way of identifying problems with a current work system's design and/or gaining insights as to what specific kind of macroergonomic intervention might be effective. This approach also can be very useful in identifying incompatibilities between the macrodesign of the work system and the microdesign of individual jobs and/or related human–machine and human–software interfaces.

A strategy I have found to be highly useful in medium and larger organizations is to begin by using a semistructured interview survey approach to identify problems in a single plant, unit, or geographical location. These results then are used to develop a questionnaire survey that I then use to determine how widespread the problems are within the entire organization.

Focus Groups

The focus group is a variation of the interview survey approach and partially overcomes the interview's disadvantages of time and money. At the same time, it retains the basic advantages of the interview method. In its basic form, the focus group approach involves bringing persons together to be interviewed as a group about one or more specific issues.

Advantages. In addition to the savings in time and money as compared with the traditional interview survey, the focus group method has another important advantage: Hearing one group member's responses may stimulate the thinking of other participants, thus yielding additional useful data that otherwise might not have come to light.

Disadvantages. Focus group participants have an even greater loss of anonymity than in an interview in that their responses also will be heard by the other participants. Consequently, participants may be reluctant to express ideas or perceptions that they perceive will not be accepted or popular with the group. In addition, there is the possibility of a "group think" phenomenon occurring, thus resulting in distorted, incomplete, or otherwise inaccurate data.

Macroergonomic Applications. A focus group approach can be highly useful in macroergonomic interventions. Persons from the specific work system of interest are brought together and interviewed by the ergonomist about specific aspects of their work system or its sociotechnical environment. Such interviews often are conducted in the form of a semistructured interview. In other cases, a work system change is simulated and the group then is interviewed together to gain their collective perceptions or opinions about specific aspects of the change. Various other data collection methods also may be used (e.g., *Monte Carlo technique* (see Glossary); *paired comparisons,* where each item, suggestion, or option is compared with all others and, each time, one is chosen over the other).

Combining Methods

Very often, two or more of the classic techniques described above are used together in carrying out a macroergonomic analysis, intervention, and/or evaluation; for example, the ergonomist might (1) conduct a field study to observe the existing work system and related performance measures; (2) do a laboratory simulation study of a proposed work system change with some employees; then follow up by (3) conducting a field experiment where a part of the work system is modified and performance effects observed; then (4) use a focus group to evaluate the field experiment; and then (5) conduct an organizational questionnaire survey after implementing the work system redesign to obtain employee responses to the change. In addition to these classic methods, the analysis phase also might involve a systematic assessment of the work systems structure and processes, as previously described in Chapters 3 and 4. As described in Chapter 2, a participatory ergonomic approach might also be used throughout the entire macroergonomic analysis, intervention, and evaluation process.

OTHER MACROERGONOMIC METHODS

In addition to the methods described above and in the three previous chapters, there also are macroergonomic methods that are targeted at specific design purposes. These methods have been developed, refined, and validated by various university groups internationally. Several methods have been developed specifically for manufacturing work systems and are described in Chapter 11. Among other methods known to the author, three appear to be particularly useful: (1) *user systems analysis,* for determining computer hardware and software requirements; (2) *ORDIT methodology,* for determining information technology (IT) system requirements, and (3) *anthropotechnology,* for ensuring that technology, and related work systems, transferred from the culture that develops the system is appropriately modified to fit the receiving culture.

User Systems Analysis

User systems analysis (USA) was first introduced to the macroergonomic community at the first International Ergonomics Association International Symposium on Human Factors in Organizational Design and Management, Honolulu, Hawaii, in August 1984 by William Glick and Rafik Beekum of the University of Texas at Austin's Department of Management (Glick & Beekum, 1984) and Leo Brady (1984) of the IBM Corporation. As described by Glick and Beekum, USA is an approach for assessing organizational needs for information processing equipment and evaluating task and organizational design. As such, it represents an integration of the sociotechnical systems approach with social network analysis. Social network analysis is a methodology from mathematical sociology (Knoke & Kuklinski, 1982). The primary unit in network analysis is the *relation* between two people. The *content* of the relation is anything that is exchanged between the two people, including work-in-progress, work-related communications, and goods (Beekun, Glick & Carsrud, 1986). In another paper at the same 1984 symposium, a colleague of Glick's and Beekum's, Alan Carsrud (1984), noted that often, organizational changes or equipment are implemented based on analyses that do not adequately consider the interrelationships of jobs and the informational needs of people in the organization. This deficiency led to a joint research project between the University of Texas at Austin and International Business Machines (IBM) to develop USA. The goal of the project was to make the workflow diagram a more useful tool in a broad range of organizational problem solving.

Brady (1984) notes that there are four basic steps to USA:

1. Development of a "USER" (i.e., work system) environment overview scenario
2. Detailed analysis of this scenario to identify key "USER" function/tasks that might be appropriate for automation
3. Development of "USER" requirements to perform these functions/tasks
4. Establishment of key system requirements (information processing hardware and software), based on "USER" needs, integrated into the environmental scenario, with development usability criteria identified.

1. *Development of a USER environment scenario.* The goal of this step is to develop an overview of how the business is conducted today—to create a general system overview diagram or scenario. This step begins by interviewing employees

and principals to obtain high-level flow diagram information describing the business as it operates today. In support of the overview, the following need to be identified: (1) the key decision-makers, (2) influences on key decision-makers (people, events), (3) the educational level of employees, (4) the analytical skills of employees, and (5) environmental pressures on key decision-makers and those who influence decision-makers. The scenario then is completed, reviewed, and critiqued by all who had input, and refined.

2. *Key "USER" task identification.* Following completion of the overview scenario, those functions involved in a specific application are identified. Additional flow diagrams detailing the supporting tasks for these selected functions then are developed. These diagrams, prepared for the scenario for the current work system, are used to identify the following: (1) specific tasks for automation, (2) interdependence of tasks, and (3) key contributors to the tasks. To support the analysis, interviews are conducted with the "USER" population (i.e., work system's employees) to gather detailed task information on such things as individual employee tasks and dependencies, existing information to support those tasks, how the information is gathered today, who uses the data, the size and organization of the data, priority of tasks, and the identified interfaces.

3. *Develop USER task requirements.* Once the key tasks for automation are identified, many data items then are gathered to support a system design implementation. For each identified task, the following information must be obtained: (1) what is to be done, (2) who does it, (3) when does it have to be done, (4) where should it be done, (5) why is it done, and (6) how is it being done now. These data then are reviewed by the employees for accuracy and completeness.

4. *Establish key system requirements.* Based on the data collected and analyzed in the first three steps, the work system's requirements for information processing and related equipment and software can be specified. The final report will specify the functions required to be performed by the hardware/software, its added value to the work system compared to how the work system currently functions, and how the hardware/software is to be acquired and serviced.

It should be noted that this is an intensely user-centered analytical process involving employee participation by all levels of the work system. As a supplement to a macroergonomics structure and process analysis of the work system, it can be a particularly useful macroergonomic methodology for determining information processing requirements for the work system and specifying specific hardware and software needs.

I personally worked with an IBM team in using this user systems analysis methodology when my colleagues (including Ogden Brown, Jr., the author of Chapter 2) and I designed a complex new university college at the University of Denver having over 30 geographically dispersed study centers in the United States and Europe. We did so in conjunction with a sociotechnical analysis to determine our work system's structure and processes. We found the methods to be highly compatible and mutually reinforcing. The opportunity occurred in the mid-1980s when a geographically dispersed master's program in systems management was transferred from the University of Southern California (USC) to the University of Denver and was used as the core program for developing a new college. I transferred with the program for three years to serve as the college's dean during its design and initial development phase.

Compared with the program's work system as it had existed at USC, this analysis enabled us to streamline the organizational structure to be more compatible with the college's sociotechnical characteristics (see Chapter 3), improve processes, better design jobs, and make more efficient use of available technology, including computers and software programs. The college realized a 23% reduction in staffing requirements and about a 25% savings in operating expenses compared with the work system as it had existed at USC. The time required for processing student registrations, grades, and other related administrative activities for the off-campus locations was reduced from an average of 3 weeks to less than 1 week. The administrative time demands on the study center managers also decreased approximately 20%, giving them more time to devote to current and prospective students.

ORDIT Method for Organizational Requirements Definition

ORDIT methodology provides a systematic macroergonomic approach for designing information technology systems. The methodology was developed under the sponsorship of the Commission of the European Communities Esprit II Programme Project No. 2301. It was a 5-year collaborative project involving MARI Computer Systems Ltd., the University of Newcastle Upon Tyne, the Work Research Centre, Ltd., and Algotech srl, in addition to the HUSAT Research Institute at Loughborough University of Technology (Olphert & Harker, 1994).

A motivation for developing this technology was the poor success rate for IT systems during the late 1970s and 1980s. Many of these failures subsequently were attributed to problems with specifying and meeting user (i.e., work system) requirements. Conventional systems analysis activities are underpinned by a sequential "waterfall" approach that largely focuses on defining technical solutions to information processing requirements rather than looking at the wider perspective of the user organization. The resulting IT systems often fail to satisfy the needs of the work system and its human operators, even though the IT systems are technically sound. The task, then, was to develop a shared framework within which user organizations and IT system developers (including ergonomists) could communicate about common issues. The resulting product was the ORDIT methodology (Olphert & Harker, 1994).

The ORDIT methodology was developed to take account of the key principle that requirements should emerge from the exploration of alternative organizational and technical futures for the work system. The methodology provides a systematic process to support organizational requirements generation and supply associated methods and tools to support the process. The basic features of the ORDIT methodology are as follows (Olphert & Harker, 1994).

1. *A sociotechnical systems approach.* The method is oriented toward achieving joint optimization of the technical and social (personnel) subsystems. It does this by systematically identifying the requirements of the social subsystem (organizational requirements) and then exploring the implications of possible technical subsystems.

2. *A user-centered approach.* Technology is treated as a tool to serve the needs of users. The method thus requires a high degree of user involvement.

3. *An iterative approach.* The method involves an iterative cycle of specification, prototyping, user testing, and redesign.

4. *Provides choices and creates new opportunities.* The method encourages scanning for opportunity and consideration of a number of different sociotechnical options to take into account a range of possible futures.

5. *Modularity.* The overall structure of the ORDIT methodology is modular. It thus takes into account the different starting points and needs of different work systems and allows choice from a variety of individual methods and techniques. Accordingly, "route guidance" showing how and why the modules might be used in different contexts is an important element of ORDIT.

6. *Enhancing communication.* ORDIT contains a set of modeling concepts that have been found effective in enabling work system personnel and IT system designers to understand and describe an organization as a set of related work roles. Structural relationships exist between role holders, and every relationship implies a conversation and the need for information. Relationships thus impose requirements in terms of information and communication structures on any IT system.

Structure of the ORDIT Methodology.   The ORDIT methodology is structured to encourage and support the requirements generation process. It focuses around the creation of models of future possibilities that take account of the organizational and technical options. The methodology consists of two elements: a set of interrelated activities, with supporting tools and techniques, and a set of navigation guidelines for moving around these activities. There are four activities within ORDIT: (1) *scoping,* (2) *requirements definition,* (3) *option generation/evaluation,* and (4) *ORDIT modeling.* These four activities are mutually supportive and are best thought of as concurrent and iterative. The work system personnel and IT system designers are likely to move rapidly and unpredictably from one activity to another.

Issues raised at some point in the conversation may illuminate issues raised in a previous conversation concerning another activity.

A full and complete description of the ORDIT methodology and it tools is beyond the scope of this text. For details on the ORDIT methodology and its tools, see the *Final Report on Project 2301—ORDIT (Deliverable 11). Report to the Commission of European Communities, December 1993,* or contact HUSAT Research Institute, Loughborough University of Technology, Loughborough, Liecs, UK.

Anthropotechnology

Based on his extensive international experience, the distinguished French ergonomist Alain Wisner (1995) noted that when planning technology transfer projects, it is necessary to study the geographic, historical and, in particular, the ethnological dimensions. Ethnology is that aspect of cultural anthropology that deals with the comparative cultures of peoples—their mores, values, belief systems, etc. To consider more adequately ethnological variables in transferring technology developed in one culture to another much different one, Wisner (1976, 1984) developed and coined the concept of *anthropotechnology.*

As described by Wisner, the orientation and goals of anthropotechnology are similar to those of ergonomics. He further noted that the general methodology also is similar, but advocated a comparative method (Wisner, 1976). Prior to the technology transfer, a study is made of the technology presently in operation in the country that developed it, or one highly similar to it. Included is an analysis of the cultural characteristics and how they impact on the design. This is done to correct its defects and highlight them in a modified design for the buyer country. The basic method advocated for this study is *ergonomic work analysis* (EWA). Secondly, an EWA is conducted of the critical points of a similar technical system operating in the buyer country. The two EWAs are compared and implications for design modifications to the system to be transferred are noted and implemented. Finally, installation of the new (modified) technological system should be done by a mixed team of managers and operators from the two countries; this should be followed up by an ergonomist who practices the necessary EWA.

Wisner (1995) noted that the factors that influence work are too numerous for a forecast to be made, from the outside, of those that constitute a determining obstacle to successful technology transfer in the particular situation. In anthropotechnology, a search for the origin of the difficulties is made; and a tree of causes is constructed that is not limited to the technical and organizational aspects that are closest to the workstation. See Wisner (1976, 1984) for a more detailed description of anthropotechnology.

Ergonomic Work Analysis (EWA)

EWA provides an exhaustive description of the activities of certain operators or users in phases of implementation of the technical system that are considered to be critical. The full value of this detailed study is revealed when it is compared with the representation the operator or user has of his or her own activities during the same period (self-confrontation) (Wisner, 1995).

Although there are numerous variations in EWA, Wisner (1995) noted that there are some common characteristics. In principle, EWA methodology includes an examination of the technical, economic, and social conditions, an analysis of the activities—the central element of the study—the diagnosis, recommendations, modifications, simulation of the work on the modified system, and evaluation of the work in the new situation. Wisner notes that the methodology is extremely cumbersome if followed in full; but that the complete work analysis process rarely is necessary. Frequently, the analysis of activities can be reduced to a few critical points and /or only to unusual ergonomic problems. See De Keyser (1991) for a more detailed description of the methodology of EWA in English.

Integrating Macroergonomics with Anthropotechnology

From the above description, it is not difficult to see how a macroergonomic approach might be used in conjunction with anthropotechnology in carrying out technology transfer projects. Top-down approach and sociotechnical analysis methods of macroergonomics provide a means for assessing not only the culture and environment of the seller country, but also of the buyer country. This comparative use of macroergonomics fits neatly with the approach of anthropotechnology; and expands macroergonomics to more adequately include systematic consideration of the full spectrum of cultural variables. In addition, it complements anthropotechnology by initially focusing on optimizing the design of the overall work system within each culture. Anthropotechnology and EWA methods provide a more thorough means of assessing the critical dimension of the personal subsystem for work system design purposes: the ethnological characteristics of the workforce.

EWA also provides a proven and thorough means for carrying the design outcome of the overall work system down to the design and/or modification of individual jobs and attendant human–machine, human–software, and human–environment interfaces. Outcomes of EWA, in turn, may identify further modifications that are needed to refine and improve the overall work system.

Taken together, using the approaches and methodologies of macroergonomics and anthropotechnology as an integrated, human-centered systems approach would appear to afford the opportunity for overcoming the traditional design and implementation problems, noted earlier, and thus increase the effectiveness of technology transfer projects.

6

Relationships Among Job Design, Macroergonomics, and Productivity

Mitsuo Nagamachi

Hiroshima International University, Japan

INTRODUCTION

Technological advancement and human development have always shared a close relationship. This is especially true in manufacturing systems with human operators. In this chapter, the harmony between the manufacturing or technology and human subsystems is emphasized using a sociotechnical systems approach. This paper first presents examples of job design in which human factors have been applied. Second, the paper explores the relationships among micro-ergonomics, macroergonomics, and the concept of self-organizing systems. Finally, the paper demonstrates how job design and macroergonomics can be integrated to create self-organizations that lead to higher productivity. It also addresses the new trend of manufacturing systems called "cell systems."

In general, the purpose of an enterprise is to maximize productivity and to optimize profit. People within these enterprises have always been involved in advancing manufacturing technology. Taylorism (1911) and the Ford assembly lines typify systems specifically designed to maximize efficiency, economies of scale, and productivity. Method engineering has been developed in the field of industrial engineering to respond to the demands of productivity and efficiency in manufacturing. More recently, information processing technology is thought to be one of the most important tools to realize higher productivity.

The trend in advanced manufacturing systems has changed jobs into ones that are monotonous and de-skilling. The advent of computer science gave impetus to manufacturing systems that emerged from numerical and programmable machine tools such as numerical control (NC), computerized numerical control (CNC), and machining center (MC). Information technology has been tied to the demand of total productivity that led to the emergence of new styles of manufacturing, such as factory automation (FA), computer-integrated manufacturing (CIM), strategic information systems (SIS), and computer-aided acquisition and logistic support (CALS).

With the transition of manufacturing and information technology, jobs and organizational structure also changed. Jobs became simple, monotonous, and repetitive; organizations became sectional and hierarchical. These trends deprive humans of outcomes such as skill, autonomy, and self-fulfillment from the organization. Recently, there is a new trend aimed at introducing CIM systems to secure higher productivity and efficiency. The new information technology will give birth to a new style of production that will enhance the human potential for achieving higher and higher productivity.

It will be possible to integrate workplace humanization and high technology by jointly considering both human and technological characteristics. Higher productivity and efficiency can be obtained by unifying both facets of the manufacturing system. Moreover, if organizations implement macroergonomics as proposed by Hendrick (1991b) and described in previous chapters in this book, this harmonized system will generate productivity greater than CIM alone.

Humans are intrinsically self-developing. The organization, as an entity of humans, also has the characteristics of a self-organizing system that potentially can permit human self-development and adaptation to the changing external environment. Higher productivity emerges as a result of both self-developing humans and organizational needs. The concept of macroergonomics has implications for integration of self-developing humans with the self-organizing functions of the organization.

HUMAN FACTORS OF JOB DESIGN

From a human factors perspective, the problem with modern manufacturing systems is that the human often is treated like a machine or tool with hands and legs. The employers want to de-skill and specialize the worker's job to obtain high productivity. The price of this de-skilled, specialized work is an imbalance in personality development. Argyris (1957) pointed out that as long as an organization does not allow humans to have personal development in a working situation, maladjustment occurs. Therefore, it is said that de-skilled and simplified jobs lead to workers' maladjustment in the workplace as well as to poor organizational functioning (Argyris, 1957; Hendrick, 1991; Hendrick & Kleiner, 2001).

Walker and Guest (1952) clarified a relation between job simplification and workers' involvement through an interview technique. Using this technique, they found that the percentage of involved workers decreased with the level of job simplification. Kornhauser (1965) interviewed workers from automotive assembly lines about their mental health. He found that the number of mentally healthy workers decreased as the degree of machine-paced and de-skilled jobs increased.

Higher Level of Brain Activation

Principle 1: **Body movement and thinking during work stimulate human brain activity that, in turn, leads to human motivation.**

One of my studies illustrates this principle. I conducted a simulated experiment using a vigilance task as an example of simple and repetitive work (Nagamachi, 1971a, 1971b). In this experiment, a subject had to check the position of a rotating pointer and push a button just before it passed the defined area in the instrument. Each subject's electroencephalogram (EEG) and heart rate were measured during the experiment. Interestingly, while performing this simple repetitive task, the distribution pattern in EEG frequency was found to be similar for all subjects.

Figure 6.1 shows a representative graph of the distribution in EEG frequency, in 10-minute intervals after the subject started the vigilance task. The horizontal axis represents the frequency bands of EEG, and the vertical axis illustrates the percentage of frequency distribution. The number in the figure indicates the time elapsed in minutes since the task started. The large percentage of EEG beta waves at the early stage of the experiment suggests that the subject concentrated his attention on the task for the first 20 to 30 minutes of work. After the first half hour, the EEG distribution changed to the alpha wave domain, indicating a drop in brain activation level. The results of the experiment suggest that a monotonous and repetitive task will force the activation level to drop and create mental fatigue.

A similar phenomenon was found by Saito (1969, 1971). Saito measured critical fusion frequency (CFF), another index of both brain activation and fatigue. In a real work situation, Saito measured the CFF of workers who were engaged in four types of tasks: table press work, assembling parts on a round table, short cycle-time repetitive work (Type I, 20 to 60 sec. cycle-time), and very short cycle-time repetitive work (Type II, 3 to 5 sec. cycle-time). The CFF results are shown in Fig. 6.2.

CFF usually decreases with longer work hours, with the decreased percentage in CFF indicating the level of fatigue. Figure 6.2 illustrates the change in CFF with working hours for the four types of jobs. The figure implies that more simplified and paced jobs produce greater fatigue. Note that the largest decrement

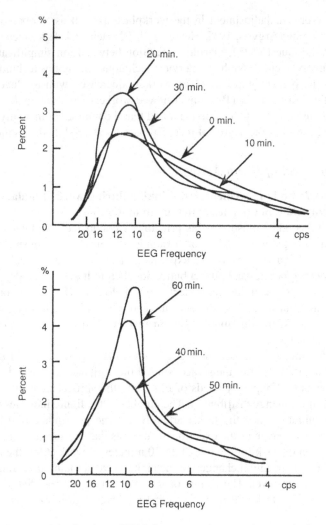

FIG. 6.1.   Distribution of EEG frequency in monotonous task.

emerged in the very short cycle-time work—the most monotonous and repetitive of these four tasks.

My colleagues and I have noted that when either the EEG or CFF indices indicate fatigue, the subjects report subjective feelings of boredom and dissatisfaction.

I have introduced a job design technology in many Japanese enterprises (Nagamachi, 1975) that originally was suggested by Davis (1966), Davis and Werling (1960), and Davis and Taylor (1972). In one case, this job design

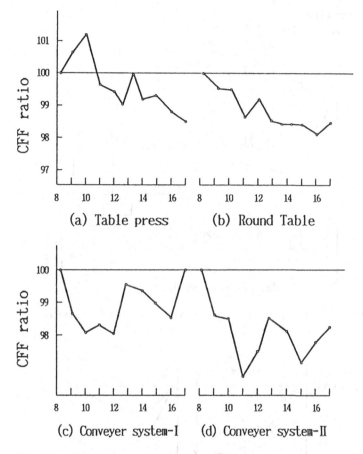

FIG. 6.2.   CFF patterns in four different jobs.

concept was introduced in a gas heater assembly line at the Nakatsugawa plant of Mitsubishi Electric Co., as shown in Fig. 6.3. The conveyor assembly system at the Mitsubishi plant was originally divided into 20 processes, as shown in the upper half of Fig. 6.3. All of these processes were combined into a one-person job for the whole assembly.

In this company, the job design is called the job enlargement system (JEL). In the JEL one-person system, a worker assembles all parts for the gas heater by her- or himself. The average job cycle-time for the 20 jobs in the original assembly line was 0.5 minutes. In contrast, the JEL one-person system took about 10 minutes for each production unit. While not increasing the total assembly time, the job redesign had a positive impact on the workers. This impact was demonstrated in a study by Kawakami and Tange (1980). They measured the

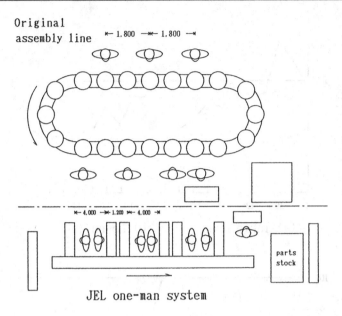

FIG. 6.3.   Original assembly line and JEL one-person system.

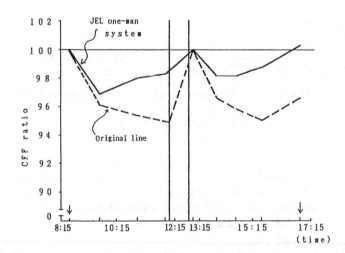

FIG. 6.4.   Comparison of CFF in original and JEL lines.

CFF for two groups of workers; one engaged in the original assembly line and another in the JEL assembly line. As shown in Fig. 6.4, the CFF values of workers engaged in the original conveyor system dropped more dramatically than the CFF values for workers in the JEL line.

The differences in CFF between both groups suggest that the JEL one-person system, constructed using the job design concept, was effective in activating the workers whose jobs were 20 times more complex than the original conveyor belt jobs.

The Work Itself

Principle 2: **A complicated job enriches a worker's skill, motivation, and satisfaction, and strengthens the worker's brain activation.**

In Herzberg's "Two Factor Theory of Motivation" (1966), the "work itself" is one of the key factors leading to the worker's motivation and satisfaction. Lawler (1969) suggested that the way jobs are designed motivates workers. Mitsubishi Electric's Fukuyama plant introduced a new JEL line for fuse-breakers to replace a traditional assembly line. As shown in Fig. 6.5, each original fuse-breaker assembly line consisted of nine processes and nine persons. The cycle-time for each process was between 40 and 62 seconds.

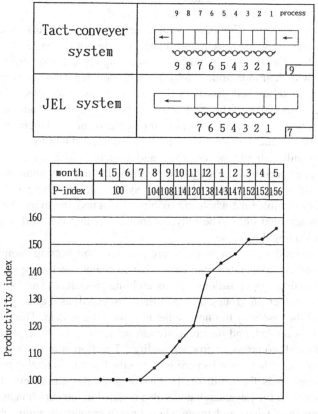

FIG. 6.5.   Increment in productivity after introducing job enlargement.

Following a participatory ergonomic approach (see Nagamachi, 1991), an industrial engineering staff interviewed the workers engaged in the original assembly lines. They found that the workers were not motivated by their work. After collecting the workers' opinions and suggestions for improving the original fuse-breaker line, the staff proposed a new assembly approach to the workers (Nagamachi, 1973a). All jobs were restructured to be enlarged in the new assembly line, and processes 2 through 7 were condensed to just one in the new line (see the middle diagram in Fig. 6.5). As a result, the new line consisted of seven persons (i.e., two persons were eliminated from the original line). In the former assembly line, the conveyor belt created a short cycle machine-paced job. In the new assembly line, the conveyor was used just as a carrier. In the new line, the workers' longest cycle-time was 4.5 minutes, or three times longer than in the old assembly line.

In comparison with the old assembly line, the new line resulted in a dramatic change in productivity. As illustrated in the lower half of Fig. 6.5, the productivity initially increased to 157% and finally to 180% of the original productivity. With the change, worker satisfaction also increased.

Fuji Heavy Industry Company, which makes passenger cars, attempted a brave challenge on work system and related job design (Nagamachi, 1977). A manager at Fuji Heavy Industry had been worrying about his subordinates' dissatisfaction with their monotonous jobs. He proposed that the workers organize a project team to solve their dissatisfaction with the assembly jobs. This project team, which was called "The New Manufacturing Steering Committee," discussed the sources of the workers' dissatisfaction and analyzed their feelings and working environment. The steering committee concluded that the job of assembling automobiles should be enlarged and enriched. The manager and the committee concluded that if each job was enlarged to the maximum, worker satisfaction level would increase. A one-man system was implemented in which each worker assembled the whole car by himself—a task involving the handling of 639 subassembled units. The only exception was that the worker got assistance when loading the engine.

In order to realize the new manufacturing system, the steering committee and assembly workers cooperated with each other to change such things as the production scheduling, tool positions, and assembling procedure. The workers were divided into six-person groups in which they discussed, as a quality circle, the efficiency of the assembly to improve the new working system. This revolutionary attempt succeeded, and the company has achieved a 100% improvement in productivity with no errors in product quality. The effort was so successful that the company was able to give its permission to the line for no final quality test.

This success was the result of the largest job enlargement effort ever attempted. Note that the job enlargements also resulted in job enrichment. It also is a good example of what can happen when a macroergonomic approach is taken to work system analysis and design.

Why Does Efficiency Increase by Introducing Ergonomic Job Design?

***Principle 3*: Flexibility and discretion in the work method result in efficiency and productivity.**

As an illustration of this principle, I carried out a simulation study in a laboratory setting (Nagamachi, 1973b, 1975). The subjects were students trained in a company like the actual workers on how to assemble a telephone. Hiroshima University contracted with a telephone company to sell the assembled telephones and was paid for the assembly. Accordingly, the subjects received an hourly wage for their work. After training, the students assembled telephones on a conveyor belt inside the laboratory.

The subjects were randomly divided into three groups: four-person teams, two-person teams, and one-person teams. Four-person teams assembled the telephone on a conveyor belt in the same way the current system operated in the real company. Two-person teams worked side by side at a round table. The one-person team had to assemble the entire telephone by himself. The students worked 8 hours each day, and the experiment continued for a month. I measured the productivity per minute and analyzed the student's micromotions through videotape recordings.

The productivity in the experiment is shown in Fig. 6.6. The figure on the left compares the productivity for the three production systems. The figures on the right show the average productivity when analyzing the micromotion study. Using the average production per minute, the one-person system was the highest, and the two-person system was second in the left figure. However, there was no statistically meaningful difference between these averages. The conveyor belt system was the lowest in productivity and was significantly lower than the other two assembly methods ($p < .05$).

It is interesting to note that a variance in productivity was quite different among the groups. As shown in Fig. 6.7, there was least variability in the

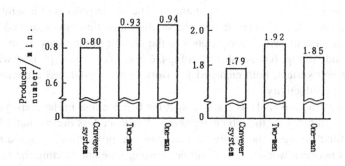

FIG. 6.6. Comparison of productivity in three simulated job systems.

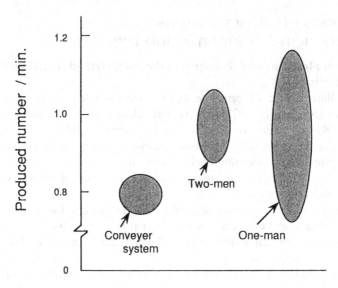

FIG. 6.7. Comparison of variation in productivity for three simulated job systems.

conveyor belt system. The variance for the two-person team was smaller than for
the one-person system, which was quite broad. From these data, we can say that
the conveyor belt system is a convenient system to maintain a consistent, but not
very high, level of productivity. The one-person system is a good technique to
achieve higher productivity, but the volume of production is uncertain and
dependent on working conditions, including a worker's skill. Among the three
production systems, the two-person system was the best job enlargement strategy. It had the appropriate flexibility required for both workers' self-fulfillment
and higher, and reasonably consistent, productivity.

The subjects' micromotions were analyzed to identify job design factors that
affected efficiency and productivity. Figure 6.8 illustrates the time required for
each of the 10 processes in decimal minutes. The whole process of assembling a
telephone consisted of 10 processes. The smallest total time was achieved by the
two-person system, which corresponds to the highest productivity as shown on
the right side in Fig. 6.6. As shown in Figs. 6.6 and 6.8, in comparison with the
conveyer belt system, both enlarged job work systems produced a greater efficiency and productivity.

The micromotion analysis clarified that the balance loss in the conveyor belt
line was eliminated by the job enlargement. A balance loss means that a loss in
process time emerges inevitably in the conveyor production by dividing the
processes between workers; for example, waiting time and starting the next job
disappeared in the enlarged jobs. More specifically, tasks 3, 5, and 8 in the four-
person system disappeared completely in the two-person system. In addition to

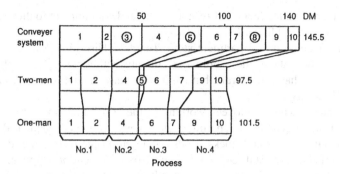

FIG. 6.8. Micromotion analysis of three simulated job systems.

the reduction of a balance loss, some other motions also were eliminated; for instance, the process time was reduced in processes 1 and 4, while processes 2 and 7 increased in time. We could not identify the eliminated motions through the film analysis. However, because of the worker's discretion in how he or she performed the job, we could imagine that two or more motions occurred simultaneously, thus reducing the total process time.

This simulation experiment shows that discretion and flexibility in operation increases worker autonomy, which, in turn, allows for the fusion of multiple motions. Thus, high efficiency and productivity result from worker autonomy.

It should be recalled that increased productivity happened in Mitsubishi Electric's Nakatsugawa and Fukuyama plants due to an introduction of job enlargement. In Fuji Heavy Industries, an even larger discretion in operation was given to the workers. Productivity improved 100% in this case. This increase in productivity resulted from the workers' self-generated ideas and suggestions that stemmed from the introduction of job enlargement and increased flexibility.

Decision Making and Responsibility

Principle 4: **The opportunity for decision making provides a worker with a psychological feeling of responsibility that leads to work motivation and satisfaction.**

When decision-making functions or authority are concentrated in the top of an organization, people in the lower levels often are unsatisfied and unmotivated. The same phenomena occur when work is divided into microtasks and people move their hands and legs like robots without thinking. In these situations, people do not participate in decision making, and they do not feel meaningfulness in their work; they cannot commit to the organization. On the other hand, if people in the organization participate in the decision making, they are able to experience utilization of their ability and judgment. As a result, this kind of situation provides people with a feeling of responsibility and commitment to the organization (Hackman & Oldham, 1976, 1980).

The labor union in a department store, Tenmaya Hiroshima Store, consulted with the author about the employees' dissatisfaction with their jobs. The Hiroshima store had low morale, but the highest sales among four stores located in Hiroshima City.

After interviewing employees working in the Hiroshima store, the situation became clear. In general, a department store has a trade management division, which annually selects and stocks goods relating to fashion, consumer demands, and taste. The division staff collects consumer preference information and makes stocking decisions accordingly. The result is good sales and organizational profit. Therefore, the authority and decision making on stocking occur at the staff level of the trade management division. The employees on the store floor feel their job is just to sell the goods that the staff selected. They thus feel little fulfillment from their work.

The solution to this problem was to provide the employees with decision-making opportunities to select and stock goods that they expected to be good sellers. This new system was named "Responsibility of Stocking Goods by Ourselves (RSO)," which is a self-order system. Because this was a very dramatic and drastic move for the employees in Tenmaya, most of the executives did not support this proposal. They were afraid that the idea could be dangerous and were anxious about losses due to the introduction of this radical proposal. They argued that not all store employees were trained as buyers. The author and the labor union proposed to the executives that the employees be trained extensively and immediately to be expert buyers. Tenmaya allowed the introduction of the RSO system only on the sixth floor of the store where a chief secretary of the union was a floor manager.

The employees were divided into small groups as in quality circles. A manager and supervisors trained them extensively on the special knowledge they would need as buyers. The workers were willing to participate in the RSO system and were eager to learn this new expertise. After the training, they were permitted to go to the wholesale stores and stock goods. They exhibited the goods in showcases that they also selected by themselves. This was the first experience for all employees in stocking goods, which represented fiscal responsibility. In the store, they welcomed the customers with a smile and tried to explain the positive features of their products. All employees were much more motivated to sell goods and were satisfied with their new jobs. They felt a strong responsibility and commitment to their jobs as well as to the organization. As a result of this effort, total sales increased dramatically and a movement for the RSO system prevailed on all floors of the Hiroshima store.

Nevertheless, the executives still were anxious about the effectiveness of the RSO system, and 3 years later the movement was completely eliminated from Tenmaya because of their misunderstanding. However, the concept of RSO was transferred to one of the largest department stores in the Tokyo area, Seibu Department Store. It attempted to introduce a variation of the RSO system. The new concept was named "Shop Master System," which implies that all management authority was provided to the master of each "small shop" inside the department store.

According to this concept, a master is responsible for managing his or her small shop, including stocking goods and selecting subordinates within the company, as well as for the profitability of the small shop. At present, the Shop Master System is one of the main department sales systems in Seibu and is prevalent in all department stores throughout Japan, ironically, including Tenmaya. The trade management division supports the Shop Master System and the overall decision-making process.

There is another case of a company in which the self-order system was introduced. Itoyokado is one of the largest supermarkets in Japan and famous for the introduction of a point-of-sales (POS) computer system in all shops. The company can access the daily sales information through the computer system. All part-time female employees are given responsibility for the self-order system. They are permitted to order goods themselves, based on hourly sales data. They manage their own shop by controlling inventory cost and by timely ordering with autonomous responsibility. Although they are part-time employees, they are not considered supplementary workers, but as key persons in the labor force. Higher organizational performance resulted from the implementation of the self-order system. The gross profit ratio in Itoyokado increased from 25.1% in 1982 to 30.3% in 1991, even though net profit already had been the highest among Japanese supermarkets. The returned goods rate dramatically dropped from 8.0 to 1.5% during this decade.

Itoyokado's introduction of the autonomous decision-making system and responsibility into the organization meant employees in this company were given the authority to decide shop control as an integral part of shop management. The organizational structure was changed to a flat hierarchy that led to easier, more effective communication. As this case demonstrated, people, if given the authority, will attempt to manage their own shops autonomously toward the desired outcome.

Self-Organization and Organizational Design

Principle 5: **The most important traits for human beings are self-development and self-realization. The organization that permits self-organization enhances the worker's self-development.**

An organization is surrounded by economic and social factors that permeate and impact it, for instance, technological progress influences the level of product development; world economic and financial conditions affect the organization's cost-cutting strategies; training and education provided to the labor force determine the development of organizations.

It is desirable that the organization solves problems on its own. The organization needs updated information about the changing factors that impact it and it needs to cultivate the people who belong to the organization to solve these problems in a timely and effective manner.

Assuming their basic physical and security needs are being met, most people in organizations want self-development and self-realization (Alderfer, 1972; Maslow, 1943). They want to know what is going on inside the organization to which they belong. They want to challenge their organizational problems. They develop on their own through their behavior in the organization and then choose a new organizational problem to solve. Self-organizing behavior focuses on solving problems originating both inside and outside the organization.

The process of organizational design can occur as a result of self-organizing change. Macroergonomics may be a special case of self-organizing design—organizational design that results from people working to solve the organization's hardware and software problems. People are able to design their organizations according to their desires and the challenging spirit and self-fulfillment that results from this activity.

In the same way, total quality control (TQC) and company-wide quality control (CWQC) can build a total quality assurance system. It can be said confidently that TQC is one way to manage the organization and guide it in the desired direction. TQC emphasizes a "market-in" philosophy to produce a quality control system. Market-in need, which reflects the customers' desires, is changeable. Accordingly, TQC must adjust to the changing direction of the customer's desire and satisfaction. Therefore, an ideal TQC structure should be a self-organizing work system design. My colleagues and I simulated the organizational characteristics of TQC through our augmented "garbage can model" (Cohen, March & Olson, 1972; Nagamachi, Kaneda & Matsubara, 1993) and discovered that both organizational performance and people's satisfaction reached to an optimal level through self-organizing behavior.

A good example of an adaptive self-organizing work system is provided by the Suntory Whisky Corporation. When Suntory decided to build its Hiroshima plant as a base for delivering its product in western Japan, the company estimated that the number of employees needed to operate this plant would be about 54. However, the plant manager estimated the number of persons needed to be 72. The plant opened with 54 employees. To meet the company's personnel requirements, the plant manager started educating and training his subordinates to strengthen their ability and, thus, increase their flexibility to accomplish the work with only 54 people. He also introduced quality circle activities based on a job design model (Nagamachi, 1973a, 1987) to enrich the employees' abilities and to expand their jobs. The quality circles discussed ways to eliminate wasteful tasks and ways to enrich their jobs to perform a larger number of tasks using a smaller number of people. In addition to this, the increased societal demand for whisky emphasized the importance of productive activity. The quality circle members solved their problems by having each worker assume several responsible roles and move to the different places as needed. They discussed ways to settle each person's multiple roles and train themselves for improving productivity in their multiple roles.

Figure 6.9 shows the change in the organizational structure of the Hiroshima plant after the self-organizing design. The chart on the left shows the original structure; the one on the right illustrates the changed structure. For instance, an electrical engineer was responsible for the maintenance of all electric equipment and systems; but now he had multiple roles, including financial, computer software, automated equipment development, and sensory test jobs. There were no distinctions between staff and workers. Each person chose the tasks that he or she felt able to do with personal motivation. Therefore, an ordinary map showing an organizational hierarchy was not applicable. The Hiroshima plant's organizational structure is complicated, as shown on the right side of Fig. 6.9.

As a result of the redesign of the work system, the original 54 employees eventually decreased to 19; but whisky production doubled. The employees were satisfied and fulfilled with the redesigned organization. I have labeled Suntory's work system an "amoeba-type organization," which implies an autonomous problem-solving group and a self-organization with the collective ability to solve problems originating from both inside and outside the firm. As noted in Chapter 3, this amoeba-type of organizational design is also called a "free form" design. The concept and activity of amoeba-type organizations has spread to all Suntory plants. As a result, these plants have achieved higher efficiency, productivity, and personnel savings, as well as increased employee satisfaction.

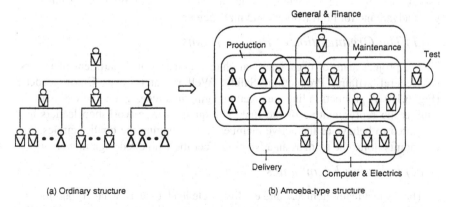

(a) Ordinary structure   (b) Amoeba-type structure

FIG. 6.9. Comparison of former and new organizational structures of Suntory Hiroshima plant.

PARTICIPATION OF WORKERS
IN JOB IMPROVEMENT

Participation of Workers in Job Redesign

Principle 6: **Workers' participation in both job and work system design covers all of Principles 1 through 5 and realizes the basis of macroergonomics in the organization.**

As also noted in Chapter 2, the most important factor for the workers is implementation of "participation" in job and work system design activities. We have dealt with the use of participation throughout this chapter. Workers are aware of what is going on in their workplaces. They know the physical and human causes related to human errors. They understand what factors lead to detrimental products. Worker participation uncovers the reasons for these incidences. This is why Japanese manufacturers use the small-group activity known as quality circle activity.

Quality circle activities begin by discussing problems in the workplace. The QC members collect the data and incidences regarding accidents, detrimental products, poor product quality, and low productivity. They look at the relationships between the factors and problematic incidences. Sometimes, a job improvement advisor or a *kaizen* consultant intervenes. They propose job improvements or job redesigns to the manager. When the manager accepts the proposal because it provides advantages to the company and workers, the manager proceeds with the job redesign with participation of the quality circles as supported by the budget.

I have introduced the formal procedure of "participatory ergonomics" to improve the whole production line in terms of organizational change (Nagamachi, 1986, 1988, 1990a, 1990b, 1990c, 1991). Quality circle activities improve a job, or group of jobs, while macroergonomic applications to work system design are useful for improving the whole production line. The combination results in a great advantage for both workers and management.

(1) The Organization of Project Teams

The project team of organizational change consists of department managers of General Affairs, Production, Safety, Welfare, and Local Union leaders (Fig. 6.10). A production line manager engaging in a redesigned line will participate as the representative of the line. The quality circle workshop leaders are permitted to behave as the active members for managing the quality circles. An ergonomist joins the team as an advisor to steer the whole redesign movement.

(2) Learning Ergonomics

The project team members and quality circle leaders learn what ergonomics is and how to redesign their production lines in terms of ergonomics, especially macroergonomics. If needed, they learn some industrial engineering to aid in surveying problems and redesigning lines in terms of micro-ergonomics.

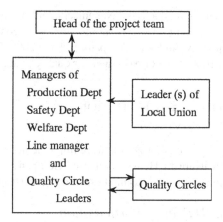

FIG. 6.10.   Organization of a project team in participatory ergonomics.

(3) Workplace Survey Activities

The project team determines the mission of the project, such as reducing accidents, enhancing product quality and production, or reducing manufacturing cost. With regard to participatory ergonomics, the mission will realize workers' welfare and higher productivity. Solution of the former usually leads to improvements in workplace safety, product quality, and productivity as well (Imada & Nagamachi, 1990). Based on the mission, and with the help of quality circles, the team surveys what is going on in the production lines and collects data about problems. The team analyzes the data in terms of macroergonomics and, where appropriate, statistical methods (e.g., multivariate analysis) to determine the macroergonomic factors affecting the work system.

(4) Ergonomic Strategy for Improving the Lines

The team chooses the ergonomic and organizational strategies to solve the factors hidden in the workplace. The quality circles are used to discuss whether the redesign strategy is the best one or if other approaches can better solve the ergonomic problems. After getting opinions from the quality circles, the project team revises the solution strategy according to macroergonomics.

(5) Ergonomic Review

After the work system redesign is conducted, the team reviews and evaluates the redesigned production lines from the viewpoint of macroergonomics and enterprise strategies. I have obtained good outcomes using participatory ergonomics and a macroergonomic approach in a number of organizations, including almost zero-level accidents in many manufacturing companies, and more than a 50% improvement in productivity (Nagamachi, 1995). An intervention using participatory ergonomics and a macroergonomic approach, conducted in an automotive company, Fuji Heavy Industry Company, led to a 200% increase in productivity and achievied a 100% level of passenger car quality (Nagamachi, 1998).

Implementation of Participation in Job Design Activities

Formal participatory ergonomic programs were implemented in the Daikin Company (Nagamachi), Mitsubishi Heavy Industries Company (Nagamachi, 1985, 1991), Nissan Oppama plant and Nissan Diesel Company (Nagamachi, 1993).

Job Redesign in Daikin by Participation

Because it is thought that workplace improvement was important for its workers' welfare, Daikin attempted to implement participatory ergonomics to improve the production lines for air conditioners. The Daikin Kanaoka plant was selected as a model plant for introducing participatory ergonomics, and a project team was established. Nagamachi's working posture measurement was selected as a tool to survey ergonomic problems. A higher workload posture was selected by this analysis and was implemented via micro-ergonomic redesign to reduce postural loading. In the second redesign stage, robots were used in the lines to improve jobs that otherwise were hard to change. This participatory ergonomics effort resulted in a more than 40% increase in productivity due to improved working postures (Nagamachi, 1985).

Change in Production Line in Nissan Diesel

Nissan Diesel is a truck maker that has a strong relationship with Nissan. A 10-ton truck assembly line in the Ageo plant was chosen for introducing participatory ergonomics and for realizing work motivation. Because the plant had quality circles, a project team was organized as shown in Fig. 6.10. First, the team surveyed the model line to find the ergonomic problems, and it interviewed the members of quality circles about their awareness of problems happening in the line. About 60 items were listed and categorized as A, B, C, in which A meant easy problems to be solved quickly; B meant problems taking several weeks for solution; and C implied big problems requiring a long time period and large investment to solve. The A problems were solved quickly by the team and the circles. B and C category items were analyzed using participatory ergonomics. One of these ergonomic problems, a propeller-shaft mounting task, was selected to be solved with regard to safety workload. A long, heavy shaft used to be attached to an engine by four workers. The assembly work forced the workers to bend their backs under the truck chassis. This work is very dangerous and four workers already had suffered back injury. The manager ordered the production technology department to devise a jig that would cause the shaft assembly to be lighter. However, because the workers did not participate in planning the jig, they did not want to use it. After organizing the project team, a design of a supporting jig for the propeller-shaft began in the quality circles with the help of the production technology department. Because introducing the new jig was thought to be insufficient to solve the ergonomic problem, the project team discussed a complete change of the truck production line. The team consulted with management to change the line and decided to use a big machine to reverse the truck frame. As it was reversed, a propeller-shaft was inserted through the frame from the overhang. After this work, the truck frame was returned to the original position with the propeller-shaft under the chassis. At the assembly line, a worker easily inserted the propeller-shaft to the engine using the new jig devised by the team.

The participation of the workers in redesigning the production line resulted in reduced accidents and increased productivity. The workers were proud of the redesigned production system because of their participation. The project team painted this area purple, so it could be easily noticed as the participation area.

NEW TREND IN PRODUCTION SYSTEMS

The Japanese worker's wage is comparatively high. Therefore, Japanese enterprises attempt to construct factories in Asia, where the workforce is easily hired at low cost. However, after a decade of this overseas cost-reduction strategy, the Japanese noticed that there was not much profit benefit and that there were chronic problems with punctual delivery. Additionally, Japanese workers desired greater satisfaction with their work and self-development through the organization. Recently, employers have endeavored to realize benefit for both management and the workers by introducing a work system design called the "cell production system." Basically, this is the new name for a one-person system described earlier.

For example, NEC Nagano (Nippon Electric Company, Nagano Company) studied the conveyor belt work of ten, five, and three person groups, and a one-person system. Experiments resulted in 150% productivity for the one-person or cell system, as compared with the baseline conveyor belt system (Nagamachi, 1996, 1997, 1998). The cell system implies that a worker assembles the whole product. In NEC Nagano, one worker assembles a whole personal computer. It is named "Yatai," which is something like a hot dog stand. In NEC Nagano, the conveyor lines were removed completely, thereby preventing the managers from returning to conveyor belt production. Many factories now have changed from production lines to the cell system. These companies include NEC Yamagata, Toshiba, Kyushu Matsushita, Alps Electric, Sony, Minolta, and Olympus. The Japanese have sought the highest efficiency of production for a long time, but now they understand the macroergonomic implication that realization of workers' motivation leads to organizational advantages as well (Fig. 6.11).

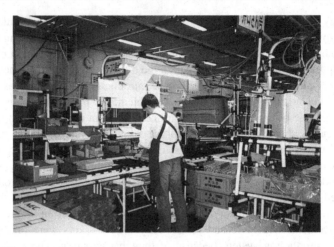

FIG. 6.11.   Scene of a "cell production system" at NEC Nagano.

FIG. 6.12.   Two-person work system at the Volvo Uddevalla plant.

Volvo in Sweden has conducted an interesting experiment in Uddevalla in which two workers assembled a passenger car together. Ellegard and her colleagues (Ellegard, Engstrom & Nilsson, 1992) investigated production with terms of 20, 10, 8, 4 and 2 workers. In Uddevalla, two workers brought the parts from the parts yard and assembled a whole car by themselves. Volvo obtained good outcomes from the redesigned production system (Fig. 6.12).

CONCLUSIONS

Production systems have a long developmental history. Initially, production was performed by only one person. But civilization and the technical revolution led to the modern assembly line system while, in turn, resulted in monotonous and repetitive work for humans.

Humans are primarily self-developed and seek autonomy. Humans want to be satisfied by work. But management's desire for efficiency and productivity forced the use of mechanistic production systems. These systems resulted in monotonous and repetitive work, often causing mental stress and physical malfunction in the workplace. One solution was to redesign the production system and related jobs using a macroergonomic perspective, approach, and methodology that followed the concept based on the six principles aforementioned.

In Japan and other countries, there are researchers and practitioners using macroergonomics in the business and social fields. One recent phenomenon is the introduction of cell production systems. The cell does not mean only a one-person system, but a small and independent or autonomous group activity. As such, it represents both job enlargement and enrichment over classical production system jobs.

We can find implementation of macroergonomics in the most advanced manufacturing systems. For example, Suzuka Fuji Xerox attempted to introduce the computer-integrated manufacturing system in its copy machine production lines. They wanted the most efficient manufacturing with low cost. However, the lead-time in this factory increased by 2 or 3 months due to malfunctions of the CIM system. Taking a macroergonomic approach, the workers proposed to management an autonomous group manufacturing system in which (1) each section of manufacturing plans the production of advanced machining and (2) the leaders of each section communicate with each other to integrate the whole production system. The redesigned work system decreased lead-time to 2 or 3 days, and the customers were satisfied.

This case demonstrates the introduction of a cell system in an organization. Each cell is an independent and autonomous part of the organization, with the cells having an integrating function as in an organism. Macroergonomics recognizes that, like organic systems, sociotechnical systems are comprised of cells. When these cells have autonomy yet act in harmony, the whole is greater than the simple sum of the parts.

Macroergonomics offers a powerful concept, approach, and technology for seeking an integration of humans with advances in manufacturing technology. Although manufacturing technology is changing and advancing at a very fast speed, a utilization of macroergonomics will give us the best solution and benefits in any stage of manufacturing and organization.

7

Laboratory and Field Research in Macroergonomics

Brian M. Kleiner

Virginia Polytechnic Institute and State University

INTRODUCTION

This chapter complements Chapter 1 and subsequent chapters that have defined and detailed macroergonomics as a discipline or subdiscipline of ergonomics. It also follows Chapter 5's treatment of various methodological approaches. Specifically, we demonstrate through several examples that empirical research that is distinct from related disciplines can, and should, be conducted to further develop our knowledge of work systems. This knowledge should lead to better design and improvement of work systems.

THEORETICAL FRAMEWORK: A GENERAL RESEARCH MODEL

Any scientific discipline should have an underlying theory base. As described in Chapter 1, much of macroergonomics's theoretical underpinnings are derived from sociotechnical systems theory. Macroergonomics is concerned with the analysis, design, and evaluation of work systems. In Chapter 1, we operationally defined work as any form of human effort or activity, including recreation and leisure-related activities. Sociotechnical systems can be simple or complex. As illustrated in Fig. 7.1, a work system is comprised of two or more persons interacting with some form of job design, hardware and/or software, an internal environment, an external environment, and an organizational design (i.e., the work system's structure and processes). Job design includes work modules, tasks, knowledge and skill requirements, etc. The hardware typically consists

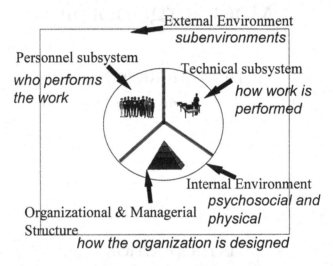

FIG. 7.1.   Basic conceptual model for a work system.

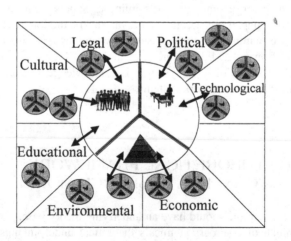

FIG. 7.2.   The work system's external environment.

of machines or tools. The internal environment consists of various physical parameters, such as temperature, humidity, illumination, noise, air quality, and vibration, and psychosocial factors. Included are political, cultural, and economic factors. The organizational design of a work system consists of its organizational structure and the processes by which the work system performs its objectives.

The external environment consists of those forces that enter the organization to which the organization must respond, as illustrated in Fig. 7.2. The work system can treat its external environment as a source of inspiration or provocation, as symbolized by the two-way arrows. In the former scenario, the work system is positively motivated by the dynamics or challenges in the environment and is proactive in response. In the latter, work systems are reactionary and negatively disposed to their environments. As a general principle, the external environment is the most influential component in the work system.

METHODOLOGICAL TRADE-OFFS
AND CONSIDERATIONS

As illustrated in Table 7.1, surveys of natural work teams or others in organizations can be beneficial. Although causal information is not obtained through survey research, correlation and other summary data are possible and useful; for example, the researcher can compare organizations and/or groups through

TABLE 7.1
Research Methods With Associated Trade-Offs

Advantages	Method	Disadvantages
• Easy to administer • Correlational statistical inquiry • Identify research variables	Survey	• Correlational data • Attitudinal data • Self-report data
• Cause and effect statistical inquiry • Access to subjects • Internal validity	Lab experiment student subjects	• External validity • Time consuming
• Cause and effect satistical inquiry • Internal validity	Lab experiment expert subjects	• Subject reliability • Time consuming
• External validity	Quasi-experiment field study	• Internal validity

administration of the same instrument. Alternatively, longitudinal affective changes can be evaluated within groups. Relative to the other methods presented, surveys are inexpensive and straightforward to administer. Disadvantages include several reliability and validity risks. The reliability and validity of self-report data can be compromised with a poorly designed or administered survey. Also, sample sizes above 50% are difficult to achieve.

In macroergonomic research, surveys have been quite useful at the front end of empirical studies to identify appropriate constructs or factors for further investigation. Also, at the back end of an empirical study, surveys can be used to poll the population to which a study is attempting to generalize in order to validate laboratory results.

In laboratory experiments, by contrast, conditions and factors can be manipulated and controlled. Subjects, representing some population, are randomly selected from a population and assigned to conditions. Independent variables are manipulated and their effects on dependent variables are measured.

Advantages include internal validity, given the researcher's ability to control the environment. This control includes the ability to design and control a task, reward structure, and manipulation and measurement of the factors of interest. Also, when students are recruited as subjects, there is typically a large pool available.

Whereas an advantage of this method is internal validity, the trade-off is with external validity or generalizability of results. Laboratory experiments are often criticized for lack of task and/or subject realism. Tasks may look nothing like the work process or task they attempt to emulate, and subjects often do not have the training or experience comparable to the workers they are representing. As illustrated in Table 7.1, an alternative is to recruit actual experts from the field or sample from the actual population of interest. This has the effect of improving external validity. However, in the experience of the author, there is a logical trade-off. Because these types of subjects are participating in the research as an extracurricular activity, their reliability is often questionable. Also related is the need to control what is measured and, thus, the construct of actual interest might not be accessible. This is certainly the case currently with many researchers attempting to measure, and thus quantify, such constructs as shared mental models within teams or groups. Also, oftentimes, especially when using within-subjects designs, relatively few subjects are employed (compared to, for example, the number of subjects accessed through a survey). In macroergonomics, researchers can build and test theories using the laboratory experiment.

A quasi-experiment in the field seeks to build on the advantages of using actual representatives from a population. Here, the researcher conducts the research in a field environment using representative subjects. This is typically a quasi-experiment because random selection is not possible. The researcher may not have control over random assignment to conditions either.

The chief advantage with this method is the high degree of external validity achieved. However, as illustrated in Table 7.1, internal validity is compromised due to lack of control. It is quite difficult to control and therefore explain the effects of extraneous variables or threats to the experiment.

In macroergonomic research, we have found it useful to conduct quasi-field experiments following laboratory experiments to validate findings in the laboratory with actual representatives of the population of interest.

Although beyond the scope of this chapter, there are other research methods used in macroergonomic research, each with its own advantages and disadvantages. Some of these were described in Chapter 5.

In macroergonomics as well as other ergonomic research, each method described can be applied, but each comes with its own trade-offs. What we have found most useful is triangulation, or combining methods so they complement one another. As a general example, one may begin with a survey of an actual work system (or by using one of the naturalistic methods presented) to ascertain hypotheses about certain factors' causes and effects. This can be followed by a laboratory study with either student or expert subjects to confirm the predicted relationships among factors. Once established in the controlled laboratory setting, these relationships can be validated in the field through a field experiment. For illustrative purposes, we describe briefly some examples of the major methods covered.

SURVEY RESEARCH FIELD STUDY EXAMPLE

Background

A study was designed to quantify empirically a construct found in the sociotechnical systems literature called "joint optimization." In theory, joint optimization suggests the need to attend to both the personnel and technological subsystems in work system design (Pasmore, 1988), but pragmatically, does this mean 50% technological and 50% personnel, or is some other ratio required?

Participants in this field study included 92 full-time, first-level managers (85 male and 7 female) in North America. Typically, these managers were foremen or supervisors, and their job duties included such titles as production lab supervisor, quality control engineer, maintenance supervisor, manufacturing supervisor, warehouse or logistics supervisor. Eleven of the organizations classified their primary function to be production, and only one company classified its primary function to be distribution. The number of managers from the companies were: American facilities ($n = 4$) and Canadian facilities ($n = 8$).

The survey utilized was derived in part from a commercially available instrument designed to assess the extent to which sociotechnical characteristics are evident in an organization. It was predicted that managers who spent 50% of their time attending to each of the personnel and technological subsystems would demonstrate higher performance than managers who did not divide their time equally.

A survey instrument was administered that measured the factors necessary to evaluate time allotment to the personnel and technological subsystems. The level of joint optimization was measured by each supervisor's perception of 20 critical sociotechnical system state characteristics such as the extent to which technology supported tasks and the extent to which personnel were important. Then, time allotment to subsystems was operationally defined and measured by each supervisor's responses to items about time spent on tasks in the technological and social subsystems. Organizational value of time use was estimated by each supervisor's perception of certain constructs related to time such as the organization's orientation toward scheduling and deadlines, autonomy of time use, awareness of time use, synchronization of tasks, and future orientation (Schriber & Gutek, 1987). Finally, department performance was evaluated by both the first-level supervisor's and plant/warehouse manager's scores on 18 items that estimated performance in such areas as departmental effectiveness, efficiency, and quality of work life.

Results

A positive relationship was observed between the level of joint optimization and department performance (see Fig. 7.3). Departments with lower joint optimization scores received lower performance scores.

Performance scores for supervisors with extremely low and extremely high scores in the data were compared to the performance scores given by their evaluating manager. First-level supervisors who divided the percentage of their time 40/60, 50/50, or 60/40 between the technological and personnel subsystems scored higher on both joint optimization and department performance than those with extreme allocations. When comparing spending more time on the personnel or technological subsystem, it was found that more time on the technological subsystem (i.e., 60/40 split) tended to result in the highest performance and joint optimization scores. Interestingly, high performing first-level supervisors who had a 50/50 time allotment received performance scores from their evaluating managers that were the most consistent with their own evaluation of departmental performance. The 50/50 ratio also resulted in the same number of departments with high performance and higher levels of joint optimization as the 60/40 split (50/50: 10 of 15 managers; 60/40: 10 of 16 managers).

In terms of differences between manager type, downstream managers in the work process reported higher levels of joint optimization than their suppliers, what were called by the researchers, "transformation managers" (Table 7.2).

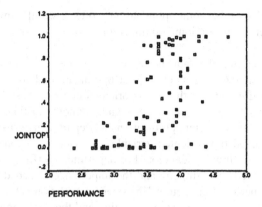

FIG. 7.3. Joint optimization and department performance (adapted from Grenille, 1997).

TABLE 7.2
One-way ANOVA of Joint Optimization Using Manager Type as the Factor

Source	SS	Df	MS	F	Sig.
Between Manager Type	.631	1	.631	4.418	.038*
Within Manager Type	12.701	89	.143		
Total	13.332	90			

* Significant at $p = .05$.

There were also no significant differences between the percentage of time spent on the personnel and technological subsystems for each level manager type.

The first-level manager's increased focus on time-oriented constructs such as time allotment to skill development and compensation and customer needs and strategic planning, in conjunction with the department's value of awareness of time use, autonomy of time use, and future orientation were significant predictors of level of joint optimization as well as departmental performance. Increased focus on time-oriented constructs such as schedules and deadlines (the importance of staying on schedule) and future orientation (importance of teamwork within time constraints) had negative relationships with department performance and level of joint optimization, respectively.

At the department level, the manager's ability to maximize sociotechnical system characteristics in the technological subsystem was a significant predictor of level of joint optimization. Managers with higher STS technological scores also tended to have higher joint optimization scores. Overall, downstream managers had higher levels of joint optimization than transformation (upstream) managers.

As illustrated earlier, the data showed a strong, positive association between department performance and level of joint optimization. This relationship was supported by the finding that three time-oriented constructs (skill development and compensation, awareness of time use, and autonomy of time use) were significant predictors of both joint optimization and department performance. Along with the three shared time-oriented variables, joint optimization was predicted by the department's future orientation. For department performance, time allotted to customer needs and strategy and the importance of schedules and deadlines completed the set of predictors. The strong correlation between department performance and the level of joint optimization and the overlapping set of time-oriented predictor variables supported the finding that time allotment can be used to operationalize and quantify the construct of joint optimization. This study illustrated the feasibility of taking theoretical constructs from the macroergonomic and sociotechnical systems literature, such as joint optimization, and empirically quantifying them. It also illustrated the feasibility and usefulness of collecting data in actual organizations. The downside of this type of research resides in the correlational nature of survey studies. As such, definitive causal relationships could not be validated.

LABORATORY RESEARCH STUDY EXAMPLE: USING STUDENTS AS SUBJECTS

Background

In agile environments, resources can be directed at technical alternatives such as the engineering process methodology used or the use of computer-support technology. Alternatively, managers can direct resources at social alternatives such as increasing the number of personnel on the project. Sociotechnical systems theory and its construct of joint optimization (Pasmore, 1988) suggests that there is an optimal combination of personnel and technological alternatives for performance. This research investigated the relationship among the important *design process*, *personnel*, and *technological* issues that must be considered by those responsible for organizations performing the design of complex work systems (Meredith, 1997). This research considered two technical considerations: (1) the overall design process methodology—concurrent engineering versus sequential engineering, and (2) whether to use or not use computer-supported collaborative work technology, and one social consideration: team size, because agility emphasizes a collaborative design process whereby engineers are integral participants in making decisions. Specifically, the study investigated whether large teams of six persons would be more effective and efficient than small teams of three persons.

The first research question asked how *design performance* is affected by engineering methodology, group size, or computer support. The second research question asked how *process time* is affected by engineering methodology, group size, or computer support. Research question 3 asked how *process cost* is affected by engineering methodology, group size, or computer support. Research question 4 asked how the *satisfaction of group members* toward the engineering design decision process is affected by engineering methodology, group size, or computer support. Research question 5 asked if there was an *optimum combination* of engineering methodology, group size, and computer support that creates the greatest technical and social outcome.

The experimental design of this research was a $2 \times 2 \times 2$ factorial, between subjects design. Large teams, consisting of six engineering students, and small teams, consisting of three engineering students, were given a set of requirements to design a transportation system that moved a payload from one point to another. Each team was asked to: (1) develop a design concept, (2) develop a detail design in the form of engineering drawings, (3) manufacture the system based on their design products (e.g., drawings and specifications) using plastic LEGOS™, and (4) test the system to determine if it met the design requirements.

Results

There were two levels of the engineering methodology variable: sequential and concurrent engineering. An ANOVA showed no significant difference between the effect of concurrent engineering and sequential engineering on design performance. There were two levels of the group size variable: small groups consisting of three people and large groups consisting of six people. The design performance of small groups was significantly greater, $F(1,32) = 13.14$, $p < 0.001$, than that of large groups, as shown in Table 7.3.

There were two levels of the computer-supported cooperative work variable: groups that used groupware in conceptual design and those that did not. An ANOVA showed no significant difference between the effect of using or not using groupware.

Given that design performance is the ratio of system effectiveness and life-cycle cost, these variables also were analyzed. While there were no main effects for the variables with respect to system effectiveness, there was an interaction among engineering methodology, group size, and the use of computer-supported cooperative work, $F(1,32) = 4.51$, $p < 0.01$, as shown in Table 7.4.

The interaction of these variables is shown in Fig. 7.4.

TABLE 7.3
Design Performance ANOVA Table

Source	Df	SS	MS	F
Eng Method	1	0.0714	0.0714	0.20
Comp support	1	0.3404	0.3404	0.96
Group Size	1	4.6717	4.6717	13.14*
EM*C	1	0.7102	0.7102	2.00
EM*GS	1	0.7981	0.7981	2.24
C*GS	1	0.0004	0.0004	0.00
EM*C*GS	1	0.2176	0.2176	0.61
Error	32	11.3808	0.3557	
Total	39	18.1906		

* Significant at $\alpha = 0.001$.

TABLE 7.4
System Effectiveness ANOVA Table

Source	Df	SS	MS	F
EM	1	18,966	18,966	0.42
C	1	6,528	6,528	0.14
GS	1	75,951	75,951	1.67
EM*C	1	116,964	116,964	2.57
EM*GS	1	13,506	13,506	0.30
C*GS	1	65,529	65,529	1.44
EM*C*GS	1	205,492	205,492	4.51*
Error	32	1,456,586	1,456,586	
Total	39	1,959,522		

* Significant at $\alpha = 0.01$.

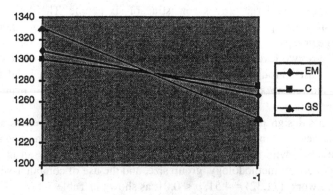

FIG. 7.4.   System effectiveness interaction.

For life-cycle cost, group size was significant, $F(1,32) = 12.44$, $p < 0.001$, as shown in Table 7.5.

There were no significant effects of engineering methodology on process time. However, the mean process time of sequential groups was greater than for concurrent groups.

There were no significant effects of group size on process time. There were no significant effects of computer-supported cooperative work on process time.

There were no significant differences between the process costs of groups using sequential versus concurrent engineering. However, the process costs for large groups was significantly higher than for small groups, $F(1,32) = 128.70$, $p < 0.001$, as shown in Table 7.6.

TABLE 7.5
Life-Cycle Cost ANOVA Table

Source	Df	SS	MS	F
EM	1	44	44	0.00
C	1	44,489	44,489	0.60
GS	1	918,090	918,090	12.44[*]
EM*C	1	180,634	180,634	2.45
EM*GS	1	55,801	55,801	0.76
C*GS	1	21,437	21,437	0.29
Error	32	2,360,781	73,774	
Total	39	3,581,275		

[*] Significant at $\alpha = 0.001$.

TABLE 7.6
Process Cost ANOVA Table

Source	Df	SS	MS	F
GS	1	1,408,501	1,408,501	128.70[*]
C	1	149,818	149,818	13.69[*]
EM	1	6,656	6,656	0.61
EM*C	1	37,088	37,088	3.39
EM*GS	1	7,290	7,290	0.67
C*GS	1	27,878	27,878	2.55
EM*C*GS	1	9,181	9,181	0.84
Error	32	350,201	10,944	
Total	39	1,996,613		

[*] Significant at $\alpha = 0.001$.

The process cost of computer-supported groups significantly exceeded the cost of noncomputer-supported groups $F(1,32)=13.69$, $p < 0.001$. Finally, member satisfaction differences failed to reach significant levels (Meredith, 1997).

This study illustrates the feasibility of simulating complex organizational processes such as concurrent engineering in the laboratory. It also illustrates the efficiency and effectiveness of training student subjects to reach a level of performance aptitude sufficient to perform in a technical experiment. In contrast, the following study illustrates the use of professional experts as subjects in a laboratory experiment.

LABORATORY RESEARCH STUDY EXAMPLE: USING EXPERTS AS SUBJECTS

Background

As stated previously, a work system is comprised of two or more people (personnel subsystem) interacting with hardware or software (technological subsystem), an internal and external environment, and within an organizational design. Communication processes are essential to the personnel subsystem. When the people in the work system are distributed, then technological support is required to create a "communication pipeline" for effective communication and decision making (Andre, Kleiner & Williges, 1998). Sometimes, as in the case of hazardous environments, technological subsystem elements do not necessarily support, but can potentially interfere with communication and the work process and thus adversely affect work system performance outcomes.

A study was conducted to understand how chemical protective equipment and clothing (e.g., mission oriented protective posture (MOPP) equipment) affects work system process and team performance. There are five levels of MOPP. The level of protection depends on the type and number of protective garments worn. Table 7.7 indicates the various clothing configurations at each MOPP level. The protective garments include overgarment, overboots, mask/hood. MOPP 0 consists of battle dress uniform (BDU) only and includes no protective clothing. Levels 0 and 4 represent minimum and maximum protection, respectively.

Team performance was operationally defined as the individual and collective behaviors of two or more individuals carried out to achieve a common goal. It is the sum of tasks performed by individuals and the coordinated actions of team members required to reach a shared goal. The measurement of team performance included both individual tasks and collective behaviors, such as communication

TABLE 7.7
MOPP Levels

MOPP Level	Overgarment	Overboots	Mask/Hood	Gloves
0	None	None	None	None
1	Worn*	Carried	Carried	Carried
2	Worn*	Worn	Carried	Carried
3	Worn*	Worn	Worn*	Carried
4	Worn closed	Worn	Worn closed	Worn

* Overgarment and/or hood worn open or closed based on the temperature.

and coordination, in order to reflect the processes involved in team performance accurately (Shiflett, Eisner, Price & Schemmer, 1985).

It was understood that the effects of MOPP on team performance can improve performance assessment, system design, and training for the military. In addition, understanding the effects of chemical protective equipment and clothing on team performance generally is critical to the performance assessment and training of many types of teams. While many team performance research studies have focused on military teams, the results of this study are particularly applicable to other civilian team-oriented operations (e.g., fire-fighting teams and chemical or nuclear disaster clean-up teams). Finally, protective clothing in uncontaminated environments, such as manufacturing "clean rooms," may impact team performance and production throughput and quality control. Studying performance outcomes alone cannot improve team process performance. This study attempted to determine which team processes, if any, were affected by MOPP during a cardiopulmonary resuscitation (CPR) task and a spinal injury management (SIM) task. Furthermore, this study investigated the performance effects of MOPP based on the role assigned (e.g., team member vs. team leader). The researchers also tried to determine which processes were related to performance outcomes and to validate previous research that has identified critical team processes that differentiated successful teams from unsuccessful teams (Grugle, 2001).

Results

The results of this study did not find a degradation of team process performance when teams wore MOPP 4 as compared to MOPP 0. The subjective ratings of performance degradation on the post-experiment questionnaire supported this result by suggesting that teams did not perceive a strong degradation in their ability to adapt to changing situations—to communicate or coordinate with one other. The significantly higher NASA TLX (workload psychophysical instrument) subjective workload ratings for MOPP 4 tasks indicated an increase in workload at the individual level, but the increased workload apparently did not affect the team's ability to communicate or coordinate. Additionally, the results of this study did not suggest an increase in the number of errors or severity of errors. Subjective ratings of overall performance were good regardless of the number of errors or error severity (Grugle, 2001).

The researchers explained these results in terms of the nature of the rescue tasks and the team performance measurement technique used in this study. In order to use the TARGETS methodology, an observational technique, all task scenario events were scripted. A less strictly controlled scenario might allow for a broader range of responses and interactions between team members. CPR and SIM required the teams to follow rigid rescue procedures, and teams were penalized with errors for deviating from the standard procedure. The levels of coordination, communication, and adaptability required to successfully perform a rescue were already designed into the training procedure. The researchers indicated it might be possible that team processes were not degraded because the teams were formally trained to respond to events and one another in certain ways. Therefore, the results of this study may not generalize to tasks that have a greater degree of complexity and a broader range of possible responses.

Task duration and task environment also might have impacted the results. The tasks might have been too short for team members to become sufficiently fatigued or frustrated enough to see a degradation in their ability to interact and coordinate actions. In addition, the controlled climate did not allow for team members to become overheated. If tension levels in individuals rise as a result of the increased heat experienced when wearing MOPP, their ability to interact with other team members may decrease.

The trade-off between experimental control of the tasks and generalizability to real-world tasks played a major role in the outcome of this study. The team performance measurement technique used and the resulting task scenarios limited the team's ability to interact at a level that might have produced significant results.

Finally, and most importantly to this study's role as an example of using experts as subjects, there were many difficulties recruiting and retaining subjects. Using experts as subjects required institutional approvals from both the university IRB (institutional review board) as well as from the external organizations supplying subjects. While managers in these organizations were quite enthusiastic about participation, getting subjects to actually appear in the research laboratory for the study was a challenge. In several cases, subjects simply did not meet their obligation and show up. Also, as compared to using students as subjects, the subjects in this study were highly trained. Nevertheless, variability in training and skill still was observed.

QUASI-EXPERIMENT FIELD STUDY EXAMPLE

Background

A research program was implemented to determine preferences for the design of distributed meeting facilities used for video-enhanced electronic meeting systems (Grenville, 2001). Although there is literature in the group communication domain, the research in this multidisciplinary area has paid minimal attention to users' environmental preferences in the design of large-scale meeting spaces. Three studies were designed and performed to explore the variables that defined the design of a distributed meeting space from the users' perspective. Twenty-five inexperienced college-enrolled participants and 25 experienced military professionals used foam-core pieces to create a design of their ideal distributed meeting space. Thirty-seven variables were used to categorize qualitative attributes of the designs. Three types of designs emerged from the sample population: V-shaped, conference (U-shaped or oval), and theater (auditorium) style. Qualitative methods were used to measure the impact of the heuristics on the users' physical design of their distributed meeting spaces.

Results

Level of experience is a distinguishing characteristic of users' preferences in the meeting environment. More experienced users assigned greater importance to auditory clarity, whereas naïve users assigned more importance to visual clarity. However, both groups shared a common set of design themes. The findings from this research study provided insight into the user's mental model of interaction in this environment. Naïve users' mental models were most concerned with the needs of virtual co-location, while the experienced users' mental models were more concerned with achieving immersion in the meeting experience despite the challenges of virtual co-location. Both nonexperts and experts found the empirically derived heuristics easy to understand. Nonexperts reported that they were slightly more difficult to apply and therefore less useful.

A macroergonomic participatory approach was attempted. However, there were issues with the amount of time required to complete this process. In actual situations, teams are unlikely to take the necessary time to plan and/or configure the facility and are not likely to spend more than a few minutes on meeting preparation. Further research is required to determine the appropriate medium for participation. Design strengths identified by participants in Study 1 were also identified by participants in Study 2. For example, themes and characteristics documented from the interviews in Study 1 were identified as design issues, strengths, and changes by participants in Study 2. This observation adds validity to the methods used in Study 1 and Study 2. The researchers observed consistent results between the field and laboratory settings.

The findings indicated that users have a shared mental model for conducting a distributed meeting. First, users' perceptions of this environment will be based on experience level. Naïve users will tend to focus on the technical goals of distributed communication. Experienced users will focus on the goals of overall communication. Naïve users will focus on the visual aspect of co-location and will focus to a lesser degree on physical co-location and aspects of a traditional meeting. Experienced users will focus on the overall communication at both meeting sites and have a heightened concern about audio quality during a meeting. Experienced users will also be more concerned about achieving the capabilities of a traditional meeting. Neither naïve nor experienced users of these sites distinguished between position and type of equipment available in the room with the exception of three items: table configuration, position of the computer, and type of computer. Results in Study 1 indicate that users' anticipated interaction with distributed technology may vary, based on experience level. Users also have different views on the need for privacy, computing power, amount of available workspace, and the ability to move machines. Visual privacy is a greater concern than auditory privacy. All of the specific issues mentioned here can be impacted by configuration, position of computer, and type of computer.

Consistent requirements emerged for the design of distributed meeting rooms. These requirements differ based on level of experience with the meeting environment. It appears from this data that the design model for the naïve user will be more focused on meeting the technical goals of distributed communication, whereas the goals of the experienced users are more focused on overall communication. A consistent set of rules emerged in the form of design themes. However, it appears that the application of the design themes also differs based on experience level. The findings about rules usage support those on requirements. More experienced users cite the usage of rules that focus on overall communication and team collaboration than naïve users. More naïve users cite the need for visual communication (large public displays) than expert users. Experienced users also tended to cite more esoteric concerns such as breakaways, flexible furniture, and lighting more often than naïve users.

Nonexperts and experts alike reported the heuristics as usable. There were differences between the two groups in their reporting of the usefulness of the heuristics. Naïve users reported them to be less useful than expert users. This may be because these heuristics were prescriptive in their approach but did not provide a step-by-step application of the heuristics to a particular room design. Instead, they required users to assimilate information and to apply the information to their design. This type of extraction of information for a set of guidelines may be more difficult for nonexperts because they lack experience in dealing with design issues related to the distributed meeting environment.

Overall, post-hoc analyses supported the notion that users do adhere to a consistent set of design rules and have a consistent set of requirements for the distributed meeting room environment. The analyses also placed emphasis on the experience level factor and further established differences between naïve and experienced users and inexperienced and expert evaluators. This study directly compared the use of laboratory and expert subjects and illustrated the complexities of performing quasi-experiments in the field with expert subjects. As was the case with bringing experts into the laboratory, described in the previous section, recruitment of subjects was a challenge. Unlike the MOPP example, in this study it was management who, on one occasion, decided to disallow the use of their personnel as subjects.

CONCLUSION

Macroergonomics concerns the analysis and design of work systems. Accordingly, macroergonomic research is needed to answer both applied and basic research questions about work systems. Several studies were presented to illustrate the feasibility of performing empirical macroergonomic research. First, we illustrated the conduct of field research in actual organizations. Here the survey methodology was employed. Though correlational in nature, this data did serve to help quantify a fuzzy sociotechnical system construct, "joint optimization." Next, we illustrated a laboratory investigation using students as subjects. In this case, we investigated the concurrent versus sequential engineering paradigms. Then, in contrast, we illustrated how actual professional experts can be brought into a laboratory setting to participate in controlled studies of macroergonomic variables. Here team process as it is affected by MOPP gear was the focus. Finally, to illustrate the application of experimentation in the field, we overviewed a study of large-scale videoconferencing.

Clearly, there are trade-offs when deciding whether to use students as subjects or experts as subjects, and whether to perform a study in the laboratory or in the field. The laboratory and use of student subjects offer more control and efficiency with respect to recruitment and conduct of the study. However, the use of specialists as subjects and actual organizational environments offers a greater degree of external validity and generalizability of results.

Together, these studies illustrate that theoretical constructs and factors from macroergonomics, and its underlying sociotechnical systems theory, can be tested and thus better understood through traditional research methods. Popular scholarly outlets for the publication of such studies is currently exemplified by such journals as *Ergonomics, Applied Ergonomics, Human Factors and Ergonomics in Manufacturing, International Journal of Industrial Ergonomics, Human Factors, Journal of Applied Psychology*, and *Human Relations*.

8

A Macroergonomic Approach to Reducing Work-Related Injuries

Andrew S. Imada

A. S. Imada & Associates

INTRODUCTION

Traditionally, occupational injuries and accidents have been viewed as singular system outcomes independent of profitability, management, or organizational context. Hendrick (1991a) proposed a macroergonomic approach that can be used to address injury prevention. This alternative view suggests that there are multiple causal factors (e.g., psychosocial, ergonomic, management, engineering) that influence safety. This approach is a major development and an overarching theme for psychological, systems, and participatory methodologies for injury reduction and mitigation. A macroergonomic approach to improve safety begs us to expand the scope of our usual interventions to reduce work-related injuries. Our traditional strategies that involve engineering, training, or administrative control may be too limiting to be effective (Nagamachi & Imada, 1992). A more robust model acknowledges that accidents and human error have multiple causal factors that extend well beyond the scene of the event. DeJoy (1990), Nagamachi (1984), and Nagamachi and Imada (1992) have presented macroergonomic models that begin to describe these mismatches in human interface. Recently, research evidence has emerged that supports these models; for example, psychosocial factors, decision latitude, and employee relationships may moderate the relationships between human interface hardware and musculo-skeletal disorders (Faucett & Rempel, 1994). Further, there is evidence that psychosocial factors related to the job and work environment play a role in work-related musculoskeletal disorders (NIOSH 1997, Chapter 7).

Traditional Approaches to Reducing Injuries

Perhaps organizations' inability to influence people's behavior and safety signifi-cantly lies in their unidimensional approach. Typically, organizations react to in-creases in frequency rates or accident events rather than manage processes. Most organizational strategies for dealing with safety issues create one or two psycho-logical consequences. The first psychological consequence is to increase workers' knowledge. The most common technique is to train people on proper techniques, machine use, or safety procedures. Other techniques for increasing knowledge in-clude creating greater awareness, alerting people to the environment, providing cautions, and introducing new information. Whatever the intervention strategy, the outcome is the same—an increased number of facts, knowledge, or ideas. The second psychological consequence that organizations create is negative affect. This negative affect is achieved through reprimands, punishment, or increased be-havior monitoring. Exhorting workers to work safely and telling them how to do their job elicits negative affect. Though this may not be the intention, these are the psychological conditions traditional programs can create. Traditional interven-tions that create increased cognitions or negative affect are insufficient to change behavior. Using training or punishment exclusively ignores the fact that the deci-sion to behave is a complex multifaceted process.

Macroergonomic Approaches to Reducing Injuries

A broader conceptualization suggests that human error or accidents are the result of an interface between workers and their environment. This human system in-terface has at least three dimensions. First, situation-based factors are related to the immediate work environment in time and space. These might include: a wet work surface, complicated workstation layout, dangerous equipment left unguarded, or uncertain environmental conditions. Second, management-based factors include: failures in communication, leadership, and awareness of risky sit-uational factors, or failure to train or warn people. Finally, human-based factors include: emotional states, attention, distractions, morale, motivation, teamwork, and stress. Traditional safety and ergonomic interventions (e.g., accident investi-gation, hazard analysis, job safety analysis, fault tree analysis, link analysis) may be useful for teasing out situational or even managerial factors. However, these traditional analyses do not capture, describe, or address human-based factors.

A macroergonomic approach to improving safety needs to address more than one set of factors. A human-centered ergonomic approach establishes counter-measures for overcoming the human-based factors and recognizes its influence on external factors. For example, when a paper roller is rotating at a high rate of speed, operators immediately recognize the danger. However, after the mainte-nance worker turns the machine off and the roller slows, this worker may place his hand on the roller to slow it down. Which interface factors alter the opera-tor's risk perception enough to encourage him to place his hand in the way of danger? No single factor is sufficient to account for this risk. Situation-based fac-tors (e.g., the absence of a guard), management-based factors (e.g., rewards that encourage workers to take risks), and human-based factors (e.g., feeling rushed) are all necessary to understand this event.

Akamiso Effects

After analyzing hundreds of industrial accidents, Nagamachi (1984) describes three levels of a humanware model. *Cognition* refers to the knowledge and skills required to do a job safely. This is typically addressed through training and trial-and-error practice. *Value judgment* refers to the operator's opinions of safety regulations and safety programs. *Affect* refers to the operator's feelings or emo-tional states or temporal moods that can cause operators to underestimate the real risk of their actions. Nagamachi (1984) describes a training method for reducing accidents by increasing workers' sensitivity to risk and understanding the role of affect in causing accidents and errors. Nagamachi Kiken Yochi (NKY) addresses these human-based factors directly and has been successful in improving safety in Japanese organizations. When operators understand sources of *akamiso* (nega-tive affect) and how these states can influence safety performance, they can begin to take control of these factors and behave safely. They also begin to un-derstand hidden sources of safety risks that were not obvious when operating under *akamiso* effects.

Management-based factors also need to be addressed in *akamiso* training. Managers need to be aware of how they contribute to hidden safety risks in the workplace. In Japan, managers are trained in NKY together with the operators. Managers then focus on: (1) identifying risks in the working environment; (2) improving these risky conditions; and (3) supporting the quality circle *kaizen* activities. These subtle and often distant changes can have dramatic effects on safety performance. The outcome is improved situation-based factors, which address the traditional ergonomic interventions.

Psychological Dimensions of Injury Reduction

Imada and Nagamachi (1990) present a similar conceptualization of attitu-dinal components. They argue that the affective component plays a large part in human error and accident events. There is evidence that links poor safety performance to affective responses such as workers' job satisfaction (Frankel, Priest & Ashford, 1980; Wolf & Pearson, 1992). This is also the component that traditional ergonomic and safety programs are least likely to address.

Psychologists have long recognized that our predisposition (our attitude) toward specific events, objects, or persons has three components—behavioral, cognitive, and affective. Traditional accident investigation and ergonomic interventions have focused on the behavioral component. If unsafe acts predispose humans to injury, we can reduce injury by altering machine design (e.g., guards, signs, barriers). The cognitive approach to influencing occupational safety and health is to introduce new information or ideas through training. The worker is taught new skills, techniques, tool usage, rules, and procedures. The third approach addresses the affective component. This approach attempts to influence feelings such as job satisfaction, stress, attitudes, temporal emotional states, and organizational climate.

While the interrelationships among these three components are not always clear, the distinction is nevertheless useful. Because few humans willfully hurt themselves, the omnipresent "human error" in accidents might be attributed to inattention or temporal affective conditions. While this appears to be common knowledge among practitioners, little is ever done to alter these affective states. This suggests that improving safety may require integrating components not typically associated with safety; for example, supervision can create emotional responses such as anger, frustration, pride, job satisfaction, and organizational identity. To the extent that these variables influence attitudes, they are important to improving safety. A macroergonomic approach considers supervisory training as important as technical training in reducing safety risks.

Risk Perception

Willingness to accept risk appears to be some function of the probability and severity of the risk, where:

$$\text{Risk Acceptance} = f(\text{Probability} \times \text{Severity})$$

Systems safety engineers have used this model to evaluate and prioritize risks. While we may not be able to define the mathematical operation of this function explicitly, the approach is nevertheless intuitively appealing. People are attracted to or repelled from events by their perceptions of the likelihood and consequences of these events. More recent research suggests that this functional relationship is not a simple linear process; for example, the introduction of new information is not additive. Existing beliefs or attitudes can influence the interpretation of incoming information, intended to change either the probability of an event or the severity of an event (Nisbett & Ross, 1980). Further, Slovic (1987) points out that the perception of risk is dependent on heuristics that humans use to filter incoming data.

Information can carry unduly heavy weight in influencing risk perception depending on existing values, attitudes, or temporary affective states; for example, events that are perceived as having potentially gruesome consequences may be intolerable despite the low probability of these events ever occurring. It may be useful to think of each of the competing factors influencing the perception of risk as dynamic forces in field theory (see Lewin, 1951). However, because human attention is limited, and we can only deal with a few pieces of information simultaneously, each factor competes for space in the central arena of consciousness. Arguably then, it is the affective component that determines which cognition takes center stage in arriving at the perceived risk. It appears that introducing new information is not a simple additive process; risk acceptance is dependent on the relationships among cognitive, affective, and behavioral components rather than cognitions alone. Both the *akamiso* and psychological models suggest that a human-centered approach to addressing safety management can make a large contribution to reducing accidents and human error. Moreover, the macroergonomic perspective presents a great advance in tying together characteristics of the human and the organizational context where work occurs.

Participatory Ergonomics

While participatory ergonomics is often viewed as an approach to improve productivity, quality, or efficiency, it has another positive effect. Involving workers in ergonomic improvements also improves their view of the work they do, ownership of the ideas, responsibility, and even the ways they see themselves. In this respect, participatory ergonomics can be viewed as a human-centered approach that can improve safety as a side effect. When people participate in improving a process, production method, or reducing costs, they generate positive affective states (Imada, 1991a). These can be observed in improved morale, teamwork, self-esteem, group identity, and commitment and ownership to ideas. According to this humanware model, positive affective states allow people to operate at a cognitive level so they are better able to identify and deal with safety hazards. (See Chapter 2 for additional discussion of participatory ergonomics.)

Nagamachi (1991) applied a participatory ergonomic approach to an automobile assembly process and achieved remarkable results. Integrating this approach into existing quality control circle (QCC) and *kaizen* (continuous improvement) processes produced results that were more effective than ergonomics would have been alone. The combination of ergonomic technology, management involvement, and human-based factors produced results far beyond the effects of these interventions individually (see Chapter 6 for additional discussion of this case).

Therefore, participatory ergonomics can be viewed as means for reducing work-related injuries, not only through the ideas it generates, but also because of its human-centered approach. Involving end-users in implementing ergonomics addresses human-based factors such as motivation and removal of negative *akamiso* effects.

Implications of Macroergonomic
Injury Reduction

Systems Thinking

There are at least two major conceptual implications for adopting a macroergonomic approach to injury reduction. First, it shifts the focus away from singular causal events to more realistic systems level thinking. This is consistent with current thinking on problem solving and quality processes that operate in other parts of the same organizations. Focusing on unidimensional interventions would be ineffective and counterintuitive to the way organizations are solving other kinds of problems. To be effective, interventions intended to affect processes as complex as accident events cannot take a functional or academic discipline as a primary approach. Instead, it should cause ergonomists, epidemiologists, and safety and health professionals to take innovative systems approaches to solve problems. The second implication of this approach is that it may cause people to look at the injury event as a chaotic process.

Chaotic Systems

Chaos is associated with the study of nonlinear dynamics, which in recent years has gained wide recognition as an alternative to current linear paradigms (see DeGreene, 1990; Gleick, 1987). Nonlinear chaos refers to a controlled randomness in systems that has disproportionate effects from stimuli. The disorderly behavior found in chaotic systems leads to a creative process that generates a rich complexity representing the spectrum of living things. Chaos can convey a feeling of harmony between order and disorder. These systems are also robust, adaptable, and can operate under a wide range of conditions enabling survival of unpredictable and changing environments. Unlike linear systems, nonlinear chaotic systems allow the disturbance to feed on itself and dissipate; the system then restabilizes. This may explain how biological systems use nonlinearity as a defense against environmental noise.

These chaotic processes often create fractal structures that defy traditional Euclidean dimensions. While fractal patterns appear random, they have an underlying characteristic called "self-similarity," which is a symmetry across scales; that is, at different levels (length and time) details of fractals may look alike. This recurring pattern within a pattern creates strength and diversity.

Unlike traditional ergonomic implementation, participation can create diversity in ergonomic strategies. The participants' varying interests, politics, education, motives, and hedonism cause ergonomics to be a fractal-like discipline rather than a linear one. Such solutions would be stronger, more adaptable, more resilient against environmental variations, and therefore able to be regenerated. Conceptually, we can envisage a myriad of solutions, which bear no relation. However, on closer examination we find that ergonomics is the theme that creates the self-similarity in these fractals (Imada, 1991b).

User involvement in ergonomic changes coupled with individual differences can replicate the strength and diversity found in fractals. These, we believe, are valuable characteristics. However, to achieve them we will be changing the way we go about our business. This will mean a paradigm shift from linear to nonlinear thinking. In its place we may have to put fractals and nonadditive systems that may not be solvable. Allowing people to participate entails shifting information from a centralized to distributive mode. This will require empowering people to be problem solvers. These changes will no doubt be stressful. However, the potential payoffs for this macroergonomic intervention to create chaos and change cannot be ignored.

A MACROERGONOMIC INTERVENTION

The Organization

This intervention involved a region of a Fortune 50 petroleum manufacturing company. This organization was one of nine geographical regions responsible for storage and transportation of bulk petroleum products to retail outlets and commercial customers. The region's geographical boundaries included southern California, southern Nevada, and southern Arizona. This product delivery system serves one of the most heavily traveled motor transportation markets in the world. This region serves the counties with the highest number of registered vehicles, licensed drivers, motor vehicle accidents, fatalities, and injuries in California. This state is the largest petroleum market in the United States and the fourth largest in the world; more than half of the gasoline stations are in this region. In 1990, there were 4,663 stations in four counties in the Los Angeles area. In 1989, drivers covered 5 million miles to deliver 1.3 billion gallons of gasoline. More than 700 deliveries are made every 24 hours, 7 days a week, with an operating budget of $38,000,000 (U.S.). The region works with three labor unions in 25 different supply sites. Currently, nine terminal managers and one area manager oversee the gasoline distribution and 235 employees.

The Problem

Historically the region's accident and lost-time rates have been higher than other regions with similar responsibilities, procedures, and functions. This should hardly be surprising given the increased risk with more motor vehicles, higher vehicle density, more accident events, and more motorists in the largest gasoline market in the world. However, the motor vehicle accident rate doubled, the industrial injury frequency rate doubled, and lost work days quadrupled between 1986 and 1987. Management's initial response was predictable; they planned to reintroduce a comprehensive training program that included technical (driver safety), ergonomic (lifting), and cognitive/emotional (stress) components. In short, the plan was to improve safety performance through training.

This is a typical reaction to increasing safety performance metrics. The results are also predictable because of the unidimensional focus on only one of the psychological dimensions. Because of the size of these changes and past failures in traditional approaches, management contacted a consultant to examine accidents and injuries as part of total organizational functioning.

A three-phased approach was proposed and approved. First, an organizational assessment was used to diagnose the current state of organization and the problem. Second, organizational and process changes followed, based on the diagnosis from the first phase. Third, monitoring and evaluation of the safety performance metrics were kept and fed back into the system for continuous corrective action and evaluation.

ORGANIZATIONAL ASSESSMENT

The Process

The first phase was an organizational assessment to review safety data, procedures, and organizational and management influences. Interviews with truck drivers and their immediate supervisors provided a large part of the data. Individual and group interviews were conducted to obtain driver perceptions of accidents and injuries and their role in the process. Interviews were conducted with regional staff, corporate headquarters staff, and internal technical specialists to learn about management practices, reward systems, incentives, disincentives, personnel policies, accident reports, and equipment design. Observations were made during meetings. Formal and informal communications were observed to understand how the organization functioned, cultural patterns, and normative behavior. Data, records, and incident reports were reviewed.

Key Organizational Features

As with most large, successful business systems, this organization had developed several strengths which, by their nature, create weaknesses. These shortcomings are the basis for the mismatch between sociotechnical system requirements and standard operating procedures. Some examples include:

- As an organization with a strong engineering culture, much of the problem solving focused on analysis. Most of the analysis focused on technical details rather than human requirements or *akamiso* solutions (Nagamachi, 1984).
- Efficiency is a highly valued feature in a strong engineering culture. At times, efficiencies came at the expense of how people may feel about the work, causing negative affect. Traditional systems were designed to maximize metrics rather than provide a human-centered approach.
- Unreliable plant and equipment made safety performance goals difficult to achieve. These working conditions also led to lower morale. Employees did not experience positive affect (e.g., pride, appreciation) when working with equipment that was perceived as being "second class." This negative affect influenced the psychological dimension in the effort to reduce injuries.
- A highly procedure-based approach to managing every detail of how people performed their jobs was developed centrally. This made employees feel like passive recipients of orders that were not well understood. Consequently, actions that were prescribed to mitigate certain risks may have been ignored because the perceived probability of the event may have altered risk perception. Finally, the highly monitored behavior led to a sense of lack of control over the environment and negative affective states.
- As an organization with long-term employees and high loyalty, a patriarchic relationship was the standard between the management and the line employees. This parent–child relationship stifled innovation in problem solving and led to unempowered employees. Success was highly dependent on leaders and less dependent on everyone else. There was little incentive for participation, no reason for buy-in, and morale was low.
- The organizational structure was a dysfunctional matrix design with well-defined silos, distinct cultures, and win–lose outcomes. Consequently, communication among organizational units was poor and cooperation was low. Success depended on personalities and individual preferences.

The engineering culture and the nature of the industry created strong, narrow areas of expertise and approaches to problems. Diversity in people, ideas, and solutions was similarly narrow. This stifled the creativity that is necessary in systems that are robust, adaptable, and operable under a wide range of conditions. Organizational design and functioning were mechanisms to maintain stability. The confusion and discomfort that come with change were dampened by lower communication, one-way communication, and hierarchy. Balance, symmetry, order, and predictability are not hallmarks of adaptive, chaotic systems. The organization needed mechanisms to infuse ergonomics and safety as part of its ongoing culture. When good ideas emerged at one level of the organization, they had no way of developing "symmetry across scales" and penetrating different levels.

Organizations need to be able to respond to their environments with changes that correspond to the environmental demand. Weick (1988) calls this ability to modulate an appropriate response *requisite variety*. Requisite variety, a well-known engineering principle, posits that to control a system, the controller must be at least as complex as the system it intends to control (Guastello, 1995, p. 122). Weick extends this logic to organizations and its ability to control itself and its environment. Lack of diversity and aversion to chaos were stifling creativity in an organizational environment that was changing dramatically; for example, long-term, loyal employees traditionally staffed the personnel subsystem. Little attention was given to identifying, selecting, and training a cadre of new employees. Regulatory and environmental requirements were being met using the same mechanisms that satisfied less stringent requirements. Costs for accidents, legal liabilities, and litigation were underestimated. In short, the organization lacked the requisite variety to function effectively in its new environment.

Recommendations

Macroergonomic concerns about these organizational features were translated into recommendations that management could use to synthesize the assessment results. The causes of safety problems were summarized in three organizational themes. First, the region was good at problem identification but unable to implement programs to solve problems. Second, the region was strong in technical analysis of safety problems but weak in managing human performance. Third, the region had potentially effective programs; however, these programs lacked integration.

THE INTERVENTION

Employee Involvement

Recommendation. Safety programs can and should be implemented through employee participation. The thrust of this recommendation was to empower employees to recognize, solve, and promote solutions to safety problems.

Increased Control Over Their Work. Drivers were allowed to control facets of their work that may not have immediate bearing on safety, but have *akamiso* effects (Imada & Nagamachi, 1990; Nagamachi, 1984) and consequently impact driver performance. An example was the sequencing of deliveries. By determining their own load sequence, drivers were able to use their knowledge and experience in reducing travel time and taking advantage of traffic patterns. This resulted in higher efficiency in the delivery system and reduced exposure to high-density traffic. A second example of increased control was scheduling working hours. At the time, a new drivers' labor contract called for a new compensation system. They went from working a traditional 40-hour week to 55 hours per week, with some drivers working as few as 44. In assigning people to different work schedules, drivers were given choices about which kind of schedule they wanted to work. Perhaps more importantly, it gave drivers greater control and commitment in their job and work assignments.

New Driver Selection.   One example of increasing the organizations' ability to respond to its environment (requisite variety) was the hiring of new product delivery drivers with safety performance as a major factor in the selection process. Prior to this intervention, local management had complete control and responsibility for identifying, screening, and training new employees. The selection decision was made on the basis of referrals, personal relationships, and a face-to-face interview with the manager. To increase the range of responses and perspectives, employees were brought into the process to create more robust perspectives on safety performance. Drivers participated in a job analysis to determine key skills for new drivers. One of the three predictors selected was "safety performance." In so doing, the drivers were looking for new drivers who were not only skilled in driving tanker trucks, but also drivers who could work injury-free because of their nondriving skills. This is in marked contrast to a system that depended on knowing the individual decision-maker and performance in a face-to-face interview alone. Job qualifications, knowledge, skills, and abilities were developed from this job analysis.

At each work location, a team of drivers was trained to screen and interview potential new employees. The training introduced them to the interviewing process (how-to, legal aspects, types of questions, purposes of the interview, mistakes to avoid, and videotaped practice sessions). The drivers made up simulations for the candidates to demonstrate their skills. A structured board interview format and behaviorally anchored expectation scales (Smith & Kendall, 1963) for each predictor and performance criteria were developed. This was an important process for a company that up to this point had very little turnover and therefore little experience hiring new employees. It was also an important message to send to the new employees; that working injury- and accident-free is an important part of the job. Drivers' participation in this selection process amplified the message.

Station Redesign.   Drivers made deliveries to locations that were poorly designed from their perspective. It is hard enough driving a 10,000-gallon vehicle in congested city traffic without built-in obstacles. Poorly designed delivery sites are the cause of frustration, stress, and other *akamiso* effects (Imada & Nagamachi, 1990; Nagamachi, 1984). This negative affect can lead to accident events. Moreover, the inability to control the design of the delivery site means that these obstacles will exist for the life of the retail outlet, often 10 or more years. Obstacles and sources of frustration included: location of storage tanks, entrance and egress requirements, parked cars for repair, location of public phones, air and water customer conveniences, landscaping, and pedestrian patterns created by station design. Drivers participated in reviewing and approving plans whenever a new gasoline station (retail outlet) was constructed or rebuilt. The drivers identified hazards, bad tank locations, unsafe places where customers are likely to park their cars, phones, air and water pumps, ornamental or decorative features, and signs that could obstruct visibility or cause an injury. Plans could not be approved without driver participation. This participation was a source of discomfort for a company with a strong engineering culture. Having people from operations participate in construction and design was a new and varied response to how the company had operated in the past. This was an example of increasing the organization's complexity to respond appropriately to its environment. Weick (1995) noted that this is not uncommon, because with greater complexity comes a greater search for and reliance on the habitual and routine (p. 87).

Reward Programs.   For ease of administration, headquarters prescribed all safety awards, and recognition. This overall strategy lacked the diversity to motivate individual drivers and locations. System stability and consistency were maintained, but individual motivations were ignored. Drivers were asked to decide on a local program that met their needs within general parameters. Several different programs were developed for the eight geographically dispersed workplaces. This gave the drivers ownership and pride in their individual safety performance. Empowering employees to select valued consequences to their behavior can be a powerful tool for altering human behavior and system performance.

Back Injury Reduction Program.   Drivers participated in a program that allowed them to select individual belts that improved lumbar support. This participation improved drivers' understanding and awareness of ergonomic problems inherent in driving. This was in sharp contrast to the top-down, mandated training that was traditionally forced on all employees.

Equipment Redesign.   Many of the ergonomic improvements came from employees' ideas for reducing injuries and accidents. Supervisors began enlisting employees' help and listening to their ideas. The region's poor safety record matched the poor productivity and unreliable plant and equipment. Not surprisingly, morale was low. At the same time, employees felt the threat from outside competition coming into the company.

Regional management began to make changes such as improving equipment and plant reliability; buying more comfortable and easy to use equipment (e.g., seats, location of equipment, performance support tools); and putting equipment (e.g., mirrors and lights) where employees said it was needed; buying tools so people could do their jobs safely; placing an emphasis on workplace safety when purchasing and replacing equipment; and following up on employee suggestions for improvements. Safety became part of the capital budget and purchasing process. While recapitalizing plant and equipment is not new, involving employees and tying equipment expenditures to safety is a novel approach. Traditional purchasing and recapitalizing focuses almost exclusively on cost, durability, and life cycle—rarely is it tied to safety.

One of the most noteworthy projects undertaken was the redesign of the hose tray on the product delivery truck. Drivers unload varying quantities of product into underground storage tanks at the stations. The gas is gravity-fed into the tanks using a plastic hose and a vapor recovery hose that returns tank vapors into the truck's compartments. The driver handles two pairs of hoses at every delivery site. Depending on the workload, drivers make anywhere from one to eight deliveries each day. This requires handling these pairs of hoses up to 16 times a day. At the end of the hose is a 14-pound (6.35 kg) fitting that is lifted each time the hose is handled. For any employee below the seventy-fifth percentile for males, this required a lift above the shoulder.

The head mechanic first noted this problem of lifting hoses at one of the vehicle maintenance shops. With no formal training on equipment design or ergonomics, he began to undertake this project singlehandedly. He worked with a tank manufacturer and came up with a prototype design. At first, this new design met a lot of resistance from the drivers.

Recognizing this resistance, we implemented a participatory ergonomic redesign strategy to make modifications to address these concerns (see Imada, 1991a). We involved three drivers and a head mechanic in several meetings over several months to discuss further improvements. They identified issues, determined criteria for solutions, brainstormed solutions, performed link analyses for movement and control, and tested different design ideas. The redesign led to several improvements without major modifications to the truck chassis, tank capacity, or safe operation. The sole purpose of the redesign was to make it easier for the driver to perform the work and reduce fatigue and injury. This was achieved by allowing drivers to stand upright and bring the load as close as possible to the body. The participatory ergonomic redesign process was useful for transferring lessons learned from this redesign. The design reduced the biomechanical requirements for the work as well as the work process. After driving on this new equipment, the drivers instructed other drivers on how work was changed and what this new work process required.

Driver College. On the initiative of two drivers, the region instituted a Chevron driver college. The concept was to have master drivers teach other drivers about safe working procedures, maintenance, pre- and post-checks, accident investigation, driving techniques, and loading and unloading procedures. The class lasted one work week and included classroom, hands-on, and driving activities. This was a process conceived, designed, taught, and evaluated by employees who were truck drivers, maintenance mechanics, master drivers, and driver instructors. All evaluations of this training initiative were positive. Drivers and driver instructors from the different work locations appreciated learning about safe work procedures and injury reduction techniques from other drivers. This college was conducted throughout southern California and Arizona.

Communication

Recommendation. Manage human performance through communication and feedback. Much of the region's work was underpublicized or perceived as unintegrated. Communication loops enabled people to see the connection between safety and other organizational functions.

Information Sharing. Management began sharing information on future programs that might impact safety. For example, part of the region was to be placed under a new incentive pay program. Safety was a point of discussion in introducing this plan. Capital programs, marketing plans, technology upgrades, and truck purchasing programs became standing agenda items at communication meetings. This was a dramatic transformation from earlier meetings that were complaint sessions and high conflict events.

Positive Feedback. The region began focusing on accomplishments when discussing safety. This positive feedback was well deserved. The majority of employees were skilled drivers operating at a high level of efficiency and safety. In an analysis of employees involved in safety incidents, we discovered that fewer than 5% of the employees had less than perfect safety performance. Yet all employees received the same negative information about safety. This positive feedback reinforced their behavior; negative feedback was reserved for offenders and low-level performers.

Communications from headquarters were once limited to a standardized form comparing regions across the country. Because this region is the most densely populated and heavily congested, it was not surprising that they had the most risk, and more safety events. The region began communicating what people wanted to know about safety: how they were doing at individual terminals; how much money was saved through safety improvements; and successes they achieved through goal setting. These milestones were celebrated through awards, social and family events, and future success was encouraged and expected.

Communication Across Levels. Communication was improved among regional management, terminal management, and line employees. In the past, any discussions about safety among these three groups were generally reactive and negative. Regional management joined in planning with the terminals and involved employees. They began talking about their successes, plans, and goals. One small but important change was the increased personal contact among managers, supervisors, and line employees. Special efforts were made to encourage contact—both organizational and casual—to facilitate communication among these groups. This made discussions about safety more proactive, positive, and participative. Finally, managers and supervisors were encouraged to thank the employees for their efforts and encourage their success. The increased contact and signs of appreciation were major milestones in improving communication across levels and throughout the organization.

Nontechnical Skills Training.   Early in the assessment process we noted that the organization was good at analyzing problems, but had difficulty implementing ideas; for example, they would conduct accident investigations, compile large amounts of facts, and come to conclusions about the causes of the accident, but fail to follow through with other employees. This was compounded by a dysfunctional matrix structure that further inhibited communication in a hierarchical, vertically differentiated organization.

We began training supervisors and managers on skills that would enable them to implement the technical aspects of their job through people, not in spite of people. Creating a more participative, open environment allowed the company to get more buy-in from its workers. Employees at the terminal were trained on communication and teamwork skills that allowed them to implement ideas as teams. Drivers were given training on stress management and how to manage *akamiso* effects. Finally, driver trainers were taught to be mentors and leaders rather than police. Office employees, dispatchers, and plant operators were trained on problem solving, interaction skills, and communication. Taken together, this human-centered skills training altered the culture to make it more receptive to participation and change.

In addition to the training, managers and supervisors had access to outside coaching and counseling. External consultants were available for advising managers on handling difficult situations, serving as a sounding board for new ideas, and acting as a confidential source for employees to pass information upward in the organization. These activities increased the quantity and quality of communication in the organization. This improved information flow increased trust and understanding across levels in the organization which made it easier to solve problems in safety and health.

As part of a company-wide effort to give management and supervisors feedback on their performance, the organization implemented an upward-feedback program in which the employees gave their perceptions about their immediate supervisor on survey questionnaires. This information was discussed among peers and fed back to the target manager in a constructive format. Similar to the coaching and counseling, this process required and built communication and trust. These were critical components for the transformed organizational culture.

An exercise in role clarity was conducted with supervisors and their direct reports. In one exercise the supervisor and subordinate described the subordinate's job, listed the job duties, and prioritized these job duties. Then each exchanged their perceptions of the job. Where there were discrepancies, both parties had to come to a consensus about what the activities were and/or why it was or was not a priority. This series of activities led to clearer performance expectations and improved teamwork. This role clarity was helpful in moving forward to addressing safety concerns. Without resolving these difficulties, there is no way the organization can cope with new and more complex problems (Weick, 1988).

Integrating Safety

Recommendation. Integrate programs through strategic planning. This recommendation makes safety an integral part of the organization throughput rather than a singular and independent outcome.

Capital Budgets. Regional management began tying capital improvements to safety. Purchasing new trucks and replacing poorly designed equipment were tied to job security and safety. These capital expenditures made a statement that management recognized the ergonomic problems as barriers to working safely and effectively.

Personnel Issues. The organization began integrating and coordinating personnel and safety functions. The assessment indicated that the region's poor safety performance could be attributed to a few drivers. Moreover, discipline and personnel matters were handled too harshly, inconsistently, or not at all. One common strategy organizations employ is the use of a policy for the entire organization to address the behavior or performance of a few individuals. Rather than creating new policies for these few individuals and imposing this on all employees, the organization focused on feedback, coaching, counseling, and discipline. Once these personnel issues were addressed, safety performance improved markedly. Enforcing personnel policies improved driver morale by affecting their sense of equity and social justice (Rousseau, 1989).

Truck Bidding. Drivers operate the same truck for one year. Naturally, all the drivers want to operate the newest equipment. As an incentive to operate newer equipment, truck bidding was tied to safety performance. Bidding on trucks for the year was based first on seniority and second on safety record. Therefore, even a senior driver could lose his position in the bidding process with a poor safety record. Conversely, it was possible for a less senior driver to improve his truck bid if he had an excellent safety performance record. This emphasized the importance of safety in the workplace and made it a priority.

Business Plans. Safety performance metrics were identified as critical success factors in the region's business plans. Safety data were translated into financial data. These plans were then presented at all the driver locations throughout the regions. Regional management talked about the goals and how drivers, supervisors, mechanics, plant operators, and office staff contributed to attaining these goals. This briefing was done at least once a year for all terminal employees. This was the first time that safety performance and feedback were tied to business plans and these business plans were presented to employees.

Truck Roadeo. Drivers were allowed to showcase their skills in a regional truck roadeo where drivers completed a written examination, an oral interview, a pre-trip inspection, and demonstrated their driving skills through a prescribed course. The course emphasized skills required to perform the driver's job safely. This included: backing, offset parking, side drop spotting, maneuvering through a narrow space, right turns, and stop lines. This roadeo was later adopted by the company in a national competition. In addition to improving morale, this activity highlighted the drivers' skills in performing work safely and productively. Inviting family and friends to the regional and national events added focus on safety as an important part of the job.

Career Development Sessions. Managers and line employees voluntarily participated in a career development session in which management openly discussed career options and opportunities. Here employees were given realistic job information about prospects for promotion, advancement, and training. At these sessions, employees were made aware of the importance of safety in career acceleration. Tying safety to performance appraisals, career mobility, and future success in the company enhanced safety and injury prevention. This was particularly true among white-collar office staff who normally did not think of accidents and injuries. It also had an impact on the terminal employees and drivers interested in career advancement. By tying these outcomes to injury prevention, the company integrated safety into the day-to-day operation of the business.

Another career development step was to remove disincentives for people to take on safety responsibilities. It had been possible to take a more responsible safety-related position (e.g., driver instructor who conducted safety check rides and certified new employees) and make less money for doing so. This sent a mixed message about the importance of safety in the organization. It also served as a barrier to attracting the most skilled, effective, and respected employees from participating in the program.

Providing Goals and a Vision. Safety was tied to the business plan in the form of goals. These goals were stated as simple, meaningful ways that employees could impact the region's overall performance. Motor vehicle safety and injuries became critical success factors (CSFs) in the region's strategic marketing plan. The goals were supplemented with feedback as business units approached goals and exceeded these targets. Incorporating operations and operating costs into the sales and marketing plan recognized the total costs of doing business. A large part of these costs were associated with motor vehicle accidents and industrial injuries.

Fostering Teamwork.   Safety was traditionally viewed as the result of individual efforts. The company began to focus on teamwork as a means of improving safety and reducing injuries. Rewards and goals focused on group as well as individual performance. The contributions of all the terminal employees were recognized. Local management emphasized how none of the jobs could be done safely and injury-free without cooperation and coordination as a team. Traditionally, truck drivers viewed their safety performance as independent of anyone else. By rotating people in and out of jobs on a voluntary, relief basis people began to understand how nondriving jobs contributed to their success. Mechanics, maintenance staff, plant operators, and driver trainers were recognized for their contribution to the overall safety effort.

Empowering People to Run a Business.   In the past, safety and injury prevention was viewed as a top-down, management-directed process; management set the goals and employees tried their best to achieve these mandates. Whether these goals were met or not met had few consequences for those doing the work. The only reliable consequence was negative information when someone did get injured. At the same time, terminal managers were not free to spend resources and money on safety projects. As in many hierarchical organizations, managers asked permission before acting. In short, employees were reactive recipients of the safety program with little discretion to address the root causes.

A major change occurred when regional management told terminals to run their injury and illness prevention programs as their own business. They were given a budget, but not told how to allocate funds, awards, or programs. Putting managers and line employees in charge of their own safety program allowed injury reduction to become part of the business rather than a separate program.

RESULTS

Safety Performance Metrics

The region made dramatic improvements in its industrial injury frequency rate, preventable motor vehicle frequency rate, and days away from work compared to the beginning of this intervention. (See Table 8.1.) The 1996 industrial injury frequency rate was 70% lower than the 1987 rate. The 1997 motor vehicle accident rate was 63% lower than the 1988 frequency rate. The motor vehicle

TABLE 8.1
Industrial Injury, Preventable Motor Vehicle Accident Rates,
and Lost Work Days by Year

Year	Industrial Injuries per 200,000 Hours	Preventable Motor Vehicle Accidents / Million miles*	Lost Work Days**
1987	8.90	—	1368
1988	5.10	4.04	369
1989	4.10	2.79	82
1990	3.20	2.25	113
1991	3.50	1.64	76
1992	3.32	2.31	156
1993	2.91	2.33	42
1994	2.00	—	
1995	1.50	1.90	
1996	2.50	2.50	
1997		1.50	

*Preventable motor vehicle accidents were not calculated in 1987, only total accidents; data for 1994 are not available. ** Because of the expansion of the geographic region, these absolute numbers are not meaningful beyond 1993.

accident frequency rate is noteworthy because a single incident in 1987 incurred $2,175,000 in actual costs, and typical motor vehicle accidents involving third-party property and injuries average $94,200 per incident. Finally, the number of lost work days was less than 5% of the 1987 figure. This represents $350,000 in direct and replacement labor costs during the first 2 years.

Overall Operating Costs

During the same time that these changes were made, a dramatic improvement was also made in reducing unit cost. The metric used to evaluate unit cost is cents per gallon (CPG). This is the total cost for transporting the gasoline from the pipeline to the customer (including terminal throughput and delivery costs) per gallon of gasoline delivered. The region delivers approximately 1.3 billion gallons of gasoline per year. Inflation adjusted CPG data for these benchmark years are shown in Table 8.2.

The inflation-adjusted operating cost for 1996 is 30% lower than in 1986. The real magnitude of this difference ($0.0068) is substantial when multiplied by the volume delivered (1.3 billion gallons). While accident and injury costs make up only part of this overall savings, they contribute directly to the overall profitability of the operation and have been conservatively estimated to contribute to one third of the overall savings.

TABLE 8.2
Actual and Inflation Adjusted Cents Per
Gallon Costs for Three Benchmark Years

Year	Actual CPG	Inflation Adjusted CPG*
1986	1.91	2.30
1990	1.75	1.75
1996	1.62	1.62

*Inflation adjustment at 2%.

CONCLUSION

These results suggest that traditional interventions alone (training, awareness, ergonomics, technical safety programs) are insufficient to reduce injuries and accidents. In this case, poor safety performance may be viewed as symptomatic of a larger organizational problem that happens to manifest itself in motor vehicle accidents and injuries. Safety must be analyzed and treated from a systems perspective. The traditional management response is to reintroduce similar programs as problems emerge. These results suggest that there are several dimensions that are directly affected by this macroergonomic intervention.

Psychological Dimensions

Human error and resulting accidents have origins in hardware (e.g., tools, equipment, design), software (e.g., management-based factors), and humanware. While traditional ergonomics is primarily concerned with hardware, it is insufficient to improve total safety performance. The willingness of a human operator to place himself or herself in a dangerous position can be better understood in terms of humanware. The operator's technical skills, way of thinking, and affect all influence the possibility of human error. The operator's interface with the hardware and software is buffered by humanware factors. Whether temporary (e.g., anger) or more enduring (e.g., loyalty to the enterprise), these humanware factors influence errors and accident events. The success of the *akamiso* and participatory interventions suggest that the human-centered approach is a viable alternative for improving safety management. These approaches suggest the need for a broader, macroergonomic approach to improving safety performance.

Management Dimensions

Managers of systems need to examine their direct and indirect influences on the human interface incompatibility. They contribute to the interface through management-based factors and indirectly through human-based factors. Negative *akamiso* effects are often a result of management's actions. Managers must be made aware of this humanware model so they can identify and eliminate risks in the workplace. They too must be aware of the factors that cause workers to take unnecessary risks intentionally. This organization's success suggests that we need to deal with these issues from an etiologic standpoint rather than a symptomatic one. Management must have the courage to entertain the possibility that it may be part of the problem, or at least in some way contributes to it. Management must be willing to assume a certain degree of risk by committing resources and capital.

Financial Dimensions

These results demonstrate the economic effect that injury reduction and mitigation can have on an organization. The savings are conservatively estimated to be 5% of the region's annual operating budget compounded annually over the past 9 years; currently about $1.5 million per year. This reduction is a serious economic incentive for organizations to consider reducing injuries and accident events. The sustained improvement over a number of years won recognition from the corporation; the region won the Chairman's Award for sustained safety performance improvement over 5 years.

As technologies become more complex, as communities become more environmentally aware, as legislation becomes more stringent, and as workers become more safety conscious, management needs to analyze and implement macrolevel changes in human factors and occupational safety. This macrolevel of analysis can be best achieved through a macroergonomic, systems approach.

9

Information and Communication Technology (ICT) and Changes in Work Life: Macroergonomic Considerations

Gunilla Bradley

The IT University, Royal Institute of Technology, Stockholm, Sweden

BACKGROUND AND INTRODUCTION

An Overview

An interdisciplinary research program on "Computer Technology and Work Life" was initiated and led by the author at Stockholm University in 1974 and continued for many years. It was followed by several similar programs in Sweden. The author developed a theoretical framework including two theoretical models, one general and one in which the concepts and their interrelationships were specified. The models were tested empirically in three large work organizations in Sweden, representing three main historical phases of computer technology. The models also were used in discussing what might be desirable goals in the *information society*. The present fifth phase, the *network phase*, is characterized by a *convergence* of three main technologies: computer technology, telecommunications technology, and media technology. Information and communication technology (ICT) is used in almost every work activity and embedded in many things around us. The author proposes a superimposed theoretical model, entitled the *convergence model*, reflecting ICT and the psychosocial life environment—a revised version of her initial model. Finally, future research is discussed with reference to the convergence model. The conclusions address major psychosocial processes, psychosocial life environments, and a call for synthesis.

My Introduction to Macroergonomics

During 1974 to 1988, I initiated and led a series of projects at Stockholm University within a research program on computerization and work life (Bradley, 1977). It was a result of an inventory study on behalf of a large labor market organization in Sweden (the Swedish Central Federation of Salaried Employees) that encompassed 80% of white-collar workers at that time. This research was presented at the Central Federation's congress in 1972 and prompted me to start cross-disciplinary research on the social and organizational impact of computerization on work life. At an early stage, this research was presented in the United States at a conference arranged by the International Committee for Occupational Health and Safety (ICOMH) in White Plains, New York, in 1977 and later was published by National Institute of Occupational Safety and Health (NIOSH) (Bradley, 1979). Some years later, the initial findings were presented and discussed at the "Quality of Work Life Conference" in Toronto 1981. On that occasion I met with leading scientists belonging to the sociotechnical school originated by the Tavistock Institute: Fred Emery, Enid Mumford, and Lou Davies. One of the leading scientists in Scandinavia was the Norwegian professor Einar Thorsrud, who arranged for me to visit some major companies in the United States that made comparatively early and intensive use of computers, including AT&T in Newark, Procter & Gamble in Cincinnati, and the automobile industry in Detroit. As a result, I returned to Sweden even more convinced that the broader view I had undertaken regarding organizational issues and the psychosocial work environment, both theoretically and empirically, was appropriate and adequate.

In 1986, I presented my major research findings—both the social and organizational impact of computerization and its impact on some health aspects—at the first WWDU (Work With Display Units) conference in Stockholm. At that conference, I met with Professor Hal Hendrick of the University of Southern California (USC), who, shortly thereafter, invited me to chair a session at the Second International Ergonomics Association (IEA) International Symposium on Human Factors in Organizational Design and Management (ODAM-II) in Vancouver in 1986 (Brown & Hendrick, 1986). As a result of my participation in that conference, I concluded that my way of thinking and the way I was viewing work life were very closely related to what was entitled "macroergonomics" or "human factors in ODAM." This awareness led to my active participation in many subsequent ODAM conferences and macroergonomic tracks at both IEA Triennial Congresses and Human Factors and Ergonomics Society Annual Meetings in the United States.

In 1987, while a visiting scholar at Stanford University, I also met with researchers belonging to the "macroergonomics community" at the University of Southern California. In 1991, I was back at Stanford to discuss with colleagues and report on my research on the social and psychological aspects of applied artificial intelligence. I thereby initiated a new research project on competence development in high-tech companies in Silicon Valley in collaboration with key persons in macroergonomics from USC. In parallel, I designed research on electronic and human communication in order to apply for financial support in Sweden.

In 1994, I chaired ODAM-IV, held in Stockholm (Bradley & Hendrick, 1994). The ODAM-IV subtitle and theme, "Development, Introduction, and Use of New Technology—Challenges for Human Organization and Human Resource Development in a Changing World," was closely related to my research. ODAM-IV took place just before the explosive development of the Internet—in retrospect, it was good timing given the conference theme. It also was the first time colleagues from Eastern European countries were significantly represented at an international ODAM conference.

Structure of This Chapter

We are living in a time of accelerated technological development, which affects us all, both in our professional and private lives and in our roles as citizens. This is true with respect to the nature of our work, the design of our organizations, our communications with others, and our leadership and managerial roles. The term "IT" or rather, "ICT" as it is known in Europe, surrounds us progressively more often. Computers and computerization are other words that sometimes are used. For me, ICT is the combination of computer technology, telecommunications technology, and media. The convergence of these three creates a very powerful force. As used herein, IT and ICT have the same meaning.

Within social psychology, we talk about human physiological and psycho-social needs (e.g., influence, social belonging, learning and development, meaningful life content, and security). When these human needs are satisfied the individual is involved in society, but when these needs are not fulfilled, we become almost "strangers." The end result is that we feel powerless, without norms, and meaningless.

What, then, are the main social and organizational impacts of computerization and the recent merger of technologies? The first part of this chapter is devoted to this question, based on my research over the years, both in terms of empirical experiences and the theoretical models developed.

The second part of this chapter summarizes the six areas that constitute my general conclusions: (1) the allocation issue, (2) transfer and growth of knowledge, power, and influence, (3) work organization and work, (4) human communication,

(5) stress, and (6) ICT and higher education. These formed my primary research areas and I have studied them during different "historical" periods of technological development—from batch systems to the current network period.

EMPIRICAL RESEARCH

This section focuses on the changes in organization, in work content, and in professional roles resulting from computerization in large, complex organizations. Various data-processing systems offer different conditions for the structure and design of work and organization. Three main types of data-processing systems were studied in the research program, Computer and Work Life (RAM). These three system types are related to three phases in the history of computer technology: (1) a *batch-processing system* at a state-owned company, (2) an *online system with display terminals* at an insurance company, and (3) a *microcomputer system* at an electronics company. Problem areas concern computerization in working life. Essential concepts derived from the theoretical model were used to study the following: (1) general questions about the work environment and work satisfaction, (2) information and participation in decision making, (3) organizational design, work content, and workload, (4) promotional and development patterns, (5) contact patterns and communications, (6) salary conditions and working hours, (7) education and training, (8) evaluation of work roles and physio-ergonomic conditions, and (9) leisure time and health.

The empirical phases included both qualitative parts with interviews and observations, etc., and quantitative phases. Indices were constructed for the quantitative part of the collection and analysis of data as follows: psychosocial work environment factors (16 indices), physio-ergonomic work environment (4 indices), democracy at work (5 indices), effects of the display/terminal information system (14 indices), effects of the personal computer system (11 indices), health problems at work (5 indices), health problems during leisure time (6 indices), the effects of work on leisure time/family life (3 indices), the weight attached to specific work environment factors and the general attitude towards work (4 indices).

From the empirical experiences, a checklist was developed to be used in the evaluation of projects for implementing computer systems, or systems in use, to give a rapid assessment of which aspects of work and working conditions were to be improved and developed. The key words were specified and were complemented with examples of areas to be selected for analysis, design, and action. These measures and tools are still relevant for studies of the social impact of ICT.

Theories, methods, and results from the RAM program are summarized in the book *Computers and the Psychosocial Work Environment* (Bradley, 1989), based on 12 RAM research reports (ISSN 0349-2015) and briefly noted in this chapter together with a discussion on actions. This section also analyzes knowledge-based systems (KBS) in an engineering environment and the organizational and psychosocial aspects of KBS introduction and use—the *fourth phase of computerization*. The KBS systems formed a kind of predecessor of what herein is labeled as the *fifth phase of computerization—the network phase*.

In the network phase, we can observe that the three main technologies, noted earlier, are merging. The summarizing term for this merger is IT (information technology) or, alternately, ICT (information and communication technology). My present research concerns the impact of ICT on changes in life conditions at work and in private life. ICT has a deep impact on our professional roles, our roles as citizens, and our private lives (summarized in my ongoing research program in Sweden, entitled "Interplay ICT—Humans—Society."

WORK ORGANIZATION

Background

As noted in Chapters 1 and 3, the structure of an organization forms the framework for changes in individual job positions or in work tasks. The design of the organization is a crucial factor in the study of the relationship between data processing and the work environment, and this same organizational design can also be seen as a direct part of the work environment.

Changes in the structure of the organization may be a direct objective of computerization. A company may want to expand some units/operations and to reduce others, to eliminate some levels via rationalizations, to steer and control others via data processing, to localize some departments or larger units within the company, or to decentralize the operations of the company. It also is possible for employees to develop data processing systems that they need on a local level, and for these employees to cooperate with other sections of the operation. In this way, increased job autonomy can be developed on a local basis. However, changes in the organizational structure are often indirect effects, unintentional effects, of the use of data processing. The organizational design already in existence before computerization often is strengthened according to our research.

Use of telecommunication technology in combination with microcomputer-based equipment of various types (for example, word-processing systems, CAD/CAM systems, personal computers) created new professional roles and patterns of communication between individuals and vocational groups—exemplified in two studies on the changing roles of secretaries and engineers (Bradley, Bergström & Lindeberg, 1984, 1988). This results indirectly in a "new look" for the organization's structure, both in terms of formal organizational charts and informal networks. Such developments should be planned carefully and be based on knowledge about people and organizations, and the interplay between them.

What changes have taken place regarding organizational structure during the computerization process? Various phases in the development of technology can be identified. In the late 1950s and, especially, in the early and mid-1960s, the dominant philosophy proclaimed that data-processing power should be centralized—it was the golden age of the mainframe computer. This was the first phase of computerization. Larger and faster computers were designed with relatively large memory capacities, and used more for administrative applications than for scientific or technical applications. Data-processing departments began to proliferate.

During the later part of the sixties, this trend was broken as minicomputers (a second phase) appeared on the scene. They had limited memory and calculation capacities, but were capable of handling numerous technical applications and tasks. Moreover, they were easier to install and handle than the larger all-purpose computers. Minicomputers increasingly resembled mainframe computers in performance, but were available at much lower prices.

Then, in a third phase of this development, the microcomputer was created. The microcomputer became the major contribution of component technology to the need in the 1970s for instruments, machines, and other equipment and, later, the *PC explosion*. Microprocessors, "the heart" of the microcomputer, can be characterized as highly complex, but adaptable to final applications by users who produce the necessary software themselves; and as very inexpensive to purchase but imparting a higher value and usefulness to the products in which they are placed.

A fourth phase in computer history can be identified when applications of AI, expert systems, or knowledge-based systems came into broader use. AI stands for artificial intelligence, a branch of research in computer science that attempts to duplicate human intellectual processes. KBS is a system of software designed to assist humans in solving complex problems that require special knowledge. Various "experts" use different definitions for the above-mentioned concepts. My colleagues and I observed extensive efforts to create practical KBS applications in both the industry and service sectors.

The AI or KBS technology was a logical predecessor to the present and ongoing fifth phase—the network phase—characterized by the merger of three major technologies: computer technology, telecommunications technology, and media technology. Thus, we increasingly are talking about information and communication technology.

Work Organization During Four Computerization Phases

In keeping with the above, the RAM project took place in four phases, each with a specific technology background.

Phase 1: The first company studied, from 1974 to 1977, the Postal Giro Service, is an example of an organization that was exclusively using mainframe computers with *batch processing*. Below is a list of some changes in the design/structure of the organization that it was possible to attribute to the computer system: (1) a decrease in the number of levels in the hierarchy, (2) the closing down of a special key-punching center in the organization, (3) expansion of the data-processing department, and (4) creation of a special department for correcting errors made during the manual handling of data.

Phase 2: The RAM project also observed the introduction and use of *display terminals* within an insurance company (Trygg-Hansa), where both mainframe and microcomputers were used, from 1976 to 1980. The following changes in the structure of the organization were observed: (1) a tendency toward eliminating a number of levels in the organization; there had been too many supervisors on the intermediate level—a result of numerous mergers, (2) enlargement of a data-processing department, especially the commissioning of a large number of outside consultants, (3) introduction of a special word-processing department. Even before the use of display terminals, the company had democratized its structure, including the introduction of leadership groups, etc., which facilitated some of the changes. Thus, a more appropriate organizational structure had been established before the arrival of the technology. The simultaneous introduction of both modern office design and display terminals resulted in the collision of these two different methods of rationalization.

Phase 3: The third phase in the development of computer technology, *micro-computerization*, was studied at Philips Electronic Industries from 1979 to 1985. The following are specific implications of the use of microcomputers: (1) the enormous growth potential of the organization created by "cheapness" and "smallness"; (2) the invisible character of "microcomputerization," as compared to "mainframe computerization"; (3) major opportunities for decentralization; (4) major opportunities for centralization, such as linkage with computer systems and databases; (5) opportunities for local initiatives—other strategies for implementation of the technology.

The question of centralization and decentralization is often discussed in connection with the impact of computerization on the design of an organization. In addition, it is usually dealt with as a special data-processing policy goal that encompasses the centralization of system development and computer operations. The development of both minicomputers and microcomputers laid the groundwork for decentralization at the same time that the technology makes effective centralized coordination possible. In integrated information systems with telecommunications technology, the opportunities for both job autonomy and centralization increase. It then becomes critical to determine in which areas job autonomy is ideal and in which centralization is ideal (e.g., see Chapter 3 with respect to decision making and other sociotechnical considerations).

Phase 4: While a guest researcher at Stanford University in 1987, I conducted a pilot study that included semistructured interviews with researchers within computer science and the behavioral sciences, people employed in high-technology organizations, and people with early experience using knowledge-based systems.

A number of assertions and hypotheses were derived from these interviews that have been used as points of departure for the empirical studies. Some of these hypotheses were "tested" in three main Swedish companies that used KBS relatively early (in the aircraft industry, banking, and a computer manufacturer).

Work Organization and Knowledge-Based Systems (KBS): Hypotheses and Some Results

- A company that learns to develop its own KBS will find it affects the design of the organizational structure. However, this requires success at the time of its introduction.
- The hierarchic structure of a company will diminish with the introduction of a KBS. When computerization takes place both from the bottom up and from the top down in the company, the middle level will grow.
- Ambiguity in organizations will be more pronounced with KBS (e.g., ambiguity related to the limits of competence of the machine and the knowledge and skills of human beings).
- The concepts of power, influence, authority, and participation in the workplace must be redefined as understanding of the psychosocial aspects of using KBS increases.
- A company that has a formal and rigid definition of work roles will have a slower rate in introducing a KBS. An unsuccessful result is more likely to result in a company where work roles are less flexible and more formal.
- To be able to cope with the KBS phase of computerization, there will be a need for cooperative patterns, reviewed salary policies, and greater use of interdisciplinary and interdepartmental contacts.
- The challenge of KBS technology is the possibility for workers to observe the impact of their work on the entire production process.
- Combining KBS with traditional computers and telematics will strengthen a company's structure if it is characterized by networking, flexibility, and a flat organization.

Cognitive issues within KBS research received the most attention. Motivational issues, on the other hand, were little studied but especially important. The first empirical findings in Sweden confirmed that successful implementation of KBS depends on motivational and psychosocial issues—and that the multidimensionality of these issues will increase. Moreover, they confirmed that earlier phases of computerization had changed the psychosocial work environment at the lower levels of organizations (see the preceding section). The use of KBS also affected the psychosocial work environment of the well educated as well, with psychosocial problems becoming dominant over physical ones.

In the companies studied, a number of individuals gained more power under the KBS without formal promotion. Broader use of the KBS technique will tend to make power within a company invisible. Influence and power are found in new channels and one less tied to positions in the company hierarchy. A KBS influences the transfer of knowledge within a company, both directly and indirectly shaping the distribution of power and influence. The empirical findings on the organizational and social impact of KBS in the three main Swedish companies are summarized in Bradley and Holm (1991).

WORK CONTENT

Work Content During Four Computerization Phases

The rationalization philosophy that for centuries has dominated both blue- and white-collar work environments has resulted in a working life that can be characterized by rigid specialization, both vertically and laterally. Work tasks that were created for the lower levels have often been routine and predetermined in detail. But the division of labor principle applies to more than one level. A strict division between planning and operations produces an exaggerated bias in work tasks in both directions.

A new rationalization aid appeared about 40 years ago, data processing, which was "perfect" for the process of removing those jobs and tasks that had been created in the Taylorian spirit. For blue-collar workers, computerization has meant freedom from a series of work environments that are dangerous to health, such as those involving lifting heavy weights, noise, and air pollution. At the same time, stress factors were identified. In the newly created jobs, such as the processing operations, psychosocial demands dominated (e.g., analytic, skills, attentiveness, and swiftness in decision making).

Computerization removed routine tasks from both blue-collar and white-collar workers. Those who performed such work tasks increasingly work with other tasks involving different demands within the same company, branch, or even in entirely different sectors. In connection with the first step of the automation process among white-collar workers, a series of new and fragmented work tasks and occupations that are predetermined (in detail) emerged, through the application of the same underlying principle of rationalization. However, in the second step of automation, these tasks were disappearing. The long-term trend was less predictable and was dependant on other labor market conditions, interventions through education and training, and a new type of technology. Use of technically advanced and integrated communication and information systems implied a fundamental change in the entire office environment. More qualified jobs were affected, first in technical and business areas (design work, budgeting, production engineering, and planning).

Apart from research strategy, it often has been possible to attribute observed changes to needs that can be related to the theory of alienation, or further developments of such a theory. However, the discussion of work content, at an early stage of computerization (phases 1 and 2), began to include new dimensions, such as: vulnerability aspects in the organization (the possibility of making and correcting errors, interruptions in operation, and waiting times); personal integrity on the job; abstraction of task and knowledge; more formalization; and more standardization. Studies on knowledge-based systems (phase 4) constituted a break in the earlier trends (Bradley & Holm, 1991). New psychosocial aspects related to work content were identified, pointing to the present stage/phase and showing new, very positive changes for the individual related to work content. Joyful experiences and involvement in work tasks were increasing and were identified in the newly created jobs.

My colleagues and I also explored various stress phenomena in workplaces during the first computerization phases. We pointed out that it is desirable that a computer system contribute to the creation of work tasks and a work situation in which the workers are not subjected to the woes of either over- or understimulation. In our various studies, we identified a risk for both over- and understimulation and a split in the labor force—a phenomena entitled "the digital divide."

Whether computerization splits the workforce into highly and poorly qualified groups, or whether an overall decrease or increase in the level of qualification takes place, will (at least in countries like Sweden) depend to a certain degree on the role given to education and training in that society. We asked if the educational system should be used to force individuals to adjust and conform, or if it should be used offensively, for example, to increase the individual's chances of gaining influence (Bradley, 1989). Some possible future directions for education were also presented (see section on action strategies).

At an early stage in the debate on computerization, the pros and cons of people versus computers regarding their capacity to handle information were discussed. People's list of characteristics, was closely linked to human intellectual functions, but people's entire social contact function and its emotional components also were brought to the fore. How work should be divided between machines and people was often discussed. A problem arises when a work task contains some subtasks that are more suited to people and some that are more suited to computers. Closer analysis of these problems calls for human factors knowledge to establish the boundaries when using technology—not the least for ethical, moral, and psychological reasons.

When more qualified work tasks of an administrative and intellectual character were first rationalized, the impact of the new technological aids on the thought processes and the opportunities for developing creativity, imagination, and emotional commitment began to be taken into consideration. In addition, the language of individuals and linguistic development were discussed. Attention has thereby been directed toward processes and conditions that are related; for example, motivation, cognition, and emotion.

The issue of people versus information and communication technology (ICT) comes into focus in certain areas in our lives. During the Internet era (see below), ICT is being used to support, not replace, human intellectual functions (e.g., cognition) and support and deepen experiential qualities (e.g., emotion).

During the fourth phase of computerization, CAD/CAM systems had become an essential part of the design process in some professional groups. For the first time, a significant impact could be noted on professional groups that previously had held qualified, creative, and secure jobs. CAD systems initially were highly mathematical and expensive which made them viable only for use on large projects within giant companies.

A special study by Bradley, Bergström and Lindeberg (1988) addressed the use of CAD technology and the role of engineers and technicians. The engineers were using CAD systems for designing integrated circuits (chips). They had at least 4 or 5 years of academic studies and were graduates in engineering. The actual drafting work was done by circuit design technicians with secondary technical school training followed by several months of specialization.

The advantages of CAD included higher performance, better overview, varied and upgraded tasks, higher product quality, less need to check details, tidier, more attractive design, and greater accuracy. Our subjects quoted such subjective improvements in their work as more influence over their work content, pace, and scheduling of work breaks, more responsibility, stimulation, and challenge in tasks.

The disadvantages of CAD included physical work environment problems, great dependence on the computer, a tendency toward isolation from colleagues, and the difficulties inherent in CAD systems. They experienced such subjective disadvantages in working with CAD as work pace dependence on the computer system, the uneven pace of the work, the demand for attention and accuracy, and company demands for efficiency.

Engineers replied that CAD was not suitable for creative thinking and problem solving, and often maintained the mind-set that "I'm the one who must solve the problem." Circuit technicians, however, wanted automation as part of their duties. They also emphasized personal contacts with colleagues when planning the work, not merely contacts via computers. They felt that their work roles had undergone deep changes due to computerization.

Work Content and Knowledge-Based Systems (KBS): Hypotheses and Results

The following hypotheses, initially formulated in 1988, were confirmed. They address skills/education/training and KBS and are closely related to work content. They still have a central role in the debate surrounding new technology.

- When a company introduces KBS, *wider participation* occurs, involving all categories of professions and persons (e.g., the end-user, the domain expert, the knowledge engineer, and the project manager).

- Six types of *educational requirements* will be very important throughout the company: learning to solve problems, plan, make decisions, work in teams (schoolwork is individual, while most work in the future will be team based), learning about women's and men's jobs, and working between cultures.
- *Skills are transferred* by being driven upward and concentrated among the experts themselves—thus the expert use of expert systems. Skills also can be driven downward, enabling workers' use of expert systems.
- Successful use of KBS requires *plans for educating and retraining* the work force, plans for job security, and *alternative careers* within the professions involved. This is valid for experts, end-users, internal knowledge engineers, and other workers.
- More extensive use of KBS *requires more generalists*, who can create new syntheses. Specialization will continue, but *interdisciplinary skills* must be developed.

Regarding work content, both the experts and the end-users are getting enhanced possibilities to develop new knowledge. The work competence level will increase for those who are responsible for updating the system. One hypothesis is that experts become better experts within their field and develop better capabilities to solve problems; end-users will increase their capability by having continual access to knowledge and expertise. Both parts of the statement are verified by our results.

Regarding professional roles, it was argued that different experts may have responsibility for different areas of knowledge, but one individual is needed who has overall responsibility.

THEORETICAL MODELS

This section presents the initial theory on computer technology and work life (Fig. 9.1) and one "application" of the theoretical framework that especially addresses "stress" (Fig. 9.2). The convergence model (Fig. 9.3), elaborated 25 years later, is also discussed. Finally, actions related to the initial theories are noted.

A theoretical framework was developed by Bradley entitled "Computer Technology and Working Environment," and was published in 1977. The framework included two theoretical models. One model was general. In the second model, the concepts and their interrelationships were specified. The models were empirically tested in three large work organizations in Sweden. Collectively, these organizations represented the three main historical periods of computer technology—from systems with batch processing to microcomputerization. Studies also were undertaken in two front-line Swedish companies that introduced

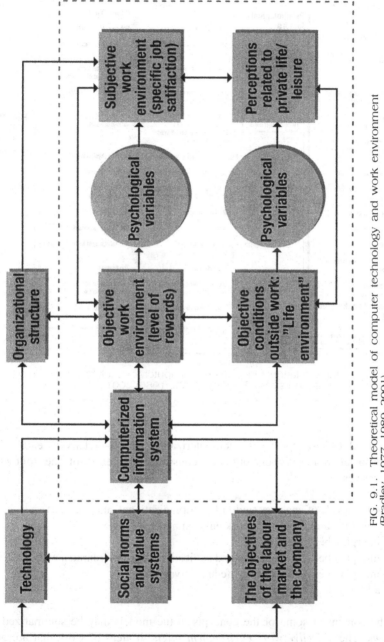

FIG. 9.1. Theoretical model of computer technology and work environment (Bradley, 1977, 1989, 2001).

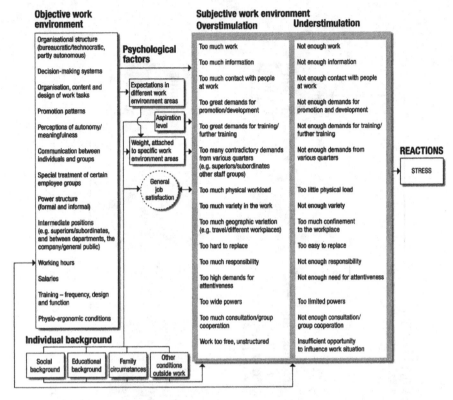

FIG. 9.2.   Relationship between computer technology, work environment, and stress (Bradley, 1977, 1986, 2001).

applied AI (KBS)—the fourth computerization period—relatively early. The
psychosocial work environment was considered in terms of the following
perspectives.

- three levels of analyses (individual, organizational, and societal)
- objective and subjective work environments
- interplay between levels
- interplay between objective and subjective work environments
- interplay between working life and private life
- a life-cycle perspective

The content of some of the concepts in the models may be summarized as
follows. The *objective work environment* refers to areas of work that are ger-
mane to large groups of employees. The *subjective work environment* consists of

perceptions and attitudes related to corresponding sets of factors in the objective work environment. The subjective work environment is closely linked to the concept of job satisfaction. The *organizational structure* includes the methods used to allocate work, the basis for decision making, and organizational aids. Organizational structure is seen as an important part of the work environment.

Psychological variables is a general term covering a number of intermediate psychologically relevant variables, such as the level of aspiration and the weight attached to specific work-environment areas. These variables are essential if one wishes to understand and explain the subjective work environment and its dynamics and, also, the perceptions of the conditions that govern our lives in general. Psychological variables are crucial in the analysis of subgroups of employees and cultural variations.

The concept of *psychosocial* refers to the process involving the interaction between the objective environment and the subjective one. *Essential concepts within the psychosocial work environment* include factors such as contact patterns and communication, organizational structure and design, work content and workload, participation in decision making, promotional and development patterns, salary conditions, and working hours. In the Anglo-Saxon use of the word, there is more emphasis on the individual and characteristics of the individual. I use the concept with the definition it is given in Sweden and in the Nordic countries in recent years in laws and agreements. I also refer to the theory supporting this definition.

The term *psychosocial work environment* is used to signify the course of events or the process that occurs when objective factors in the environment are reflected in the individual's perception (either positive or negative) of work and conditions of work. Its essence is the interaction between the environment and the individual. Thus work environment factors exist at different levels—the level of society, the level of the company, and the level of the individual—and they interact with one another. But there also is a distinction between the objective and the subjective work environments, which also interact (Bradley, 1989).

Objective conditions outside work refers to behavior and consumption, the conditions that prevail during the hours spent away from work. These may be affected by change related to the use of computerization at work. Certain attitudes, values, and experiences related to private life/leisure and family life also were analyzed against the background of the introduction of computer technology into the individual's work. This was in the initial phase of our research. Now, 25 years later, we are directly targeting the use of ICT in the home environment.

The theoretical models used in the RAM program also were used as models in discussing what *structure* a computerized society should have (see the two-way arrows in Fig. 9.2) and in specifying *desirable goals*. This was done in a special chapter in Bradley (1989) and also in the book, *Computers and Society* (Bradley, 1993).

The Convergence Model:
ICT and Life Environment

The fifth historical period of computerization I refer to as the "network period." It is very much based on the convergence and integration of three main technologies: computer, telecommunication, and media. ICT is increasingly being used in almost every activity and embedded in most of things around us. Thus, the graphical representations in the models need to be changed.

Comments on the model:

- A convergence of computer technology, telecommunication technology, and media is occurring.
- Professional roles (work life), private roles (private life), and citizen's roles also converge.
- Work environment and home environment are converging to become life environment.
- Effects on the individual are becoming more multifaceted and complex.
- Technology, norms and values, and the labor market interact in the globalization process.
- There is a new emphasis on certain dimensions in the psychosocial environment.
- New dimensions are appearing in the psychosocial environment. Openness to unforeseen implications is required.

Some comments are in order regarding the convergence model and informatics and other IT-related disciplines. A discussion about focus is taking place; both analysis and design need to address not only the work process and management of production life, but also people's life environment—moving from a work system to a life system perspective. Our roles as citizens and private persons are crucial. Community research in a broad sense has come to the fore with respect to both physical and virtual communities. Analysis and design of ICT and societal systems, both at the local level and globally, are now important. The labor market parties in the Nordic countries used to play a role in the discussion of system development and the orientation of research in informatics and ICT-related disciplines, emphasizing the participatory part. A renewal of the structures and focus within these organizations currently is taking place in Sweden and in Europe. There also is a need for new and additional actors in the integration of ICT in society (e.g., children, elderly, and consumer organizations).

The design perspective is broadening. Systems design, organizational design, and role design are increasingly accompanied by regional planning and societal design, and are key issues for the emerging IT-related disciplines. These design perspectives can be represented by convergent circles in a three-dimensional "circle of action." Of course, more traditional fields like architecture, industrial

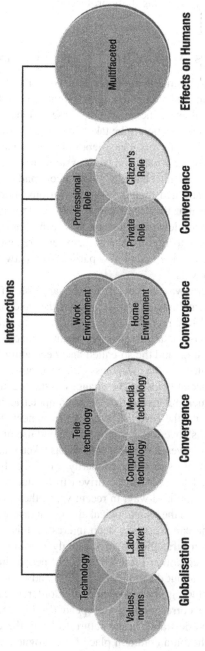

FIG. 9.3. Convergence model—ICT and psychosocial life environment (Bradley, 2000, 2001).

189

design, and graphic design are involved and, digital technology also interacts at the various levels of analysis among these professions.

Action Strategies According to the Theory

According to the main and initial theoretical framework (Fig. 9.1), people are affected by factors at the national, company/organizational, and individual levels. Individuals, in turn, can influence these factors. These different interaction levels also may feature discussion of action to be taken in computerization. What actions do the various players—the state, company, labor market, organizations, and individuals—undertake to achieve a balanced work system and a balanced society, to use the classical term from sociotechnical systems theory and macroergonomics?

An action tree was patterned that showed the various actions undertaken in Sweden during the time that the RAM project was underway (Bradley, 1989, 1993). Referring to the overview of actions on new technology and work life, the most challenging task was to find a balance between strategies at the different levels or "branches in a tree of action"—particularly between efforts at the national and individual levels.

For many years, there was success from the application of the so-called "Swedish model" or "Scandinavian model" within the field of new technology and work organization. More recently, that model has been debated and successively revised. These revisions are related to both the economic development in the country and globalization and the so-called "new economy."

The RAM project contributed an awareness of the various interacting factors of computerization, with special focus on the changes in the organizational and psychosocial work environment related to the use of computer technology. The term psychosocial was built into the Swedish Work Environment Act in which there also are certain paragraphs covering the introduction and use of technology. Laws cannot easily be changed—even if political change has taken place. Hence the psychosocial work environment issue, which primarily was driven by trade unions and their central federations, later come to be driven from the employers' side and, moreover, in a more decentralized way. In recent years, there has been a focus on technology—to keep up with the technological armament race worldwide.

At the same time, one can see that a deep interest in this broader type of research and related action strategies is increasingly coming from informatics, computer science, other computer professionals, and people involved in organizational change. Visions about desirable workplaces and societies are a focus of many conferences, including the International Conferences on Information Systems (ICIS) and the corresponding European (ECIS) and Asian (PACIS) conferences. It would be desirable in the future to bring the ergonomic people and the IT-people together in a common place for dialogue and mutual support. In this dialogue human computer interaction (HCI) and computer human interaction (CHI) societies are essential partners.

SUMMARY OF ICT AND CHANGES
IN WORK LIFE

I will cover six areas in my summary of the impact of ICT on work life: (1) the allocation issue, (2) transfer and growth of knowledge/power and influence, (3) work organization and work content, (4) human communication, (5) stress, and (6) ICT and higher education. They all relate to my main research areas and I have studied them during the different "historical" periods in the technological development of computerization.

The Allocation Issue and ICT Have
Come to the Fore

Twenty-five years ago, at a time when ICT was entitled EDP (electronic data processing), I used to close my speeches by arguing that computerization is really an issue of allocation. It has very much to do with allocating the so-called "good life."

- Allocation of (1) work and leisure time, (2) citizens' services (paid/unpaid), and (3) production and reproduction
- Allocation between cities and rural areas
- Allocation of profit between (1) sectors within a country, (2) industrialized countries, and (3) industrialized countries and the third world

Both physical power/strength (muscular) and "thought power" are being replaced. However, the amount of human life that is devoted to work—what we today entitle work—does not seem to have diminished. We have achieved a subdivision that has created one group that is overworked and one group that is excluded from the workforce. This is not necessary. More and more people could have access to a good life. The allocation question and ICT is per se an important research area. Furthermore, ICT can support the "weak" in society—those people who have various kinds of linguistic, physical, or intellectual handicaps. There is a need for contributions from all humans (Hirst & Thompson, 1996).

Transfer/Growth of Knowledge
and Influence

What we call "knowledge transfer" is an important phenomenon in the present and future knowledge society. This is valid for knowledge transfer between urban and rural areas, Sweden and Europe, and the center and the periphery, as well as globally.

The technology now used to transfer knowledge to people earlier was available only to experts, and access to knowledge has strengthened people—both in their professional roles, and their roles as citizens. Learning, in itself, is changing. We now can be coworkers and colleagues when shaping new knowledge. Distance education is becoming more and more common. The teacher's role and the student's role are also changing, and we are learning more and more together.

Hence, ICT involves a transfer of power connected with knowledge. A decomposition of the traditional hierarchical structure is occurring. There is an embryo for renewal. Long distance work, distance education, and long-distance services already provide new prerequisites for the role of regions (Bradley & Bradley, 1998).

An old lace cloth is, to my mind, the best model for how the world might work—what social systems, organizations, and official authorities will look like in the future. I think of a lace that was actually crocheted by my grandmother. The network era has arrived and seems to be here to stay. It is possible to crochet all the time; each new loop is connected to another through the same yarn. The distribution of power is now possible in a profound sense. Competence is transferred to the periphery, down to the line.

Work Content and Work Organization Are Changing in Depth

During the industrial era, which was characterized by automation of work tasks, research showed an increased alienation: Workers felt estranged from their work tasks and the results of their work, as well as in their personal relations. Work primarily was a source for earning money. During the early periods of computerization, however, new phenomena in work life arose (e.g., vulnerability and threat to personal integrity) and development of more abstract and formalized work tasks were focused in research on work life environment and in the discussion of the new technology.

What is happening now with work content and work organization? We have achieved more flexible work processes regarding both the professional role itself and leadership. Further, the professional role, the learning role, and the role of citizen are becoming more and more integrated. Repetitive jobs and physically strenuous jobs, including routine work, are beginning to vanish and a total upgrading of qualifications is occurring. In parallel, the organization is becoming flattened. The type of organizational structure that has become progressively more common is networks. From an international perspective, more work tasks are becoming similar because software programs are sold worldwide and the work tasks are carried out in a similar way.

The hierarchical structures of companies mirrored industrialization and industrial technology during the mainframe period of the computerization era. What characteristics does the network organization have and how are people affected by this new structure? Some examples include: (1) direct communication between the various levels of the organization, (2) barriers between idea and execution are disappearing, (3) reallocation of power in the organization, (4) continuous changing of structure and roles, and (5) openness to the surrounding world.

How Does the Workforce in the Flexible Company Function?

At the center there is a core workforce of permanent, full-time employees who enjoy a wide range of employment rights and benefits; however, the core workforce is decreasing. The other, increasing part, is called the peripheral workforce (Barnatt, 1997). This consists of part-time staff, self-employed consultants, subcontracted and outsourced workers, and temporary and agency employees. Some of these "knowledge workers" are key resources, while others are exchangeable. Through the network organizational structure, they might have very strong positions in the company because of their expertise or social contact, although this is invisible. Power is invisible in these new forms of organizations. Power has no outward manifestation and is not reflected to the same extent as before in properties and gadgets linked to leadership. One common trait is, however, that the peripheral workers are so-called free agents: They take care of their own security, skills development, and personal marketing. They are very loosely, if at all, tied into the welfare system. They are strong when health and good times are present, but are in a high-risk situation if health and family problems sap their energy and motivation.

Hence, we find increasingly *organic* organizational structures with a focus on flexible work processes, including dynamic networks for capital and human resources (compare the network organizations above). Economic systems are being created in which the present boundaries are increasingly eroded. Advanced ICT support is used for various forms of collaboration. More and more individuals function as self-governing company units.

ICT Has an Impact on the Social Contact and Communication Between People

There is a change occurring in the structure and quantity of the contacts and collaboration between people in work life, in private life (e.g., in homes), and in the community as a whole. Electronic networks, electronic subcultures, and electronic communities are emerging in work and private life with various functions.

If technology is used in a proper way, it can give us more time for human contacts. Unfortunately, in many cases, it has produced the opposite effect. Many people are working at a computer screen the whole day and only interacting with the computer—they do not meet with living human beings. The dialogue between people is running the risk of getting thinner. However, a new world is opening when we think in terms of the virtual company, the virtual classroom, and the virtual living room. It seems like these new opportunities provide us with an insight into the value of meeting in person—its importance for listening, for trust, and emotional support and safety. It is important that human emotional development can find its place. Sometimes people can hide behind the computer screen to avoid conflicts, or avoid difficult meetings with other persons. There are experiences that, in the long run, can contribute to personal growth and development. Children and young persons may lose an important time in their identity development if they are exposed too early and too intensely to electronic communication.

However, the feeling of belonging when meeting in person can be facilitated and ennobled with the support of electronic communication. Electronic meetings also can be deepened by follow-up contacts in the form of face-to-face meetings. This ought to be an important goal. "Electronic solitude," that is structural loneliness forced on a person, and which exists today, has to be prevented in the ICT society, or at least combated and counteracted.

We now have new collaborative structures—and possibilities for a deepening of democracy as well as international understanding. Today there are real prerequisites for peace.

Communication between people with an increased use of ICT makes it clear to us that communication has different purposes: a knowledge function, a social function, a control function and, not least, an expressive function. Qualitative aspects of communication (e.g., trust, confidence, interest, listening, and emotional engagement) are now essential. New dimensions in the quality of communication will occur.

With ICT Our Tempo Is Increasing: A Stress Factor

Our perception of time and space is changing. New opportunities for flexible work (telework), to work and learn independent of location, have changed our perception of space. Our requirements for ourselves regarding pace and tempo are becoming higher all the time—there is a change in the level of expectations (aspiration). An adaptation to the machine occurs. The pace of the machine affects us in an unconscious way. The words "slowly" and "fast" have new meanings. The same is true for the words "close" and "far away."

There is a basic level of stress in our technological environments in large cities. It used to be entitled "technostress," a phenomenon on the societal level. Technostress is an accelerated tempo which, to a large extent, is a consequence of effectiveness and efficiency. Perhaps technology fits better into a societal structure on the macro level where small scale, closeness to the environment, to nature, to the woods, lakes, and the sea, exists. One could talk about overstimulation, often in the big cities, and understimulation, often in rural areas, as promoting stress. This could be balanced.

Certain ICT stress is related to the fact that we have an increased dependency on computers, electronic networks, and on the equipment functioning reliably and properly. With the Internet, we have a super network that dominates electronic communication in business and private life. This dependency is accentuated when there is a demand for access to information. Stress phenomena in the Internet world are information overload, contact overload, demand for availability, lack of organizational filters, hard to separate "noise" from essentials, changing level of expectations, and changed perception of time and space in general.

The phenomenon called ICT stress can be characterized by too much or too little of various things, such as workload, information, contact with people, flexibility, opportunities for development, and training (Bradley & Robertson, 1991). Perhaps we could even revise plans and develop a "society of moderation."

ICT Has a Deep Impact on Learning and Education

What are the important competencies (capacities) that we will need in the future? My colleagues and I have studied both environments where ICT is being developed—the Silicon Valley in California and the corresponding area in Sweden, Kista, which is close to Stockholm—and environments where ICT is used extensively. In all of these environments, the following seem to be needed: (1) leadership rather than management, (2) creativity, (3) problem-solving capacity, (4) social competence and communication skills (human to human) (5) computer competency, (6) ability to work and function in multicultural environments, and (7) ability to cope with stress and psychological strain.

During the last two years, I have attended many conferences dealing with stress and ICT. What strikes me is the prevailing resignation: The focus is not primarily on preventing stress in the workplace, but rather, how to cope with the stress that we necessarily have to live with. This is a challenge for research. Some of my computer science students have started to become stress researchers—focusing on the preventive aspect. What can technology do? How can society contribute? What actions can be undertaken in both the close and more remote environment?

Jacques Delors' report, *Learning—The Treasure Within* (UNESCO), brings forward four pillars of education that I think are very central for all learning in the future:

- Learning to know (for knowing)
- Learning to do: (1) from skill to competence, (2) from intellectual to behavioral skills, and (3) innovation and creativity in a local context

- Learning to live to together: (1) discovering others and (2) working towards common objectives
- Learning to *be*.

IMPORTANT RESEARCH
AREAS FOR THE FUTURE

The theoretical models illustrate a view of working life and its processes. However, there is a risk that the same phenomena that occurs in working life in general—the fragmentation of knowledge—will also occur in the field of data-processing research and research into the consequences of computerization. Integration is easily lost, and it becomes increasingly difficult to gain an overview of research results in terms of computer technology and various combinations of new technology. Parts of the reality can be separated out and described in great detail—effects can be shown—but the force of these effects in the interplay with other parts and "fragments" is often uncertain. Atomization versus the holistic approach has always been one of the dilemmas of research, but becomes a more serious consideration in some disciplines—particularly when theories about humans are applied and there is a need for interdisciplinary approaches.

Focus on Psychosocial Life Environment/Quality
of Life and Well Being

I think that when we have been embedded in a society deeply and broadly affected by the new technology, it might be harder to identify both the risks and opportunities. A new generation has grown up in the digital environment. There are reasons today to go back to classics in research. There are no more work environments in the traditional sense—they are dissolving; but the phenomena identified in work life research (where Scandinavia was very active for many years) have to be reviewed with a new perspective. How are human needs for influence, belonging, and meaning being met in the new structures? In the development of "the home of the future," there is a growing market for all kinds of ICT applications—services and equipment. We need to analyze and design "the good home" (Bradley, Andersson & Bradley, 2000).

Some examples of desirable research focusing on the individual are: How is ICT changing our identity and self-perception, social competence, creativity, integrity, and trust? How is ICT changing:

- Being connected and not being connected
- Things and services
- Balance between emotional and rational components

- Balance between female and male
- Involvement and alienation
- Individual and collective.

Focus on Some Major Psychosocial Processes as Policy Statements

One way to summarize the discussion on the IT society and the individual is to address psychosocial processes. They could be formulated as policy statements or viewed as research questions. By now there are reasons to start up normative research that, later on, could be discussed across wider cultures. It concerns the classic question: Is research value-free?

- *Integration or isolation:* Normatively, ICT should contribute to an enrichment in social contact between people and should be used to prevent social isolation and facilitate integration.
- *Autonomy or control:* ICT should contribute to greater autonomy for the individual. Control or freedom is a classic issue, often described in terms of "privacy-integrity."
- *Overstimulation or understimulation*: ICT should facilitate information access for all and support individual learning, but at the same time prohibit various kinds of overload (e.g., information overload, contact overload).
- *Dehumanization or humanization:* ICT should contribute to the deepening and development of true human qualities and be used to provide time for people to develop themselves as human beings.
- *E-conflicts (wars) or E-cooperation (peace):* E-conflicts must be avoided. E-co-operation and peace are major concerns.

Focus on Synthesis

We need to focus on syntheses. Again I think there is a need for normative research in the next step where we place human welfare and quality of life for all as crucial societal goals.

Some main areas for research would be to study ICT and various cultures, ICT and democratic processes, and ICT and leadership—on a governmental level as well as in work organizations and schools, analyses of life styles and values, and new organizational models of work and life in general. ICT itself serves as a tool for both quantitative and qualitative analyses.

Research program/projects during the last few years in which I was directly involved are: (1) interactive creative environments (2) networking—organizational and psychosocial communication, (3) towards a global learning space, (4) the home of the future, societal and psychosocial challenges, and (5) ICT—comparisons between a rural community and a suburban community.

CONCLUDING REMARKS

In the summer of 2000, I was honored to serve as one of the keynote speakers at the IEA/HFES Triennial Congress in San Diego (Bradley, 2000). I closed my speech by using some animations with seeds and sunflowers. My comments were as follows:

> New applications of information and communication technology are like seeds. They pop up all the time. Some of them will not survive and cannot grow. Others will stay and grow to beautiful plants and flowers. This is true for those applications that fulfill human needs and requirements: Influence on one's own life conditions and society, social belonging, a meaningful life content, learning and developing oneself. We will hopefully have a field of flowers that will broaden and deepen quality of life for all.

As I pointed out at the beginning of this chapter, the perspective in my research is very similar to Chapter 1, "An Overview of Macroergonomics", even if the language is somewhat different. Using my initial theory and the convergence model as a basis for analysis, the integrated design criteria in macroergonomics are most helpful in the design, not only of our workplaces, but also for our homes which are becoming centers for a lot of activities related to the professional role and the citizen's role, in addition to the private role—a pronounced communication sphere. Hence, in the design of the ICT society the following macroergonomic design criteria have a key role:

- *Joint design:* Human or personnel subsystems are considered jointly with the technological subsystems, with a built-in process of participation and influence by people.
- *Humanized approach:* The starting point in design should be people and humans needs, capabilities, competencies. Later, various allocation issues can be considered.
- *Integration of the organization's sociotechnical characteristics into the design:* A continuous evaluation process must be done at various levels of analysis (organization, individual, societal) as a basis for integrating the crucial psychosocial factors into the design—to lead towards well being and quality of life for all, or a balanced society.

10

Macroergonomics
of Hazard Management

Thomas J. Smith
University of Minnesota

INTRODUCTION

The control of hazards by biological systems to avoid danger and secure survival
is as old as life itself. Emergence of organized systems of work, more lethal
weapons, and increasingly complex technology has prompted human appreciation of hazard control as a key to safety, security, and productivity. This suggests
that management of hazards should represent an integral aspect of macroergonomics. Nevertheless, the recognition of safety as a legitimate management
function is relatively recent, emerging in the United States gradually only over
the past five decades or so. Further, there is by no means a consensus within the
safety community regarding what it is about safety that should be managed.
Indeed, "loss control," "accident prevention," and "safety" itself are cited far
more often than hazards as the proper management focus.

Accordingly, this chapter first provides definitions, scope, rationale, and background for a macroergonomics of hazard management (HM). Subsequent sections introduce a behavioral cybernetic model of HM, and go on to address
macroergonomic keys to HM embodied in the model.

Definitions and Scope

In this chapter, the terms *safety management* and *safety program* refer to any organizational function or program with a general focus on safety and accident prevention, whereas the term *hazard management* refers to a safety program with a specific focus on detecting, evaluating, and abating hazards. *Hazard* refers, in a general sense, to any work design factor that elevates the risk of detrimental performance by a worker (employee or manager) or an organizational system. *Macroergonomic* refers to the organizational design and management (ODAM) characteristics of a safety or HM function, program, or system. *Safety performance*, in a broad sense, refers to the integrated performance of all organizational and individual entities whose activities affect safety.

Heinrich (1959, Appendices I and II) and Montgomery (1956) provide a chronology and historical background of the safety movement, from the classical to the modern era. Pope (1990) provides an extensive set of safety definitions. *Safety* refers to freedom/security from danger, injury, or damage. Hazard is derived from Old French and Middle English terms of the same spelling referring to games of chance—the term thus connotes chance or risk.

Even today, the meaning of the term safety is not uniformly understood within the safety community and elsewhere. Pope (1990) asserts that the term lacks absolute definition and is not acceptable for precise administrative language. Petersen (1971, p. 26) suggests that "safety is not a resource: it is not an influence: it is not a procedure: and it is certainly not a 'program.' Rather, safety is a state of mind, an atmosphere that must become an integral part of each and every procedure that the company has." Zaidel (1991), with reference to driving safety, suggests that, "safety . . . represents behaviors, situations, or conditions that . . . are associated with either higher or lower probability of accidents."

The latter definition invokes a long-standing and cherished assumption in the safety field that the primary concern of safety is dealing with accidents. Indeed, the first text of the modern era on occupational safety is entitled, *The Prevention of Accidents* (Calder, 1899). Webster's definition of "accident" is instructive: "an event that takes place without one's foresight or expectation; an event which proceeds from an unknown cause, or is an unusual effect of a known cause, and therefore not expected; chance." In a similar vein, Pope (1990, p. 107) offers a series of definitions of the term, all of which refer to the unexpected, unplanned, and uncontrolled nature of accidents.

Some in the safety community observe that given their unpredictable and uncontrollable nature, a focus on accidents is incompatible with a philosophy of safety management. Guarnieri (1992) notes that the word accident has almost disappeared from safety science and engineering use.

The safety field is huge with an extensive literature; Pope (1990, pp. 71–75) cites some 163 texts and monographs relevant to the field published since 1899, over half published since 1975. The primary purpose of this chapter therefore is to develop and address major themes, concepts, principles, and practical considerations relevant to the macroergonomics of managing hazards. The primary focus will be on occupational safety at the organizational and workplace levels, with an emphasis on U.S. experience. Chapter 16 deals with the macroergonomics of major disasters.

Rationale and Background

Accidents by their very nature are inherently unmanageable. Safety management has emerged as a major focus of the safety field. However, the imprecise meaning of safety leaves unclear what it is precisely that is being managed. Connotations of the term loss control may go beyond safety-related loss, and hazard control arguably represents an essential prerequisite for loss control.

The term *hazard management* avoids these shortcomings. According to Webster, the term management refers to, "the executive function of planning, organizing, coordinating, directing, controlling and supervising." Hazards represent concrete, recognizable entities whose existence can be characterized and quantified. With adequate data, informed judgments and predictions can be made about possible outcomes associated with the prevalence and persistence of particular hazards. Reduction in risk associated with hazard modification or elimination thus can be estimated. In other words, hazard conditions represent a highly eligible target for management. Moreover, reference to other chapters in this text suggests that HM represents a prominent target of macroergonomics generally.

Today, hazard control appears to represent the prevailing focus of the safety management profession. Grimaldi and Simonds (1989, p. 16) illustrate this point:

it seems clear that the pursuit of safety must be pointed toward the identification of hazards, determination of their significance, evaluation of the available correctives, and selection of the optimal remedies. When this path is followed, it ends with the control of unwanted events at an irreducible minimum. "Accidents" in this case are not the target. Instead, it is the *hazard* that causes the unwanted harmful event, which is to be eliminated. This is the primary concern of safety management.

Recognition of safety as a legitimate management function and target for macroergonomic analysis is relatively recent. An early statement of this viewpoint is that of Agricola (1912) in 1556, who notes in a discussion of mine accidents that, "often, indeed, the negligence of the foreman is to blame." More recently, a broad consensus has emerged in the safety community. According to the compilation of Pope (1990, pp. 71–75), from 1956 through 1990 a total of 18 texts and monographs with the term safety management in the title were published. The National Safety Management Society was founded by Creswell and Pope in 1966 (Pope, 1990, p. 56). Support for the management approach to safety is provided by Borowka (1986, 1988), Creswell (1988), Cutter and Wilkenson (1959), DeJoy

(1994), Marcum (1984), Miller (1982), Morris (1984), Nuernberger (1956), Pope (1981), Pope and Cresswell (1965), Veltri (1991), and Wangler (1986).

One consequence of the emergence of the concept of safety as a management function is an increased emphasis on management responsibility for safety problems. Indeed, some advocates of safety management assume that any and every safety problem represents a management defect. The first principle of loss control cited by Petersen (1971, p. 19) illustrates this point of view: "an unsafe act, an unsafe condition, an accident: all these are symptoms of something wrong in the management system." This perspective provides a strong rationale for the application of macroergonomic principles and practices to improve the design of safety management systems.

What kind of hazards are we referring to as the object of HM? An ongoing debate in the safety community contrasts the role of hazardous conditions versus unsafe behavior in accident causation. This debate originates with Heinrich (1931, 1959), who in the later publication argues that 88% of 75,000 industrial accident cases reviewed were attributable to what he calls "unsafe acts of persons." The prevailing view is that accidents have multiple, complex causes related to both physical-environmental hazards and behavioral variability (Petersen, 1971).

The rationale for the present chapter rests in part on this debate because the prominent focus of ergonomics is on design variability, as opposed to behavioral variability. The model in the next section presents a behavioral cybernetic perspective on multiple causation theory, in which safety performance of the organizational system is viewed as a process of interaction between the human and design elements of the system, and hazards are viewed as manifestations of flaws in workplace design subject to organizational management and control.

More specifically, the nature of risk associated with hazards can be understood from a human factors and safety performance perspective with the central concept that across many different categories and forms, a hazard represents a job or workplace factor that gives rise to a fundamental mismatch between worker behavior and workplace conditions (K.U. Smith, 1973).

K.U. Smith is one of the pioneers in adoption and development of HM as a human factors–based systems theory and process for safety improvement (K.U. Smith, 1975a, 1975b, 1979, 1988, 1990). With a series of HM projects and programs in the Behavioral Cybernetics Laboratory at the University of Wisconsin-Madison, Smith elaborated the theory, principles, and applications of HM in the 1960s and 1970s as an early macroergonomic formulation of worker–management participation and union–executive cooperation in operational management of safety and production efficiency in the workplace. The HM initiatives of Smith are among the first specific instances in management science in which macroergonomic principles were applied to safety management of complex systems. As such, they can rightly be viewed as early pioneering efforts in organizational macroergonomics generally and safety macroergonomics particularly.

BEHAVIORAL CYBERNETIC MODEL
OF HAZARD MANAGEMENT

The term *ergonomics* refers to the natural laws (*nomos*) of work (*erg*) (Konz, 1995, p. 12). In the ergonomics/human factors (E/HF) community, broad agreement has emerged that such "laws" refer to design factors, conditions, or strategies that accommodate human capabilities and limitations, and that E/HF science is intimately linked to design (Chapanis, 1991; Meister, 1989).

Given the fundamental focus of the field on design, it may be argued that the basic premise underlying E/HF science is that variability in the behavioral outcomes of human performance and work—such as safety, health, quality, productivity, or user acceptance—is critically influenced by design attributes of the environment in which performance occurs (T.J. Smith, 1993, 1994, 1998, 1999; T.J. Smith, Henning & Smith, 1994). This premise forms the basis for the HM model presented in this section, which assumes a closed-loop or cybernetic relationship between workplace design conditions (including hazards) and safety performance of organizational systems (behavior) to account for differences observed in the safety record of different systems in different contexts.

This model is based on behavioral cybernetic theory (K.U. Smith, 1972; K.U. Smith & Smith, 1962; T.J. Smith & Smith, 1987), which maintains that human behavior is guided as a closed-loop, feedback-controlled process. A major implication of this theory is that to a substantial degree, variability in the control system (behavior) should be referenced to variability in what is being controlled (sensory feedback). The cybernetic model of behavior thus establishes a biological basis for the interdependence of performance and design, supported by an extensive body of empirical evidence (Flach & Hancock, 1992; Meister, 1989; T.J. Smith, 1993, 1994, 1998; T.J. Smith, Henning & Smith, 1994).

Figure 10.1 applies behavioral cybernetic concepts to the particular case of HM. Elements of the hazard control system specified in the shaded region of the figure are: (1) a hazard detection system (the sensor), which provides a means of monitoring sensory feedback from workplace hazards (input); (2) a safety reporting system (the controller), which provides a means of referencing actual with desired safety performance; (3) actual safety performance (output) of the organizational system (effector); and (4) system safety performance targets and goals (the reference). The model assumes that effective management control of safety is mediated by comparison of feedback from actual system safety performance with safety performance targets and goals. Error between actual and desired performance is used to guide management decision making directed at tightening system control over workplace hazards (hazard feedback control), thereby reducing system performance error and improving accident and injury prevention. Many years ago, Juran (1964) introduced a similar servomechanism model to characterize the management of system quality.

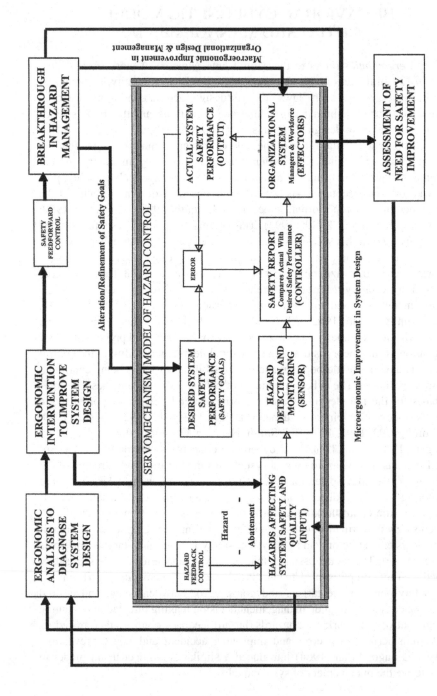

FIG. 10.1.   Closed-loop control model of hazard management, with ergonomic analysis and intervention specified as keys to breakthrough in system performance in a feedforward control manner.

It is the area of desired performance—safety goals and standards—that distinguishes the macroergonomics of HM as a matter of particular complexity in both the design and the operation of the organizational system. This is because of the number of players with a direct, vested interest in influencing workplace safety goals. In addition to managers, the workforce, and customers, these also include insurance companies, government regulators, and labor unions. By comparison, the latter three parties typically are less immediately involved in setting targets for such organizational performance attributes as quality or productivity.

One of the implications of the model in Fig. 10.1 is that it points to a role for ergonomics in facilitating breakthrough in safety performance. The key to maintaining and sustaining performance of a system at any given level is the operation of system *control*, defined by Juran (1995, p. 1) as "staying on course, adherence to standard, prevention of change." As he points out (Juran, 1995), "under complete control, nothing would change—we would be in a static, quiescent world." However, Juran (1995, p. 3) goes on to observe that: "control can be a cruel hoax, a built-in procedure for avoiding progress—we can become so preoccupied with *meeting* targets that we fail to challenge *the target itself*—this brings us to a consideration of breakthrough."

Juran (1995, p. 3) defines managerial breakthrough as "change, a dynamic, decisive movement to new, higher levels of performance." The model in Fig. 10.1 conceives of breakthrough as a closed-loop process involving *feedforward* rather than feedback (i.e., servomechanism) control (T.J. Smith, 1999). Feedforward control is ubiquitous among biological systems. It enables them to rely on sensory feedback from present conditions to project their behavior into the future.

Similarly, the impetus for breakthrough in organizational performance is some feedback indicator suggesting that the current level of performance is no longer adequate. This prompts the system to initiate breakthrough by feedforward projecting its behavior into the future to achieve a new, improved performance level. It is likely that often, if not always, the root cause of inadequate performance at any given level is some sort of design flaw, either in microergonomic design of the work process or environment, or in macroergonomic design of the organizational system, or both. In the safety field, these design flaws are termed hazards (Fig. 10.1).

This is where ergonomics comes into play. As denoted by the system control elements outside the shaded region in Fig. 10.1, ergonomic analysis can be used to detect system hazards. Ergonomic intervention then can be used to abate hazards and improve system design in order to facilitate the breakthrough process. Application of ergonomics can thus serve as a key breakthrough strategy by means of which the system elevates its safety performance from one control level to the next. In this manner, the organizational design incorporates ergonomic systems provisions for alteration of the safety goals themselves to deal with altered circumstances (new regulations, new technology, etc.).

The basic premise in Fig. 10.1 is that an occupational ergonomics program meets this requisite system design need for initiating and managing breakthrough.

Methods of ergonomics analysis are admirably tailored for hazard assessment; results of this analysis therefore can serve as early warning sentinels for the need for initiating a breakthrough process. Once design problems have been identified, methods of ergonomics intervention then are admirably suited for contributing to problem resolution, through microergonomic improvement in system design features, including hazard abatement, macroergonomics improvement in organizational design features, and alteration or refinement of safety targets and goals.

A second implication of the model in Fig. 10.1 is that variability in safety performance of the organizational system should be prominently influenced by design characteristics of the particular set of hazards at which the HM process is directed. In other words, the model predicts context specificity in organizational HM performance analogous to that observed with individual performance (T.J. Smith, 1993, 1998; T.J. Smith, Henning & Smith, 1994). This in turn suggests that there should be no "magic bullet" or completely generalized scheme for macroergonomic design of HM systems. Rather, the design of each HM system must be customized and specialized in relation to the particular mix of hazards being addressed. Moreover, as or when the design of the hazard environment is modified (through abatement of existing hazards, introduction of new ones, or both), it may be necessary to in turn modify the macroergonomic design of the HM system itself to deal most effectively with the new hazard mix.

MACROERGONOMIC KEYS
TO HAZARD MANAGEMENT

The remainder of this chapter is devoted to a discussion of various macroergonomic strategies whose adoption might support breakthrough in HM success specified in the figure. Rather than reviewing widely accepted safety management strategies (Grimaldi & Simonds, 1989; Heinrich, 1959; Petersen, 1971; Pope, 1990), the emphasis is on innovative macroergonomic approaches to safety that may not be generally familiar to the safety community, yet whose benefits have been reasonably well documented and whose effects on safety therefore should synergistically complement those of more widely recognized strategies.

Organizational "Best Practices"
for Safety Management

One obvious source of inspiration for HM breakthrough is the safety management practices of companies with commendable safety records. Cohen (1977) and M.J. Smith and colleagues (1978) have investigated factors and characteristics that differentiate highly successful from less successful occupational safety programs. Both studies found that relative to high accident rate companies and sites, low accident rate companies and sites are distinguished by the following ODAM approaches to safety (the order does not represent an importance ranking).

1. Greater management concern and involvement in safety matters, reflected in regular inclusion of safety issues in plant meeting agendas, and routine, personal inspections of work areas by top plant officials.
2. Greater management skill in managing both material and human resources.
3. Full-time safety director who reports to top management.
4. Greater emphasis on hazard control, with tidier work areas, more orderly plant operations, better ventilation and lighting, and lower noise levels.
5. More open, informal communication between workers and management, with frequent contacts between workers and supervisors on safety and other matters.
6. More stable workforce, with older, more experienced workers, and less absenteeism and turnover.
7. Well-defined employee selection, placement, and job advancement procedures.
8. More flexibility in discipline for safety violations, such as safety counseling in addition to suspensions and dismissals.
9. Greater availability of recreational facilities for worker use during off-job hours.
10. Greater effort to involve worker families in safety campaigns.
11. Frequent use of safety incentives and promotions.
12. Frequent use of accident investigations and formal accident reporting systems.
13. Formal safety training for employees and supervisors.

Both studies conclude that management commitment to safety is the most critical key to HM success, and also report that customary use of conventional safety practices, such as safety promotion, safety training, and accident investigations (Points 11–13), did not differentiate as strongly between low versus high accident rate companies.

In its emphasis on the link between safety performance and management of the organizational system, and on the importance of hazard control, the model in Fig. 10.1 is aligned with findings cited above. More broadly, the significance of macroergonomics to success in safety and HM is supported by the following conclusion of Cohen (1977, p. 177).

Overall, this review would suggest that the more distinguishable elements of successful safety performance rest largely on psychological or human factor considerations. Management commitment, aspects of interpersonal communications and interaction, early safety indoctrination and follow-up training, workforce stability, and personnel development and support programs all fall in the psychological domain. While not belittling the importance of engineering approaches to accident prevention, the evidence here would argue for increased emphasis on nonengineering measures.

M.J. Smith and Beringer (1987) echo this theme in advocating four basic E/HF approaches that can be used to control workplace hazards: (1) applying ergonomic principles to the design of workplaces and jobs in order to provide work situations and environments that capitalize on worker skills and capabilities; (2) designing organizational structures that encourage safe working behaviors; (3) training workers in hazard recognition and in employing effective work behaviors for dealing with these hazards; and (4) improving safe work behavior through work practices improvement.

Worker Self-Regulation of Hazard Management

Probably no component of the safety system modeled in Fig. 10.1 has received greater attention than that of hazard detection and monitoring. As noted earlier, understanding in the safety community of what a hazard is has been dominated by the thinking of Heinrich (1931, 1959), who focuses on unsafe acts (behavioral hazards) and unsafe conditions (physical hazards) as the hazard categories of greatest concern.

A fundamental problem with the characterization of accident and injury risk in terms of behavioral and physical-environmental hazards is that HM strategies directed at abating these hazards do not reliably and predictably result in accident and injury prevention. For example, Jones in 1973 documented the limitations of dependence on engineering controls and related enforcement measures in bringing down occupational injury rates to acceptable levels. Gill and Martin (1976) echo this finding with the claim that the engineering approach to safety management is not sufficient to prevent all accidents, and that innovative hazard control strategies such as performance standards and worker participation should be considered to achieve further improvement in accident prevention. M.J. Smith and colleagues (1971) and Gottlieb and Coleman (1977) report that inspections carried out by OSHA are relatively ineffective in preventing accidents because most hazards cannot be identified by traditional workplace inspections. Results from these and other studies suggest that, in general, only 5 to 25% of accidents can be avoided by rigorous compliance with conventional safety standards (Ellis, 1975; K.U. Smith, 1979). OSHA itself has noted that, "OSHA's own statistics appear to indicate that 70–80% of all deaths and injuries each year are not attributable to a violation of any OSHA specification standard" (Occupational Safety and Health Reporter, 1976, p. 684).

These considerations suggest that along with physical and behavioral hazards, there is an entire additional domain of hazards related to macroergonomic design defects in the management of safety. The analyses of Cohen (1977), Ellis (1975), and M.J. Smith and colleagues (1978) suggest that in terms of their adverse impact on safety, this class of hazards may be more significant than either behavioral or physical-environmental hazards. This perspective has prompted the development and testing of a range of macroergonomic strategies that emphasize the role of workers in self-regulating their interaction with workplace hazards and with the occupational safety system. Various aspects of this topic have been reviewed by Cohen, Smith, and Anger (1979), Coleman and Sauter (1978), K.U. Smith (1979, 1988, 1990), M.J. Smith (1994), M.J. Smith and Beringer (1987), and T.J. Smith, Lockhart, and Smith (1983).

The concept of worker self-regulation of HM is based on the premise that compared with managers and safety professionals, workers often know as much or more about their jobs, about job-related hazards that they encounter on a daily basis, and about how to reduce or eliminate those hazards. Four macroergonomic strategies based on this premise, three of which explicitly involve the participatory approach (Chapter 2), are summarized in sections below: (1) the worker hazard survey; (2) measures that encourage worker self-protection against workplace hazards; (3) worker involvement in safety program decision making; and (4) ergonomic intervention to improve HM.

Worker Hazard Survey

The purpose of the worker hazard survey, developed by K.U. Smith and colleagues (K.U. Smith, 1979, 1988, 1994; M.J. Smith, 1994; M.J. Smith & Beringer, 1987; T.J. Smith, Lockhart & Smith, 1983), is to obtain information from workers about the identity and control of job-related hazards. The survey therefore represents one alternative for the hazard detection and monitoring stage of HM, as depicted in the model in Fig. 10.1. Survey information is acquired through short interviews conducted by unbiased interviewers.

The major purpose of the worker hazard survey is to discover what workers know about job hazards, and to establish what knowledge, skills, techniques, and actions workers employ in detecting and controlling hazards. To this end, an interview with a worker should focus on: (1) identification of job hazards perceived by the worker, itemized in order of severity; (2) worker rating of each hazard (high, moderate, or low risk for causing accident or injury); (3) methods used by worker to detect and control identified hazards; (4) worker recommendations for reducing or eliminating hazards; (5) worker ideas and recommendations for improving performance skills required to control identified hazards; and (6) description of near, minor, or major accidents associated with identified hazards.

Figure 10.2 illustrates a worker hazard survey form developed for this purpose (Kaplan, Knutson & Coleman, 1976). M.J. Smith and Beringer (1987, Fig. 7.1.5, p. 779) illustrate an alternative form developed for the State of Wisconsin Department of Industry, Labor and Human Relations.

The worker hazard survey process can be used as a base, or foundation, for introducing an entire HM program. In particular, once worker hazard survey information has been collected, a series of organizational follow-up steps are recommended (K.U. Smith, 1979, 1988, 1994) to develop and establish a defined HM program. These are summarized in Fig. 10.3 in flowchart form.

WORKER HAZARD SURVEY FORM

Organization _____ Division _____ Department _____

Job _____ Shift _____ Interviewer _____

Date _____ Telephone _____ Fax _____

List of hazards in order of severity. For each hazard, describe specific job operation (related to situation, tool, equipment, materials, process, or organizational factor) associated with hazard.	Description of specific tool, physical machine or equipment, work station, and/or interface associated with hazard.	Description of techniques/methods/procedures used for detecting, monitoring and controlling the hazard specified.
1. (most severe)		
2.		
3.		
4.		
5.		
6.		
7.		
8.		
9.		
10.		

Description of prior accidents with this job associated with any specific operations and hazards listed above:

FIG. 10.2. Worker hazard survey form.

Hazard Management Benefits
of Worker Hazard Survey

That workers can recognize hazards, and have knowledge and skill in self-controlling them to improve safety performance, represents the fundamental premise of the worker hazard survey. This premise has been tested, with positive results, in a series of cross-industry studies conducted within the conceptual framework of the University of Wisconsin HM program (Cleveland, 1976; Coleman & Sauter, 1978; Coleman & Smith, 1976; Gottlieb, 1976; Kaplan & Coleman, 1976; Kaplan, Knutson & Colemen, 1976; Richardson, 1973; M.J. Smith, 1994).

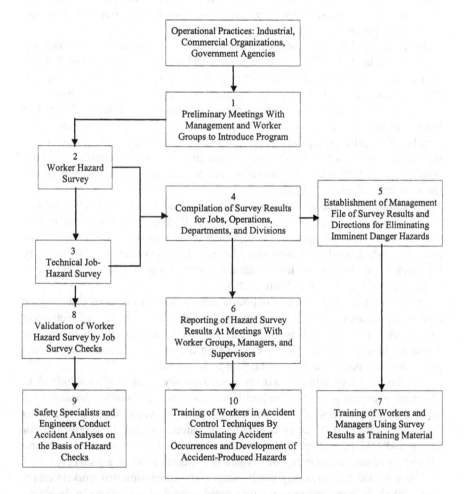

FIG. 10.3. Process flowchart for installation of a hazard management program.

Generally, the worker hazard survey results reveal that employees are one of the best sources of information about day-to-day hazards of their jobs. Major conclusions supported by this work can be summarized as follows:

Workers generally identify hazards that most directly affect their personal safety. Unsafe conditions and operating procedures often are specified, but it is not uncommon for possibly hazardous actions or behaviors to be candidly cited.

Worker-identified hazards largely comprise those of immediate self-concern. Whereas government standards are specification-based, tending to focus on physical-environmental hazards, workers tend to identify a broader range of hazards, often related to job performance, to dynamic operational circumstances, and/or to intermittent or transitory work situations or conditions, for which no physical standards or standard operating procedures exist. Hazards of this type are termed *operational* hazards (K.U. Smith, 1979, 1988, 1994).

Based on worker hazard survey results from six Wisconsin metal processing plants (Cleveland, 1976; Coleman & Sauter, 1978), employee-identified hazards bear a significantly closer resemblance to actual causes of accidents than do those designated by government specification standards or state inspections.

Experience with the worker hazard survey method suggests that for every 100 hazards reported by employees, 1 is very serious and should be addressed immediately, 24 require prompt attention to avert a potential accident, 50 represent less serious hazards that may require some minor action to improve working conditions, and 25 relate not to safety but to personal employee complaints (M.J. Smith & Beringer, 1987).

Experience with the worker hazard survey approach in the public sector was not uniformly successful (Richardson, 1973). Follow-up by management to deal with identified hazards appeared to represent the key to success in public agencies; in agencies in which management failed to take action and to provide safety feedback to employees, the hazard identification programs failed due to lack of employee involvement (M.J. Smith, 1994).

Workers typically can clearly delineate their actions in avoiding, reducing, or controlling hazards. Workers tend to identify health hazards in terms of their acute toxic effects, rather than their long-term potential for causing illness or disease. There is substantial awareness of engineering defects in ventilation or personal protection that may cause increased risk of toxic exposure.

Employers cooperating in worker hazard surveys generally conclude that: (1) employees can provide more and better information about job hazards than other available sources; (2) workers often identify hazards that management is totally unaware of; (3) survey results are of value in establishing priorities for safety and HM procedures; (4) the surveys do not tend to disrupt work; and (5) employee awareness of hazards typically increases following a survey.

These worker hazard survey results support the conclusion that workers represent a relatively untapped resource for hazard control. A few others in the safety community have reached this same conclusion. Swain (1974) refers to workers as subject-matter experts on their work situations, whose input to management can serve as a key contributor to safety performance. He notes that worker hazard data represent a rare but effective source of information about near accidents and potentially unsafe situations. Hammer (1972, 1976) likewise presents evidence showing that workers have high awareness of job-related hazards and that asking them directly represents the best way of obtaining this information. Both authors note marked discrepancies between worker-provided hazard data versus that provided by more conventional specialist reports and postaccident investigations.

Worker Self-Protection Against Workplace Hazards

Given findings cited above that workers readily can identify and evaluate risk associated with workplace hazards, it seems logical to suggest that macroergonomic initiatives directed at encouraging more effective behavioral strategies on the part of workers in interacting with and controlling hazards might also yield safety benefits. Results from a number of such initiatives, variously termed *worker self-protection* or *behavioral safety* programs and built primarily around provision of safety performance feedback, training, and safety incentives, support this assumption. Worker self-protection programs therefore represent a viable option for the hazard feedback control stage of HM, as depicted in the model in Fig. 10.1.

A conceptual analysis by Cohen, Smith, and Anger (1979) of strategies that promote worker self-protective behaviors targets use of safety training, incentives contingent on worker safety performance, "selling" and "telling" safety communication, and tangible expressions of management commitment to safety.

These recommendations are supported by the review by McAfee and Winn (1989) of a series of 24 field studies, dating from 1971 to 1987, that examined use of safety incentives and/or safety performance feedback to enhance workplace safety. The major finding was that every study reviewed, bar none, found that use of safety incentives and/or feedback enhanced workplace safety and/or reduced accidents, at least in the short term. However, the analysis revealed that not all safety-related behaviors improved in every study, suggesting that there are limitations to the use of safety incentives and feedback for enhancing worker self-protective behavior.

Because studies reviewed by McAfee and Winn (1989) employed different types of safety incentives and/or feedback in different combinations, accompanied in some cases by safety training, the authors could not address the relative effectiveness of different behavioral safety strategies in promoting worker self-protective behaviors. In an attempt to address this shortcoming, Ray, Bishop, and Wang (1997) conducted a safety study in a General Motors automobile parts plant which comprised: (1) two groups of experienced workers (experimental and control) performing similar operations; (2) introduction of a behavioral safety program to the experimental group; and (3) no intervention for control group. Results from the study indicate that safety performance: (1) does not improve with training alone; (2) improves significantly when training is accompanied by feedback; and (3) improves even more when feedback is accompanied by goal setting.

Participation by Workers in Hazard Management Decision Making

The thesis that workers should be actively involved in the process of making decisions about matters governing their own safety in the workplace is an implicit feature of the closed-loop model of HM in Fig. 10.1. Evidence cited above demonstrates that workers have capabilities for both recognizing sensory feedback from hazards and developing behavioral proficiency in hazard feedback control. From a cybernetic perspective (Juran, 1964; T.J. Smith & Smith, 1987), therefore, it is unlikely that safety will be effectively served by a process that confines safety-related decisions exclusively to management, with no opportunity for worker input. As Viaene (1980, cited in Bryce, 1981, p. 1) puts it, "one of the most important trends that has emerged from the recent reforms is that of the transfer of power and responsibility on an ever-growing scale to those who are, fundamentally, the people who have to face these hazards."

Bryce (1981) and the U.S. Department of Labor (1989) note that the joint labor–management occupational safety and health committee represents one of the most effective avenues for worker influence over safety decision making. Generally speaking, these committees (to varying degrees in different countries) are aimed at advancing the right of the worker to: (1) participate in occupational safety and health programs and in setting safety guidelines and standards; (2) have access to job-related safety and health information and relevant training; and (3) refuse to participate in unsafe work and to avoid penalty if this right is exercised (LaBar, 1990). The U.S. Department of Labor survey (1989) notes that there are thousands of such committees in the United States (most established after passage of the 1970 OSHA Act), and describes six case studies that document the relative success of these committees in achieving labor–management cooperation in dealing with safety problems.

Ergonomic Intervention to Improve Hazard Management

One of the premises of the model in Fig. 10.1 is that ergonomic analysis and intervention can support breakthrough improvement in system safety by allowing the system to project the trajectory of its safety performance into the future, in a feedforward manner, through use of ergonomic analysis to identify and evaluate workplace and system hazards, followed by ergonomic intervention to abate targeted hazards. Behavioral cybernetic theory assumes that variability in safety performance should be influenced by the design of the performance environment with which the performing system interacts, just as is observed with individual and organizational performance generally (T.J. Smith, 1993, 1994, 1998; T.J. Smith,

Henning & Smith, 1994). Ergonomic intervention strategies aimed at improving work design deficits therefore should beneficially impact variability in safety performance in a manner that complements and extends the impact of traditional safety programs. Various lines of evidence support this viewpoint.

While other chapters deal extensively with various facets of macroergonomic intervention, this section will confine itself to evidence pertaining to the impact of ergonomic intervention on safety and HM. Cohen and colleagues (1997, pp. 114–18) review 24 ergonomic intervention studies reported since 1979, all of which were directed at reducing the incidence of musculoskeletal injuries and discomfort through implementation of microergonomic or macroergonomic redesign strategies. Tangible improvement in various measures of safety performance was cited in every study.

In a similar but somewhat more objective vein, Westgaard and Winkel (1997) review findings from 92 ergonomic intervention studies published between 1969 and 1995, all directed at improving musculoskeletal health. The studies are categorized into interventions to reduce musculoskeletal injuries using one of three strategies: (1) redesign of workstations, work methods, tasks, jobs, or work places to reduce exposure to mechanical hazards; (2) redesign of production systems, often with a dual aim of increasing productivity, and frequently involving participatory ergonomics; and (3) training/treatment methods, encompassing physiotherapy, health education, exercise, relaxation training, work technique training, or combinations thereof. Seventy-four of the 92 studies reviewed report reduction of musculoskeletal injuries and/or discomfort attributable to the intervention.

Possibly the most persuasive evidence regarding the distinctive benefits that ergonomic intervention can bring to improving safety performance is that provided by Guastello (1993). He reviews findings from studies of 48 different intervention programs, each of which employed one of eight different intervention strategies for accident prevention: personnel selection, poster campaign, exercise/stress management, safety audit system, quality circles, technological intervention, behavior modification (safety training, incentives, goal setting), and comprehensive ergonomic programs. The latter were most effective in accident reduction (51.6% average reduction in accidents), subsequently confirmed in a follow-up study of Shannon and Guastello (1997).

Synergistic Effects of Hazard Management and Quality Management on Safety Performance

A variety of observations suggest that successful safety and quality management (QM) programs share a number of parallel macroergonomic design features (T.J. Smith, 1999; T.J. Smith & Larson, 1991). These include management commitment to the program, employee participation in program decision making, defined methods for tracking program performance, and a closed-loop management control system (as in Fig. 10.1) that links program input (e.g., hazards, quality defects) to program performance (e.g., hazard control, quality control). Social cybernetic theory (T.J. Smith, 1990; T.J. Smith, Henning & Smith, 1994, 1995; T.J. Smith & Larson, 1991; T.J. Smith & Smith, 1987) suggests that if design features of different interacting operational programs in an organizational system are closely comparable, we may anticipate effective social tracking and mutual performance synergism between them.

Evidence from both conceptual analyses and field experience can be cited to support these hypotheses. Salazar (1989), Krause (1993), and Millar (1993) advocate the application of principles of QM to improve safety system performance. Winn (1994) points out that acceptance of QM principles has encountered skepticism and resistance in the safety community, but that nevertheless there is evidence that application of the QM approach affects safety positively. Companies that report positive safety performance benefits from adoption of QM principles in safety management include Unocal Chemical and Minerals Division (Minter, 1991), Inco Manitoba Division (Scott, 1991), PPG Glass Group (Durbin, 1993), and Georgia Gulf's Delaware City operation (DiPiero, Davis & Krause, 1993).

T.J. Smith (T.J. Smith, 1999; T.J. Smith & Larson, 1991) reports similar experience for a manufacturer of floor finishing equipment. Over 15 years after the company had implemented a progressive safety program that included elements of management commitment, safety performance tracking, and employee involvement, a QM program was introduced. Over the ensuing 10 years, performance in both quality and safety improved virtually in lockstep with one another, such that improvement in the latter was highly correlated with improvement in the former.

A plausible explanation for observations of synergism between safety and quality performance is that individual responsibility for careful workmanship in producing defect-free products and in participating in quality program decision making, embodied in the QM approach, naturally carries over to working safely and to participating in HM. The more general behavioral cybernetic interpretation is that emphasis placed on employee involvement in both programs introduces intimate behavioral feedback links between the safety and the quality of work performance that leads inevitably to the pattern of program interaction observed. The linchpin of both quality and safety is worker performance, and management commitment to support and encourage self-responsibility in the effective execution of work can be expected to benefit results in both areas.

CONCLUSIONS

A detached look at occupational safety today encounters a field replete with uncertainty, a situation that seems not to have changed much throughout the modern era. For example, over 20 years ago Ellis (1975) surveyed the efficacy of five alternative approaches to occupational accident and injury prevention: (1) inspection for compliance with government safety standards; (2) employee safety training; (3) accident statistics feedback; (4) management-sponsored safety programs; and (5) economic incentives, including workers' compensation (WC). Conclusions reached are that: (1) sound evidence supporting the safety effectiveness of inspections is lacking; (2) complete abatement of all hazards identified by inspection would prevent only about one fourth of all occupational injuries; (3) evidence regarding the safety benefits of training is mixed; (4) safety benefits of accident statistics feedback is uncertain; (5) accident and injury prevention outcomes of safety management programs are consistently positive, despite a diversity of approaches; and (6) no solid evidence exists that WC yields safety benefits. Point 5 supports the macroergonomic approach to HM.

Many other observations and lines of evidence support the pattern of inconsistency and contradiction inherent in the findings of Ellis (1975); for example, there have been dramatic improvements in workplace safety over the course of the century, yet job-related accidents and injuries, particularly of the musculoskeletal variety, continue to extract a formidable personal and financial toll. There is widespread agreement that safety should be treated as an integral management function. Yet the philosophy of safety engineering, pioneered by Heinrich almost 70 years ago, continues to dominate the field (Pope, 1990). A key premise of this philosophy, the unsafe acts/unsafe conditions model of accident causation, has attracted both praise (Heinrich, 1956) and criticism (Blake, 1943, 1956; Petersen, 1971). Engineering controls are emphasized as a key factor in the success of both ergonomic intervention and safety programs (Cohen et al., 1997; M.J. Smith & Beringer, 1987), yet various observers conclude that engineering controls, in and of themselves, do not dramatically improve safety performance (Cohen, 1977; Gill & Martin, 1976; Jones, 1973). Behavioral modification through safety incentive programs (promotions, slogans, and/or performance feedback), as a key strategy for improving safety performance, encounters praise (Bruening, 1989, 1990a; Kendall, 1986, 1987; Krause, Hidley & Hodson, 1990; McAfee & Winn, 1989; Moretz, 1988; Topf & Preston, 1991), conditional support (LaBar, 1989), or skepticism (Bruening, 1990b; Kohn, 1993); an objective survey of the efficacy of various accident prevention strategies shows a mixed record for different behavioral modification approaches (Guastello, 1993; Shannon & Guastello, 1997). Safety training is lauded as a key to effective safety performance (Cohen, 1977; Reid, 1987; M.J. Smith & Beringer, 1987; M.J. Smith et al., 1978), despite evidence showing that training does not compare favorably with other macroergonomic strategies for safety improvement (Ray, Bishop & Wang, 1997). It's enough to make dedicated safety professionals tear their hair.

From a behavioral cybernetic perspective, these contradictions and inconsistencies are strongly suggestive that safety performance, like individual, social, and organizational performance (T.J. Smith, 1998; T.J. Smith, Henning & Smith, 1994,

1995), is context specific (section 2), specialized in relation to design features of particular work environments. That is, variability in safety performance of different operations should be prominently influenced by distinctive design contexts of their physical, operational, and organizational environments. Interventions directed at altering the design of the work environment and/or the work system for purposes of hazard abatement thus should differentially affect safety performance of different operations in different ways. This is exactly what is observed.

Section 3 discusses a series of macroergonomic strategies that are likely to benefit safety performance: best practices analysis, worker hazard survey, behavioral safety analysis, participatory HM, ergonomic intervention, and integrating QM with HM. Although evidence points to the safety benefits of each of these strategies, not all have achieved widespread adoption. The analysis of HM offered here supports the following macroergonomic recommendations for safety program design and management that offer potential for bringing about tangible gains in occupational hazard abatement, safety performance, and accident and injury loss control.

1. *Manage safety hazards, not safety behaviors.* Behavior is self-controlled, not directly susceptible to outside manipulation. However, variability in behavioral performance is strongly influenced by design factors in the performance environment. Given the closed-loop coupling of design and performance, redesign of the workplace through macroergonomic intervention to abate hazards is highly likely to yield safety benefits by reducing unwanted variability in safety behavior.

2. *Close the loop between hazard analysis and hazard control.* The most effective system of hazard assessment is of little use if not accompanied by an equally effective system of hazard control, which rests in turn on management commitment to safety.

3. *Validate program content and effectiveness.* The effectiveness of hazard control is predicated on the validity of the HM program, demonstrated by face, criterion, and economic validation procedures plus program reliability assessment. Traditional approaches to validation of a safety program tend to focus largely, if not exclusively, on standard measures of safety performance, such as accident and injury statistics. However, no entirely satisfactory approach to safety performance assessment currently exists (Grimaldi, 1970; Grimaldi & Simonds, 1989, Chap. 7). This consideration supports the following recommendations: (1) consider a multidimensional approach to program validation; (2) criterion validation of the program should focus on hazard control efforts, because both identification and abatement of hazards can be objectively assessed, and the beneficial impact of hazard abatement on safety performance is well documented; (3) to satisfy regulatory reality, high priority organizational need, and acceptance by both managers and workers, criterion validation of safety performance itself should focus on standard lost time incidence measures of

performance, with an emphasis on the elimination of severe events (Grimaldi, 1970; Grimaldi & Simonds, 1989, Chap. 7); and (4) pursue research to develop feasible, meaningful, and acceptable measures of safety performance.

4. *Adopt a systems perspective on hazards.* From a systems perspective, occupational hazards represent incidents or conditions that exist or arise through the variable, sometimes transitory, and typically synergistic convergence of organizational, behavioral, operational, and physical-environmental factors specific to particular tasks, job operations, and sociotechnical systems. The complexity of this interaction mocks simplistic attempts to pigeonhole hazards into a limited number of categories (e.g., physical versus behavioral), defeats efforts to attribute occupational accidents, injuries, and safety problems to one type of hazard or another, and stymies abatement strategies based on dissective assumptions about hazards. In today's workplace, operational hazards cause the most accidents and injuries. Typically, these are not addressed by government standards, are not detected by outside inspectors, and are not abated by remedial approaches prompted by standardized accident and injury investigations. Further, operational hazards are context specific to particular tasks, jobs, and operations. From the standpoint of both organizational and personal self-interest, therefore, guided by dictates of the Pareto principle (Juran, 1964, 1995), the HM program should focus on operational hazards as a major priority for hazard control.

5. *Exploit parallels between proficiency in safety performance and motor performance.* Safety performance by workers in the workplace represents, in essence, a manifestation of motor behavioral performance. Skill in safety performance therefore rests, in large part, on acquisition of motor skill. From the motor behavioral literature, we know that there are three fundamental keys to the acquisition of motor skill (Schmidt & Lee, 1998): provision of feedback, repetition, and patterning of movement. We know from the safety literature that provision of safety performance feedback, worker experience (repetition), and safety training (patterning) benefit safety, although findings regarding training are mixed (Cohen, 1977; Ellis, 1975; Guastello, 1993; McAfee & Winn, 1989; Shannon & Guastello, 1997; M.J. Smith & Beringer, 1987; M.J. Smith et al., 1978). These parallels are not accidental. Macroergonomic strategies that promote the acquisition of motor behavioral skill in safety performance may be deemed a key prerequisite for a successful HM program.

6. *Utilize workers as a resource.* As noted earlier, front line workers: (1) are acutely aware of hazards that directly threaten their health and well being; (2) can recognize a broad array of operational hazards that standards and inspections do not address; (3) can devise methods for abating these hazards; and (4) can beneficially participate in safety decision making. Utilization of workers as an integral contributor to HM represents a macroergonomic strategy whose widespread adoption is long overdue.

7. *Integrate ergonomics into HM.* Among different options for HM, ergonomic programs have proven effectiveness for the identification, evaluation, and

abatement of operational hazards, particularly those associated with risk of musculoskeletal injury.

8. *Integrate safety with production.* A number of arguments can be advanced for this recommendation, advocated by a number of observers (Petersen, 1971; Schlueter, 1965; T.J. Smith & Smith, 1983). First, it unifies, rather than playing off, the mission of the production system with that of the safety system. Second, evidence shows that it is cost effective, by reducing both production and safety costs. Finally, if quality management has been incorporated into the production system, there is likely to be a synergistic and beneficial influence on performance of the safety system (T.J. Smith, 1997, 1999; T.J. Smith & Larson, 1991).

9. *Consider performance standards.* For dealing with operational hazards that are not adequately addressed by specification standards, performance standards (K.U. Smith, 1973) offer a complementary approach that may benefit HM. A further advantage is that safety performance can be made a defined performance criterion for both employees and managers.

10. *Adopt a no-fault philosophy.* From the systems perspective of a hazard as comprising a complex and variable interaction between a design flaw (hazard) and worker or organizational behavior, it makes little sense to attempt to assign blame for a particular incident to one system component or another. Nevertheless, blame-mongering continues to plague the safety field—pervasive use of "human error" as code for "operator error" represents a prime example. Conversely, the term "systems error" implies that the performance-design system as a whole is at fault. Various lines of evidence indicate that negligence-based WC or inspection-based approaches to safety are of marginal effectiveness in improving safety performance. These considerations suggest that a no-fault approach to safety performance appraisal (K.U. Smith, 1973; T.J. Smith & Smith, 1983) may benefit HM. Operationally, what this means is that hazard abatement should pursue the dual approach of design improvement coupled with macroergonomic interventions to improve proficiency in safety performance.

11. *Aim for continuous improvement.* Humans are imperfect control systems. Human behavior always has been and always will be highly individualistic, and somewhat variable and unpredictable. Safety behavior is not immune from these biological realities. Therefore, achieving perfection in safety performance (Pope, 1990) represents a naive and unattainable goal. A more realistic and attainable objective is to strive for continuous improvement in safety performance (DiPiero, Davis & Krause, 1993; Salazar, 1989). This philosophy has paid well-substantiated dividends in the arena of quality management and quality performance (Deming, 1986). There is every reason to believe that safety performance will benefit from a similar approach.

12. *Eschew safety nostrums and quick fixes.* All evidence suggests that variability in safety performance, like that observed in individual performance and other domains of organizational performance, is highly specialized in relation to distinctive physical, organizational, and operational design characteristics

of different organizations and operations. From this perspective, one operation's strategy for safety salvation may not necessarily generalize to other operations. For effective HM of a particular operation, the key is to target those macroergonomic strategies that are likely to positively influence safety performance of that operation, to implement those strategies in an evolutionary rather than revolutionary manner, to track their effectiveness (Point 3 above), and to modify and redesign the approach if necessary in an iterative manner. Usability testing (Point 13) offers a model for this kind of process.

13. *Consider usability testing as a model for fine-tuning safety performance.* The process of usability testing first was widely employed in the computer industry (Gould, 1990), based on recognition that user acceptance of new technological designs could not necessarily be predicted in advance. The process is iterative— user acceptance of a new design first is assessed, and the design iterated based on user feedback until an acceptable level of acceptance is achieved. This process appears to offer lessons for the safety field. Effects of new macroergonomic HM design strategies on safety performance may not be entirely predictable. Performance consequences and user acceptance therefore both should be assessed, and HM strategic designs iterated until acceptable levels of both performance and acceptance are achieved.

14. *New technology—new hazards—new hazard management demands.* M.J. Smith (1994) notes that emergence of new technology has introduced a new constellation of hazards which pose new challenges for HM. For example, rising complaints of work-related musculoskeletal disorders have been attributed to the growth of video display terminal work (Bammer, 1987; Hendrick, 1994b). Thus, as with business generally, the only constant in the safety field is change. Workers and safety managers alike must be constantly alert to the possibility of new hazards associated with new technology, and must be prepared to consider new macroergonomic strategies for managing these hazards.

ACKNOWLEDGMENT

This chapter is dedicated to my father, Karl U. Smith, whose seminal scientific contributions to conceptual, empirical, and academic research and development of the field of hazard management inspired many of the ideas presented here.

11

Macroergonomic Aspects of Manufacturing

Waldemar Karwowski, Jussi Kantola,
and David Rodrick

University of Louisville

Gavriel Salvendy

Purdue University

INTRODUCTION

Contemporary enterprises apply information-based technology to automate different aspects of design, planning, and operation of manufacturing systems including, for example, engineering, production planning and control, management of human resources, marketing, purchasing, and finance. All these functions must also be integrated to improve competitive advantage in view of changing market requirements. This is because effectiveness of such systems depends on abilities of the people to compensate for the limitations of technology. Therefore, organizational and related human issues, which are the focus of the macroergonomic discipline, are critical to the success of technology, which is greatly dependent on the nontechnical characteristics of a corporation. As discussed by the Manufacturing Studies Board (1986a), achieving the full benefits of information-based manufacturing requires interrelated changes in many areas of human resource practices, including: plant culture; organization; job design; compensation and appraisal; selection, training, and education; and labor–management relations. Other studies on the fusion of manufacturing systems and information technologies in Europe (Brödner, 1987) also concluded that organization is the key element of success, which should be valued and appreciated at a level equal to information technology; for example, human resource utilization is considered essential for assuring efficient manufacturing and high quality products under conditions of market uncertainty (Jaikumar, 1986; Kember & Murray, 1988; Majchrzak, 1988; Wall, Clegg & Kemp, 1987).

Manufacturing companies are continually exposed to changing market demands, rapid progress in technology, evolving legal provisions, and concurrent social changes. These demands produce complex and dynamic changes in the business environment, and greatly influence the requirements for organization and management of human work. Because contemporary manufacturing facilities are social, technological, and informational systems, a vision for integrating people, organization, technology, and information systems has recently gained significant attention (Corbett, 1988; Goldman, Nagel & Preiss, 1995; Karwowski & Salvendy, 1994; Kidd, 1994; Majchrzak, 1988, 1995; Wall et al., 1987). The design of enabling information technologies often focuses on the system's organizational architecture in order to enhance human communication and to support cooperative work between people (Savage, 1991). Under this framework, people are knowledge contributors who also have specific personal needs. Such needs affecting work design requirements are the social needs (i.e., desire for significant social relationships) and the growth needs (i.e., desire for personal accomplishment, learning, and development) (Wall et al., 1987).

As discussed by Karwowski et al. (1994), developments in information-based manufacturing systems have had profound impact on work design (i.e., on how the tasks are grouped into jobs and allocated among different work groups). Two key technological features that affect work design are *technical interdependence*, or the extent to which the technology requires cooperation among employees to produce a product or service, and *technical uncertainty*, or the amount of information processing and decision making workers must exercise during task execution (Wall et al., 1987). Work design also is affected by task environment. When task environments are relatively stable, the jobs can be programmed and standardized; when the environment is relatively dynamic, the jobs must be managed adaptively (see Chapter 3). Implementation of information-based manufacturing requires nontraditional work designs and significant modifications in the work context. The new forms of technology tend to increase technical interdependence, uncertainty, and environmental dynamics. In addition to assessment of skills, training, and *human–machine* interfaces, effective design of manufacturing systems includes examination of the system of reward of human skills, direction of technological innovation, and human operator control over manufacturing operations (Majchrzak, 1995).

Potential contributions of macroergonomics to modern manufacturing reflect the variety of human components of an industrial enterprise and include the following broad areas of application: (1) development of business practices, (2) design of enabling (information-based) technologies, and (3) management of new technologies and business practices. This chapter presents an overview of some of the main macroergonomic issues in information-based manufacturing systems, and discusses selected applications of macroergonomic principles applicable to the design and operation of such systems.

EARLY MANUFACTURING CONCEPTS

The first organizational structures in manufacturing utilized narrow and precise definitions of tasks leading to job specification and tasks grouping (or departmentalization). For example, Charles Babbage (1832) advocated an increase in work efficiency from job specialization achieved through decreased learning time and increased skill due to repetition, fewer tool changes, and decreased waste. The horizontal differentiation extended through the number of required specialties and the level of training for specific jobs (Hendrick, 1986). Taylor (1911) developed the principles of scientific management (i.e., vertical work specialization based on the division of work and allocation of responsibility between the management and workers (see Chapter 3 for additional explanation). The creation of management science, focusing on one optimal method for definition of work task and determining all of its details, formed the basis of Taylor's system. This system was characterized by division and simplification of work and work standards, and led to a separation of planning, controlling, and monitoring activities.

Weber's (1946) ideal bureaucracy reinforced and extended the Tayloristic design principles based on the "one best way" by focusing on organizational structure that emphasized efficiency, stability and control, and division of labor through well-defined, simple and routine tasks performed by the workers. The widespread adoption of Taylor's (1911) principles of scientific management on the shop floor led to drastic job fragmentation. From a purely technological standpoint, the one best way of doing the job, matching people to tasks, supervision, incentive schemes, and creation of staff groups to plan and control of workers, was considered progress toward creation of an effective factory organization (Hayes et al., 1988). The main weakness of this approach, however, was undesirable social consequences (MacKenzie & Wajcman, 1985; Rosenbrock, 1983).

Separation of managers and staff from workers, disregard of the need for continued improvement of workers' skills and organizational capabilities, and job fragmentation proved to be serious deficiencies of many systems. The technology-driven or technocentric approach to design and management of information-based manufacturing was in principle no different from the Tayloristic method (Brodner, 1987; Cooley, 1986; Corbett, 1985; Kidd, 1987). As such, the quest for information-based manufacturing as practiced in the recent past further enhanced vertical division of work, and aimed to reduce human involvement in the manufacturing processes.

The emergence of an information-based global economy, as well as technical requirements of flexible and mass-customized production, stimulated introduction of the new paradigm in manufacturing philosophy, characterized by the following assumptions: (1) producing is a service, (2) flexibility ensures productivity, (3) productivity in information management increasingly influences the productivity of material processing (utilization of work, capital, material, energy), (4) more productivity is gained by uniting information processing, decision making and realization in small, fast control loops or work groups, (5) the workforce becomes more important, and (6) manufacturing systems have to be designed as more use-oriented than cost-oriented units. In order to realize these objectives and requirements, a set of new ideas and new ways of thinking have been developed (Warnecke, 1993). Majchrzak (1995) concluded that information-based manufacturing implementation is more likely to be successful when the technology, organization, and human resource/people issues have been designed to complement and integrate with each other. Many of these failures in implementation efforts could be attributed to managers and technology, as well as designers lacking an understanding of the required organizational and human resource changes.

CONTEMPORARY MANUFACTURING CONCEPTS

The contemporary manufacturing paradigm stipulates the systems approach to integration of technology, people, and organization (Goldman et al., 1995; Karwowski et al., 1994; Karwowski & Salvendy, 1994; Kidd, 1994; Majchrzak & Wang, 1994). Appropriate technology needs to be implemented with due consideration to the requirements for selection and training of human resources, allocation of decision making, and control over technology by the people charged with its use. Organizational structures allowing persons to manage the available resources and adapt quickly to a changing marketplace are also needed. While many different concepts were developed in the recent past to realize this new manufacturing paradigm, they all attempt to capture critical requirements of the global economy and changing market conditions (i.e., cost, quality, flexibility, and time) (Itoh, 1991). These concepts are briefly reviewed below.

Lean Manufacturing

The concept of lean manufacturing Womack, Jones, and Roos (1990), which replaces the traditional mass production paradigm, considers the organizational structures and human resources, alongside of information-based technology, as important factors for generating expected productivity and quality results. Lean production combines the advantages of both craft and mass production, including high volume and diversified production using flexible automation, multiskilled teamwork, and an overall corporate culture of continuous improvement and high quality. Wobbe and Charles (1994) identified five general principles for lean production. First, the more complex products become, the more quality is dependent on upgrading of all stages of manufacturing and demands the full dedication of employees at all levels. Second, the more sophisticated manufacturing technology becomes, the more it is vulnerable and dependent on human skills for control and maintenance. Third, the more customized productions are, the more human intervention is necessary with regard to change-over, setting up machines, adaptation, adjustment, and control. Fourth, if products demand a high service input, and aftersales service and maintenance, skilled people are required to accomplish these tasks. Finally, the shorter the life cycle of products becomes, the more innovation comes into play; take-off phases occur more frequently, and their mastery is dependent on experienced personnel with formal knowledge to overcome new challenges connected with the start of a new product.

Concurrent Engineering

Concurrent engineering (CE) aims for near simultaneous design of new products and processes (Sinclair et al., 1995) through identifying the effects of possible design features on each other as well as the intended performance criteria—including management practices such as organizational design, job design, and training. King and Majchrzak (1995) discussed different management practices dependent on the key design decisions regarding stand-alone versus integrated manufacturing systems, and application of information-based technology. The basic assumptions about concurrent engineering, with respect to communication and (lateral) coordination between disciplines, data representation for designers, and human–computer interaction of individuals, were found to be inconsistent. Therefore, the following complementary strategies were recommended. The first strategy advocated the use of a user-centered design approach for organizations committed to change over several design cycles. A second strategy stated that CE tools should integrate human aspects of design with the technical elements, and allow sharing of cognitive models as well as technologies to accommodate the variety of perspectives that team members bring to a CE design process. A third strategy advocated sharing of ideas with appropriate people at the closeout of each session, rather than demanding it using standardized project management techniques.

Agile Manufacturing

A more recent manufacturing philosophy and business approach is the concept of agile manufacturing (Goldman et al., 1995; Kidd, 1994; Kidd & Karwowski, 1994). The term agile manufacturing was first introduced with the publication of the *21st Century Manufacturing Enterprise Strategy* (Iacocca Institute, 1991), which made three important points in an effort to define agility manufacturing. These key points are as follows:

1. A new competitive environment is emerging, which is acting as a driving force for change in manufacturing.
2. Competitive advantage will accrue to those enterprises that develop the capability to rapidly respond to the demand for high quality, highly customized products.
3. To achieve the agility that is required to respond to these driving forces and to develop the required capability, it is necessary to integrate flexible technologies with a highly skilled, knowledgeable, motivated, and empowered workforce. This must be done within organization and management structures that stimulate cooperation both within and between firms.

As discussed by Kidd (1994), lean manufacturing and agile manufacturing are two different concepts. Agility, which implies a quickly moving, nimble, and active system, is not the same as manufacturing flexibility, which means adaptability and versatility. While flexibility is a necessary requirement for competitive corporations, it does not constitute agility; nor does the concept of lean manufacturing (i.e., doing everything with less) (Jones, 1992). Although lean manufacturing is necessary for agility, it is not sufficient. Similarly, computer-integrated manufacturing and computer-integrated enterprises, even though they assure rapid communication, exchange and reuse of data, are not necessarily agile. However, computers that link across applications, functions, and enterprises may provide for the necessary conditions for agile manufacturing (Goldman ct al., 1995). An agile company aims to manage change and uncertainty through the entrepreneurial approach, combined with a flexible organizational structure and distributed managerial decision-making authority, which allows for rapid reconfiguration of the human and technological resources in a quest for meeting globally changing market requirements (Karwowski et al., 1994).

According to the Iacocca Institute (1991), the agile manufacturing enterprise exhibits the following characteristics: (1) concurrency in all activities, (2) continuing education for all employees, (3) customer responsiveness, (4) dynamic multiventuring capabilities, (5) employees valued as vital assets, (6) empowered individuals working in teams, (7) environmental concern and proactive approach, (8) accessible and usable information, (9) skilled and knowledgeable employees, (10) open system architectures, (11) right first time designs, (12) total quality philosophy, (13) short cycle times, (14) technology awareness and leadership, (15) enterprise integration, and (16) vision-based management.

Agile manufacturing can also be considered as a framework, within which a company can develop its own business strategies (Kidd, 1994) supported by three primary resources: innovative management structures and organization;

a skill base of knowledgeable and empowered people; and flexible and intelligent technologies. Agility can be achieved through the integration of these three resources into a coordinated and interdependent system that aims to achieve cooperation and innovation in response to the need to supply customers with high-quality customized products. The first principle of agile manufacturing is that an enterprise should be built on the competitive foundations of continuous change, rapid response, quality improvement, and social responsibility in terms of environment and employees (Kidd, 1994).

Implementation of the agile manufacturing principles may often require extensive changes in the existing business practices, or even a complete overhaul of such practices (Goldman et al., 1995; Greiss, 1993). Because success may depend on rapid development and delivery of quality products, modern corporations aim to maximize their ability to capture and utilize corporate experience in product design through concurrent engineering practices (Goldman & Priess, 1992). Also, an implementation of collaborative design efforts is required to assure fast-paced design decisions in a competitive environment with no error tolerance (Forsythe & Ashby, 1994).

Consideration of macroergonomic factors affecting decision making within dynamic markets is very much needed for development of manufacturing agility. This includes knowledge of team dynamics, individual information requirements, information management and utilization, and monitoring and assessment of the status of complex and dynamic systems (Forsythe & Karwowski, 1995). The appropriate organizational structure and infrastructure support for corporate communication and information transfer also are needed, including system and software compatibility. In agile manufacturing enterprises, integration and networking occurs at all levels of the organizational structure. Therefore, the needs of a complex infrastructure and the numerous human issues should be addressed (Haney, Reece, Wilhelmsen & Romero, 1994).

Virtual Enterprise and Human Networking

Agile manufacturing can be viewed as the business concept that brings together many ideas in order to develop an appropriate response to global market opportunities (Goldman et al., 1995; Kidd, 1994), and which can be realized through virtual enterprises. The agility comes from integration of organization, people, and information-based technology into a coordinated interdependent system. A framework of generic features of agile manufacturing may be comprised of the following elements (Kidd, 1994): (1) integrated enterprise, (2) human networking, (3) enterprise based on natural groups, (4) increased competencies of all people, (5) focus on the core enterprise competencies, (6) virtual corporation, (7) an environment supportive of experimentation, learning, and innovation, (8) multiskilled and flexible people, (9) team working, (10) empowering of all the people in the enterprise, (11) knowledge management, (12) skill and knowledge enhancing technologies, (13) continuing improvement involving all people, and (14) change and risk management.

The two important building blocks of manufacturing agility are human networking and employee empowerment. Under the concept of human networking (Naisbitt, 1984; Savage, 1990) workers are networked together by interactions that occur between all people in the organization. Traditional, information-based networks also can be used to support human networking.

Human Networking and Manufacturing Effectiveness

The manufacturing enterprise is effective when it accomplishes its stated goals, makes maximum use of its resources, and satisfies its strategic constituencies (Savage, 1991). The goal attainment approach seeks to satisfy the goals of an organization. However, there can be long- and short-term goals, and individual and organization goals. By making use of the principles of human networking in modern manufacturing, it is possible to build cross-functional teams with common goals. Peer-to-peer interaction and an integrated structure can ensure timely flow of information between teams and efficient sharing of knowledge and skills.

Savage (1991) proposed that the organizational functioning of modern manufacturing be based on the strategic constituencies approach, which holds that any organization must satisfy its strategic constituencies, owners, employees, customers, suppliers, unions, and government to survive. To satisfy all these groups the organization must develop a network connecting all groups. Employees, both blue and white collar, are critical to the success of an organization. Under the contemporary manufacturing concept, work has been divided into parts and subparts with one worker doing only a small part of the whole work. Furthermore, work is seen as a process, and the worker is intimately connected with product through this process. The worker has the vision about the product and the knowledge of the process to work toward the product.

HUMAN-CENTERED MANUFACTURING

The concept of human-centered manufacturing focuses on five basic principles of work design (Corbett, 1990; Murphy, 1989); (1) retention and enhancement of existing skills, (2) extension of operator choice and control (humans control the technology, not vice versa), (3) minimal subdivision of work (unites planning, execution, and monitoring), (4) maximization of human operator knowledge of the whole production process (encourages social communication and interaction), and (5) consideration of ergonomic factors (layout of equipment, design of keyboards, software, etc.). The above allows us to extend traditional human factors considerations through the design of manufacturing architecture within adaptive organizational structures.

Under the above, modern manufacturing enterprises are viewed as socio-informational systems, with focus on the following issues: (1) parallel design; a cyclical and iterative process of simultaneous planning of both human and technical aspects of new systems, (2) the complementarities of technical and social components of the new systems; design of equipment hardware and software as a set of tools to perform routine transfer and control tasks and enhance the ability of humans to make choices in situations of uncertainty and deal with unpredictable events, and (3) new system design techniques created to overcome designers' lack of understanding of contemporary manufacturing operating requirements (Badham & Schallock, 1991).

According to Corbett (1988), the human-centered computer integrated manufacturing (HCCIM) system satisfies several basic requirements. In general, the more degrees of freedom open to users to shape their own working behavior and objectives, the more human-centered is the manufacturing system. The HCCIM aims to unite the planning, execution, and monitoring component of work in order to minimize the division of labor common to present day practices. HCCIM encourages social communication (both formal and informal) between users, preserving face-to-face interaction in favor of electronically transmitted data exchange, and accepts the present skills of the user and allows them to develop. This approach is contrary to conventional design, which tends to incorporate the skills into the machine. Finally, it provides a healthy, safe, and efficient work environment.

The design of human-centered components of the manufacturing systems extends beyond providing computer-supported group work on the shop floor. For example, Badham and Schallock (1991) outlined a human-centered work-oriented model for advanced manufacturing based on the strategy of integrated group manufacturing. First, operators' tasks were widened as far as possible. Second, computer-aided planning facilities were located at the shop floor level rather than at the planning department level. Third, as much as possible, planning and scheduling functions were supported at the production island rather than the foreman/area control level. The above model was adapted to conditions of smaller firms and batch sizes, high quality and customized markets, relatively high skill levels, unionization, and "high trust" production cultures (Badham & Schallock, 1991).

This approach has been developed within the ESPRIT research program (Kidd, 1990a). By splitting orders instead of dividing labor, job shop manufacturing was changed into group manufacturing where part families were manufactured in their entirety (Brödner, 1986). Such integration enabling a tool facilitated the exchange of ideas and allowed for interactive learning between the computer and designers. Another application is the use of human-centered concepts in designing an automated manufacturing cell. The operator runs the manufacturing cell with the aid of powerful software tools, including creation of machine programs using high-level software tools, optimization of these programs using his or her skills and experience to minimize the cutting time, performing machine scheduling, and programming the work-handler to load/unload, and doing the leftover tasks.

Open System Design in Manufacturing

The human- and computer-integrated manufacturing systems require flexibility, adaptation, improved responsiveness, and the need to motivate people and make better use of their skills, judgment, and experience (Kidd, 1992). The above implies that organizational structures, work practices, and technologies should be developed in such a way as to allow people at all levels in the company to adapt their work strategies to the variety of system control situations. Therefore, they should be designed and developed as open systems. The term *open system* is used to describe a system that receives inputs from and sends outputs to the system's environment (Kidd, 1992). This term was associated, in the context of modern manufacturing, with the system architectures based on the International Standards Organization Open Systems Interconnection model (ESPRIT Consortium AMICE, 1989). The idea can be applied not only to system architectures and organizational structures, but also to work practices, human–computer interfaces, and the relationship between people and technologies, such as scheduling, control systems, decision support systems, etc. (Kidd, 1990a, 1990b).

An open manufacturing system allows people a large degree of freedom to define the mode of operating the system and adapting to its environment. For example, Kidd (1990b) has demonstrated the concept of an open system for the human–computer interface (HCI) in the workshop-oriented computer numerical controlled (CNC) cell. The HCI system allows the human operator to customize the interface to his or her own personal preferences by changing the dialogue, the screen layout, etc. In a closed manufacturing system, actions of the user are determined by system designers through hardware, software, or performance constraints, and the user–computer relationship is predefined. An example of a closed system is the design of the human–computer interface of CNC machine tools, which are predetermined by the manufacturing designers. A closed system limits the user's freedom of action, or forces the user to use the manufacturing system in a particular way. A closed system approach aims to automate a particular task, but when the manufacturing process fails, it leaves the user without the necessary information-based decision support.

An open manufacturing system does not restrict what the human operator can do by allowing the designer to make use of any of the infinite number of possibilities between the two extremes—the computer acting alone or the user acting alone. In the open system, the relationship between the user and the computer is determined by the user, not by the designer. The role of designer is to create a system that will satisfy the user's personal preferences and allow the users to work in a way that they find most appropriate, making the system more adaptable and agile. Wall, Corbett, Martin, Clegg, and Jackson (1990) examined two work design styles used to manage and operate CNC stand-alone systems. The work designs of interest were the specialist control and operator-centered control designs. In the specialist control mode, engineers and computer specialists maintain, repair, write, and fine-tune the programs, while the operator has minimal involvement. In the operator-centered control mode, the operator is responsible for monitoring and maintenance and programming of problems as they occur. The results showed that introduction of an enhanced operator control over CNC assembly machines led to reduction in downtime for high variance machines, and suggested that work redesign improved intrinsic job satisfaction and reduced feelings of job pressure among operators.

HUMAN ROLES IN INFORMATION-BASED MANUFACTURING

Complementarities of Cognitive and Macroergonomic Approaches

With increasing automation, the nature of tasks in modern manufacturing systems shifted from those that mostly require perceptual-motor skills to cognitive activities, such as problem solving and decision making in supervisory control tasks (Goodstein, Anderson & Olsen, 1988). Traditional approaches to design and management of information-based manufacturing consider workers as deterministic input-output systems, and tend to disregard the teleological nature of human behavior (i.e., the goal-oriented behavior relying on seeking relevant information and active selection of goals) (Rasmussen, 1983). However, to be successful, the design and management of advanced manufacturing systems needs to incorporate descriptions of human mental functions required for a specific task.

Sanderson (1989) identified two complementary approaches to defining the role of people in contemporary manufacturing systems: (1) cognitive engineering, based on cognitive and motivational psychology and focusing on the individual, and (2) macroergonomics, based on a combination of industrial/organizational psychology, social psychology, and systems theory, which focuses on the entire organization. These two approaches are complementary because the actions of the individual operators are restricted by the policies and "culture" of the organization. On the other hand, the functioning of the entire manufacturing organization is reflected in the actions of the people who supervise the system's operation.

The application of information technology affects the human roles in contemporary manufacturing systems in two ways: hierarchically and horizontally (Sanderson, 1989). The hierarchical effect of information technology manifests itself through the process of automation, and leads to human operators acting as supervisors of the artificially intelligent manufacturing processes. The horizontal effect of information technology can be described in information processing terms, and illustrates the situation in which the human operators have access to information about all aspects of the manufacturing system. The cognitive engineering approach (Hollnagel & Woods, 1983) emphasizes appropriate goals for the human roles, a need to understand the details of the particular system, and system complexity. Human knowledge and skill are considered as an inherent part of system design requirements. The interaction between the technical and human parts of the system is to be designed to make the best use of the knowledge and skills of both parts of the system.

System Complexity and Change Process

Management of change is the systematic application of a change management methodology that includes the politics of management, development of strategies, and treating human resource management as a business issue. It also includes analysis of work system design, design of technology, and quality. The support for a change process can be established through such actions as: (1) development of skills and training for new forms of team work, (2) implementation of the reward systems for participation in teams, (3) design of a comprehensive communication system that incorporates a knowledge of change, and (4) establishment of procedures that allow for experimentation, creating a culture of change, and providing facilities to support change. The knowledge and tools of the macroergonomic discipline can be very useful in this process.

Contemporary manufacturing enterprises are by their nature very complex and uncertain systems. The reduction of such complexity can be achieved by comprehensive use of the unique worker skills and on the basis of human-oriented organizational schemes (Kidd, 1994). Human skills can be developed and utilized by taking into consideration all the factors that define the working tasks and management procedures. These procedures, in turn, have to be appropriately designed with respect to the organization, use of technology, and corporate policies. An appropriate framework needs to be developed to deal with change, risk, and complexity. For example, an integration of human resources across departments and at different hierarchical levels in developing the strategic vision should be followed by a strategy to realize appropriate changes within a new organizational structure.

HUMAN- AND COMPUTER-INTEGRATED MANUFACTURING

Human- and computer-integrated manufacturing refers to developing integrated systems of people, organization, and technology (Kidd, 1991). Three main types of system integration are: (1) integration of people, by assuring effective communication between people, (2) human–computer integration, by designing suitable interfaces and interaction between people and computers, and (3) technological integration, by assuring effective interfacing and interactions between machines. One of the main ingredients of integrated manufacturing is the concept of skill-based technology—a strategy that involves people who work within appropriate organizational structures and are supported by information-based technology. The above-defined concept takes its origins from the early ideas of the human-centered approach to design of advanced manufacturing technology (Rosenbrock, 1989).

Organizational Design Issues

Contemporary manufacturing systems must use the process of differentiation in order to divide its staff, functions, and processes into distinguishable units (Majchrzak, 1988). At the same time, however, these units must also be closely coordinated through the process of (computer) integration. Maintaining the balance between integration and differentiation is a central dilemma for organizing manufacturing systems. Majchrzak (1988) proposed a model describing six major components of the human infrastructure in contemporary manufacturing systems. This model is arranged into four components as stages to signify the fact that earlier decisions affect later options. These stages include:

1. The consideration of equipment features (parameters) and selection of equipment,
2. effects that decisions about equipment features could have on jobs (first-order effects),
3. effects that the decisions about jobs can have on personnel training, selection, and other related policies, and on organizational structure (second-order effects), and
4. the ultimate significance of the equipment for the production process, human infrastructure benefits, and organizational survival.

The two remaining components of the above model are the planned change process for implementing the decisions and constraints on the human infrastructure decisions (i.e., workforce, control over resources, predictability of marketplace, and management of human resources priorities). The choices about human infrastructure must be compatible with equipment parameters to achieve optimal use of the new technology.

Socio-Informational View of the Manufacturing Enterprise

Many corporations made an effort to develop and introduce new work structures that would be compatible with different branches of modern manufacturing systems (Womack, Jones & Roos, 1990). Significant improvements were made by structuring manufacturing into functional blocks (products in assemblies, production in work groups, and manufacturing cells), leading to greater system transparency of internal processes as well as to greater system flexibility. In such a decentralized system, with people who are adequately skilled, production planning and control can be shifted to the operational (factory floor) level, which leads to shorter and faster control loops and smaller organizational hierarchy. Organizational flexibility also was enhanced by including other support activities in the change, such as programming, maintenance, servicing, and tasks connected with quality assurance. The improved use of the system (availability) compensated for additional costs for personnel caused by higher qualification. Finally, utilization of the teamwork concept allowed for increased flexibility concerning quantity and personnel. Recently, planning and controlling tasks are increasingly being transferred to task groups, so that there is not only an increase in autonomy but also in responsibility.

In order to account for the multitude of factors encompassed in human, organization, and technology issues, manufacturing enterprises focus on the interactions or congruence of these factors (Majchrzak, 1988). The manner in which such factors match one another is more important than any single factor alone, and it is the degree of appropriate match that the organization should promote. Some of the important tools designed to assure congruence between different manufacturing factors were developed by Majchrzak and Gasser (1992) and Majchrzak (1995). These tools, such as HITOP and ACTION, incorporate expert system techniques and knowledge bases in order to provide managers with structured information about organizational structures, given the technology and human attributes of the manufacturing facility.

TOOLS FOR MANUFACTURING INTEGRATION

Integration Process

Computer-integrated manufacturing (CIM) aims to integrate technology, organization, information systems, and people to improve the industrial performance of a company. The objectives of CIM are to achieve multiple and changing business goals, long-term competitiveness, and productivity. Unfortunately, many such systems fail to achieve their intended objectives. While most of them are technically successful, only 14% show significant improvements in the competitive business position of the enterprise (Jaikumar, 1986). These failures have both short-term and long-term effects. The short-term losses include wasting of valuable resources and failing business objectives. The long-term losses are unfulfilled strategies; lower confidence in leadership; increased resistance to change, and threats to organizational survival. Similar consequences can be expected from failing to adapt the CIM architecture to changing internal and external future needs.

Many of the current CIM reference architectures, which are the models used as a benchmarks against which to assess both the structure and performance of a particular CIM facility, utilize out-of-date design criteria that lack a comprehensive consideration of macro- and micro-ergonomics issues. Other problems with these architectures are complexity and difficulty in transferring abstract design into an implementation plan. Simple but comprehensive quantitative tools that would help in designing successful CIM systems are needed (Kantola & Karwowski, 1999). Such tools would improve the success rate of CIM implementation efforts and help to determine requirements and resources needed for CIM implementation.

According to Gasser and Majchrzak (1994), one needs to address the following set of issues in preparation for using the manufacturing integration tools in practice:

1. Are the management and engineering staffs aware of the technology failure rates internal to the organization?
2. Have there been any efforts in the organization to document learning from the failures to confirm that the failures have been attributable to lack of planning of how the technology would be integrated with the people and organization? Have these lessons learned been disseminated to managers and engineers?
3. Have there been any efforts to examine in the organization how technology design, especially as it relates to organizational and people issues, is conducted today? Has there been any effort to benchmark how the organization does technology design with how other companies do it?
4. Have there been any efforts to examine how tools could be used to help facilitate the technology-organization-people (TOP) integrative design process? Has a list of criteria for a worthwhile tool been generated?
5. Have there been any efforts to search for available tools that have been or could be used to help facilitate a technology-organization-people (TOP) integrative design process?
6. Have there been any efforts, albeit small steps, to involve shop floor workers in continuous process improvement efforts in which they work collaboratively with engineers?

CIM Implementation and Complexity

Computer-integrated manufacturing differs from nonintegrated systems with respect to technical complexity and the scope of operations (Majchrzak, 1995). Complexity means that with integrated manufacturing, the equipment often serves multiple and flexibly interchangeable functions, and significant information typically is needed by the human operator to interpret, evaluate, and diagnose events. Such a situation may lead to technical problems that are more difficult to diagnose than those that can occur in the nonintegrated manufacturing systems. The scope of operations means that under integrated manufacturing systems the equipment performs many more operations than before, making disturbance removal more difficult (i.e., a solution to a problem at one machine needs to be considered in relation to other machines). Therefore, appropriate management strategies must be developed to allow workers to cope effectively with the greater complexity and scope of integrated manufacturing.

Some of the ways to support workers in managing complexity include the following (Majchrzak, 1988; Majchrzak, 1992; and Liker et al., 1993):

1. broadening manufacturing operators' job responsibilities to include machine repair, process improvements, and inspection
2. enlarging maintenance workers' job responsibilities to include teaching, ordering parts, scheduling, and machine operations
3. extending supervisory job responsibilities to include working with other departments to resolve problems
4. more maintenance people to compensate for increasing equipment unpredictability
5. increased use of work teams to provide a coordinated response to broad problems
6. operator selection based more on human relations skills than seniority to ensure necessary communication and coordination capabilities
7. increased training in problem solving and how the various manufacturing processes function to handle the increased scope of problems

Majchrzak (1995) also postulated that enterprises that successfully utilize integrated manufacturing technology exhibit more of the management practices discussed above than the less successful plants. In addition to implementing the integrated manufacturing, the successful plants also have these management practices in place. For corporations with nonintegrated manufacturing, the kinds of management practices needed for integrated manufacturing are not nearly as critical for the plant to perform successfully.

With flexible manufacturing automation, workers need to be able to respond to a greater number of issues than with manual operations. They thus need to have the appropriate authority, technical skills, latitude in work procedures, and motivating rewards for quick and effective response. Therefore, design of TOP integration should include the following requirements (Majchrzak, 1995):

1. a broadly diffused expertise among workers, engineers, technicians, and managers about different TOP integration options and their impacts on each other
2. different disciplines and stakeholders working together as a team to create a comprehensive picture of the as-is and to-be TOP aspects of the enterprise

3. modeling techniques that allow the team to explore and assess the consequences of different integration options so that intuitive predictions about effects of technological or organizational change can be tested against best-practice models
4. documentation of designs with the expectation that they will evolve over time in response to ongoing learning and adjustment; so as workers gain experience with new technical and organizational designs, they can improve their designs.

The HITOP System

HITOP, or high integration of technology, organization, and people (Majchrzak, 1990), is a methodology that allows a design team to conduct an analysis of a TOP system. A HITOP analysis involves the design team completing a series of checklists and forms that describe their organization and current technology plans, and then helps the design team to identify the implications of those plans with respect to organizational and people issues. However, the HITOP analysis contains no knowledge base, nor does it recommend specific design options.

The ACTION System

ACTION is an interactive software system and methodology that embodies an extensive knowledge base about relationships among technical, organizational, and strategic features of manufacturing corporations (Majchrzak, 1995). The system allows one to model the impacts of different organizational, technology, and strategy choices. This can be done, for example, by specifying the characteristics of the technology, best practice skills, information, performance measures, rewards and norms, and empowerment needs for the comprehensive data sets related to different activities, process variance control strategies, and business strategies (including new product development flexibility, minimizing throughput time, and maximizing process quality). A profile of the ideal organization can then be invoked from the knowledge base in terms of the required activities, information, skills, technologies, etc.

According to Majchrzak (1995), the identified system integration-related problems (gaps) and alternative priorities for solving such problems, which are based on different prioritization criteria, can be in analyzed in ACTION and presented to the user. The prioritization is calculated based on a probability model in which a high probability of accomplishing a particular business strategy is calculated based on the number of problem areas meeting particular criteria of successful organizational designs, minimized coordination needs, maximizing unit capabilities, and motivating workers through appropriate performance metrics and rewards. ACTION promotes the modeling of alternative TOP integration designs by a computer-accessible knowledge depository of best practices. Furthermore, the system uses a simple mode of changing inputs in order to describe alternative ideas, and allows for an arbitrary order for input of data, design scenarios, information processing, and output.

Although the ACTION theory and software tool is considered to be the most powerful tool available for implementing socio-informational systems, because it allows one to align social and technical subsystems in manufacturing organizations, it has also some shortcomings. ACTION does not cover all critical design aspects of a CIM system and the reference architectures and ACTION are qualitative tools only.

The CIMOP System

Computer-integrated manufacturing facilities or enterprises are complex systems that require a high level of integration between its four subsystems, i.e., technology, organization, information system, and human resources. Knowledge about the relationships during these subsystems is essential for achieving the multiple and changing objectives of the entire CIM system. While each of the subsystems represents different design aspects, the whole system design must be optimized with respect to a large number of relevant variables. It should also be noted that the design aspects of a CIM have different levels of importance, and these decisions must be made based on the system hierarchy.

Recently, Kantola and Karwowski (1999) developed the CIMOP method for evaluating computer-integrated manufacturing, organization, and people system design (Fig. 11.1). The critical design aspects are quantified and are called the design factors (DFs). These factors relate a CIM system design to practice and allow for quantitative: (1) evaluation of a CIM system design, (2) evaluation of an existing CIM system, (3) comparison between CIM system designs, and (4) "what-if" type analyses of the effect of alternative system design improvements. The ability

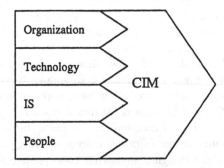

FIG. 11.1. Organization, technology, information systems and people are integral parts of a CIM system.

to evaluate CIM subsystems individually (from their integration point of view), in addition to the whole system, helps to pinpoint problem areas in the design.

CIMOP utilizes 75 design factors representing critical design aspects, including: (1) organization subsystem (12 DFs), (2) technology subsystem (17 DFs), (3) information systems (IS) subsystem (22 DFs), and (4) people subsystem (24 DFs). These design factors were defined based on the latest knowledge from the following disciplines: human-centered CIM (HC-CIM); manufacturing organization design; manufacturing system design; information systems in manufacturing; flexibility, reliability, availability, and maintainability (RAM); human–computer interaction (HCI); and human–machine system design (HMS).

The CIMOP method allows users to select specific DFs for inclusion in the evaluation criteria. While it is strongly recommended to include all DFs in every CIM system evaluation, limited evaluation criteria (with less than 75 DFs) can be used to evaluate systems with a lower level of integration. DFs are arranged in a top-down manner to a functional hierarchy, which shows their primacy and subordination to each other. The design factors and their grouping are shown in Table 11.1. The evaluation of DFs indicates the degree to which specific objectives of a CIM system design are met. DFs are not independent, and the interaction between them allows for trade-offs through compensatory effects of different inputs related to the specific output. The DFs were chosen to be universal in a quest to minimize the need for different definitions under varied social environments.

In CIMOP, five evaluation modes are used to determine the overall system design quality: (1) CIM system design, (2) organization subsystem, (3) technology subsystem, (4) IS subsystem, and (5) people subsystem. The greatest advantage from such an evaluation is the ability to evaluate CIM structure prior to its implementation so as to avoid possible design failures. Costs at the design stage are relatively low compared to the costs resulting from a CIM system failure. Figure 11.2 illustrates the hierarchical structure of CIM system design evaluation.

The DFs, DF groups, CIM subsystems, and overall CIM design quality are defined as fuzzy linguistic variables. Three linguistic states, LOW, MEDIUM, and HIGH, are used to describe the state of DFs, and also the state of inputs (intermediate results) to the next level of the evaluation process. The value of HIGH describes the desired state of DFs. LOW denotes an unacceptable state of DF, which is a generally known obstacle to a successful CIM system. The MEDIUM state describes the value of DF, which is acceptable, but needs improvement. Generally, MEDIUM is estimated halfway between LOW and HIGH. Five linguistic states: LOW, LTM, MEDIUM, LTH, and HIGH are used to describe the design quality of DF groups (intermediate result), the design quality of CIM subsystems, and the design quality of the CIM system.

CIMOP allows the one to account for the DFs imprecision, i.e., the uncertainty in evaluation of DFs. Such uncertainty is present due to the following reasons: (1) DFs are qualitative and they are imprecisely defined; (2) the assessment of DFs is subjective; (3) partial ignorance about the state of DFs can

TABLE 11.1
Design Factors and Their Grouping

CIM Subsystem	Design Factor Group	Design Factors
Organization	Organizational design	Allocation of operations, allocation of controls, information flows, organizational design process
	Functional structure	Organizational levels, recursivity of cells, central coordination, system dependence
	Organizational learning	Intracell learning, intercell learning, system-level learning, organizational self-renewal
Technology	Components	Machine tools, tools and fixtures, intracell material handling system, intercell material handling system, quality assurance (QA)
	Features	Layout, safety, possibility to intervene, complexity, upgrades
	Flexibility	Basic, process, routing, product, production
	RAM	Reliability, maintenance
IS	Implementation	Networks, compatibility, standardization, IS management, upgrades
	IS components	PPC, CAD/CAE, CA(P)P, CAM, CAQ*
	Information	Information types, data access, quality of the information, data entry, databases
	HCI	Use of IS, feedback, flexibility, design of human–computer interaction, interfaces
	RAM	Reliability, maintenance
People	(Top) management	Support, role, openness to new ideas, performance measures, environmentally conscious manufacturing
	Teams	Determination of skill requirements, norms, skill requirements, structure, hiring
	Tasks	Task requirements, continuation, task allocation between man and machine, task allocation between team members, workload
	Communication	Customer involvement, top management, horizontal communication, vertical communication
	Motivation	Working environment, salary, monetary compensations, nonmonetary compensations, flexible work time

*PPC   Production Planning and Control
CAD/CAE Computer-aided Design and Engineering
CA(P)P Computer-aided Process Planning
CAM   Computer-aided Manufacturing
CAQ   Computer-aided Quality Control

be involved in the evaluation; (4) the evaluation done by the members of an evaluation group is not uniform. The linguistic values, which are utilized for design quality evaluation, are modeled in CIMOP by fuzzy sets, which represent the CIM system design quality better than crisp sets. This is because the predicates in propositions representing CIM system design quality do not have crisp

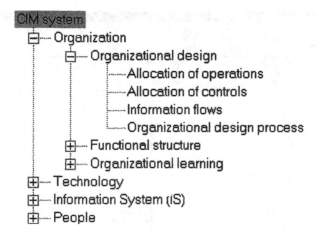

FIG. 11.2.   Example of the hierarchical structure of a CIM system design evaluation.

denotations, and their explicit and implicit quantifiers are intrinsically fuzzy. CIM system design quality is expressed through a degree of membership for a fuzzy set representing different levels of design quality.

The reasoning scheme of the CIMOP Method utilizes a fuzzy logic control (FLC). FLC is applied to model the imprecise, complex, and abstract aspects of CIM system design. In CIMOP, the FLC has a hierarchical structure, which resembles the hierarchical structure of a CIM system design. A FLC consists of four modules: (1) fuzzy rule base, (2) fuzzy engine, (3) fuzzification module, and (4) de-fuzzification module. The CIMOP method was programmed with Microsoft Visual Basic 5.0 along with MS Access database (Visual Basic, 1987–1997). Database queries are done with the Structured Query Language (SQL). The CIMOP method operates on a PC in the Windows 95 operating system. The method has an online help including the following help topics: the CIMOP method, DFs, Evaluation, Comparison, Analysis, Results, and Output scale. Projects can be printed. Printouts include general information, evaluation inputs, and graphical results.

The Evaluation screen allows for the quantitative evaluation of a CIM system design. This is done by evaluating the state of DFs of the system design (Fig. 11.3). DFs are arranged as a hierarchy, as described earlier, at the left side of the screen. The tree can be expanded or collapsed by clicking the nodes. The rightmost branches are the DFs. Predefined levels of the DFs appear on the textbox on the right side of tree. After all DFs have been evaluated, results can be viewed. Results are presented graphically on the Results screen. Another way to view results is to make printouts.

The CIMOP method has two alternative output scales for the results: Numerical and Verbal. The output scale can be selected by clicking on the Output scale listbox above the tree. Both scales are linear. The range of the numerical scale is

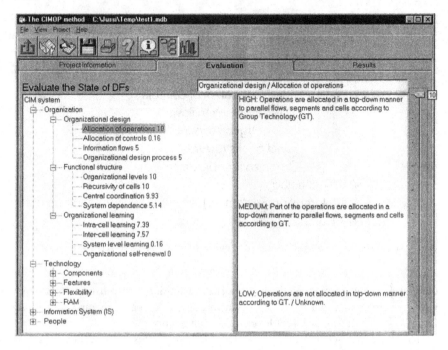

FIG. 11.3.   Evaluation screen of the CIMOP method.

from 0 to 10 scaled from the defuzzified values. The suggested verbal scale has 3 levels: Design is acceptable, Design needs improvement, and Design is unacceptable. Results can be reviewed for the overall CIM system design, organization subsystem, technology subsystem, IS subsystem, and people subsystem (Fig. 11.4). Results indicate the level of design quality for the whole system, and indicate how different subsystems contribute to the CIM design quality, and how different DF groups contribute to the quality of subsystem design.

The evaluation in CIMOP is to be conducted by an evaluation team representing the most reliable information about the CIM system design. The team should have representatives at least from (top) management/factory level, human resources, IS management, and manufacturing cells/shop floor level. The evaluation team then compares the state of each DF against the predefined levels (Fig. 11.2). After the state of each DF in the project has been estimated, graphical results representing the design quality can be examined. The graphical results indicate the design quality of the CIM system, its subsystems, and DF groups (Fig. 11.4). Result of the evaluation can aid in determining whether a particular system design should be implemented or improved.

Based on the outcome, an implementation plan can be determined in each particular case. Intended users of the CIMOP method are companies with an

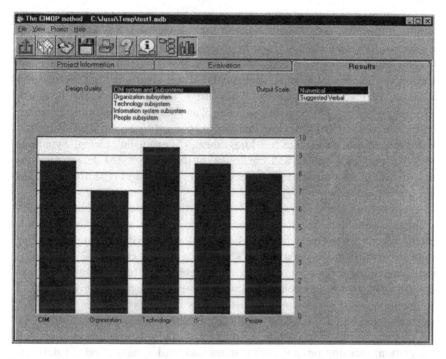

FIG. 11.4. Design quality of a CIM system and its subsystems on the numerical scale.

existing CIM system, companies designing, redesigning, or implementing a new CIM system, consultants, researchers, or any other parties engaged with CIM system design and evaluation.

CONCLUSIONS

Recent changes in the global economy induce a need to implement highly adaptive organizational structures, and adopt people-oriented, rather than technology-centered, approaches to design and operation of information-based manufacturing systems. Some of the existing organizational structures are too rigid to cope with external complexity and rapid changes in products and processes suitable to meet market demands (see Chapter 3). Such functional structures underutilize specific skills of human operators (i.e., workers' ability to cope with the system's uncertainties and emergencies by modifying rules when appropriate). Information-based manufacturing companies are learning to create organizational structures and procedures that would enhance communication and cooperation between different organizational units. Such structures also should enable people to do what they can do best, through an improvement of their technical qualifications, a shift of decision-making authority, utilization of advanced tools for information handling, and feedback systems.

The macroergonomic discipline can offer specialized knowledge on many aspects of the above paradigm, specifically on the management of change with respect to: (1) work organization, (2) job design, new ways of organizing manufacturing processes, (3) skill-oriented control and responsibility, (4) management of change processes by assessing the critical change factors and developing systems, procedures, and tactics to address them, (5) evaluating change by aiming to make problems visible and create energy for change, (6) determining the cost/benefit of solution alternatives based on the degree and type of change, and (7) specific tools, techniques, and methods.

As discussed by Karwowski et al. (1994), the knowledge of change implementation should be translated into a comprehensive and practically usable methodology. Macroergonomics can help to improve and develop manufacturing industry capabilities in the areas of: (1) management of change, (2) technology deployment, (3) integration of organization, people, and technology into a coordinated system aimed at delivering competitive advantage, (4) development of information-based and knowledge-based tools to support analysis of complex organizations, based on holistic methodologies, (5) development of skills and competencies in the area of systems strategy, systems architecting, and systems integration, focusing on both soft systems and hard systems, and (6) development of technologies that leverage the skills and knowledge of users.

Particular emphases of macroergonomics efforts are being directed toward facilitating development of agile manufacturing enterprises. There is a need for new methods and tools to support an interdisciplinary approach, with special emphasis on combining top-down and bottom-up methods, holistic approaches, organizational simulation, appropriate and selective use of technology, increased user involvement, rapid prototyping, etc. The macroergonomic discipline has an important role to play in both advanced manufacturing technology development and in defining various technology systems and their usage (Forsythe & Karwowski, 1995; Karwowski et al., 1994). For example, a challenge of agile manufacturing is socio-informational in nature (Forsythe & Ashby, 1996), as it imposes new demands on managers and floor workers. Empowerment of product development teams and increased openness of information sharing may lead to decreased power on the part of management. At the factory floor level, there is an increased responsibility on the part of the workers who are brought into the product development and decision-making processes, and the threat posed by computerization and automation of fabrication and assembly tasks.

Contemporary manufacturing faces increasingly competitive global markets of the 21st century by adopting new management and technology concepts, including the ability to develop and make new products rapidly (Forsythe & Karwowski, 1995; Kovac, 1993; Nagel & Dove, 1992). In recent years, the manufacturing industry experienced an infusion of information-based technologies and pronounced changes in work culture and work practices. Macroergonomics plays an important role in assuring that the technical and social objectives of information-based manufacturing are met, and in shaping the new manufacturing paradigms of the future.

12

Macroergonomics of Training Systems Development

Michelle M. Robertson
Liberty Mutual Research Center for Safety and Health

INTRODUCTION

Macroergonomics is concerned with human–organization interface (HOI) technology, which includes the systematic consideration of the personnel subsystem (see Chapters 1 and 3). As described in Chapter 3, macroergonomics involves designing the work system to be compatible with the psychosocial and skill characteristics of the people who will constitute the personnel portion of the system. Accordingly, the personnel subsystem also involves the analysis, design, and evaluation of training systems to ensure that they meet the personnel subsystem requirements of the work system. Today, many organizations are developing training programs to meet the demands of both complex external environments and new technologies being introduced into the workplace.

Training is an integral part of a larger work system, with the individual or end-user in the center. It is important that the entire work system be taken into account, including the balance of the interrelated system elements. These work system elements include: job design, environmental design, technology, and the work system's structure and processes. All of these integrally related system elements should be considered to enhance productivity effectively and minimize the negative health and safety effects arising from poorly designed work systems.

249

These training elements, processes, and models are applicable to a broad spectrum of work environments and conditions, such as aviation maintenance (Robertson, 1998), manufacturing, telecommunications, government, military, utility, and service organizations, among others. The effectiveness of these training programs varies, depending on several issues such as the organization's culture.

When a macroergonomic approach is taken, training is part of a comprehensive, systematic approach to enabling knowledge within an organization. This approach enables training to play a key role in linking an organization's corporate goals with organizational effectiveness and facilitating required change processes. Training programs can modify how people work together, solve work-related problems, and actively fulfill their role in implementing workplace solutions. For instance, when a successful office ergonomic training program is implemented, the result is an increased ability for the worker to change his or her work environment, reduce exposure to work-related risk factors, and promote healthy work practices. High-quality training also incorporates a "participatory" approach, in which end-users, managers, designers, support staff, and others are involved. A participatory approach assures that each employee develops the knowledge, skills, and motivation to feed back useful suggestions for improvements to the organization, including its work system design (see Chapter 2). For example, it is this participatory aspect, along with the ergonomic training, that forms the basis for creating an improved work environment and continual work system improvement within the organization.

To illustrate macroergonomic aspects of the instructional systems design (ISD) process, this chapter will provide an overview of a macroergonomic training approach to designing, implementing, and evaluating ergonomic training programs. A case study then will be presented to illustrate the macroergonomic ISD model. The case study demonstrates the applicability of this macroergonomics training model in the occupational context of aviation maintenance operations. The goals of the training programs were to reduce human error and improve safety in aviation maintenance environments. Elements necessary for successful programs and training models for designing and evaluating the effectiveness of training programs are also presented.

MACROERGONOMIC COMPONENTS
OF A SUCCESSFUL TRAINING PROGRAM

Successful training programs consistently incorporate several critical components. These include creating a responsive organizational environment, supporting active participation, developing active learning experiences, creating continuous learning and improvement pathways, providing continuous feedback, training supervisors and middle managers, having senior management support and commitment, and using an instructional systems design model. Each component is discussed below along with examples.

Creating a Responsive Organizational Environment

Training of managers and supervisors also is necessary in order to provide a responsive environment in which employees are encouraged to utilize their training through reinforcement and reward. Because supervisors have more influence on the daily performance of individual employees, their participation in the training process is essential for the success of the ergonomic training program. Luopajarvi (1987) found that supervisors responded best to training that emphasized situations over which they had some measure of control, and that such training made them more cooperative and supportive of change. This type of training assures that supervisors learn to respond effectively to suggestions regarding office ergonomics given by employees.

Supporting Active Participation

Participating in the creation, development, and implementation of a training program stimulates a feeling of individual ownership. Active involvement creates a sense of commitment to supporting the training program goals and a willingness to engage in the required cultural change process. Being a member of a team that is designing and implementing an office ergonomic training program is motivating, rewarding, and beneficial to the individual and the organization. Working together on a cross-functional, interdisciplinary team provides a unique strength in designing and developing a training program. If there is a lack of active worker participation in the training program, workers' motivation for, and understanding of, the material presented is low and their resistance to change is high (Luopajarvi, 1987).

Developing Active Learning Experiences

More effective instructional methods, sometimes called "inquiry" or "discover" learning, emphasize the involvement of learners. These methods include trainees by having them participate in problem-solving activities and group discussions. The strength of this approach is that the use of group exercises and related case studies promotes an active learning environment. It creates an interactive, highly motivating approach, since the trainees are doing more than just passively receiving the information—they are actively applying and using the concepts and skills. To further strengthen this approach, training courses can be cofacilitated with the trainer by trained workers who are knowledgeable in specific work requirements and processes. These facilitators can encourage students to participate by bringing "real world" experiences into the classroom. Often, these employee facilitators are viewed as relevant and valid experts. If the class makeup consists of individuals from various job positions in the organization, active and interactive discussions can occur providing an ideal opportunity to discuss "real" problems together.

Continuous Learning and Improvement

Every work system changes over time. For example, in the systems approach, office ergonomic training and practice must be viewed as part of the overall health, safety, and ergonomic program. As such, the training program must be adapted to other kinds of changes that occur in the work system—such as the introduction of new technology or new work procedures—as well as to changes in health and safety practices. The idea of continuous change and adaptation is fundamental to making any work system responsive to the needs of the workers, as well as to changes in the organization's technology or relevant external environment. Financial and organizational resources must be committed to actively support the change process. For example, with respect to changes in the organization's safety culture, this includes the commitment of human resources within the company, such as administrators, trainers, and curriculum developers, and media and computer application developers. Continuous improvement is not a short-term activity. Rather it requires long-term management commitment to continuously adapt and improve the program.

Providing Continuous Feedback

Performance improves more quickly when people are given feedback (Hackman & Oldham, 1976, 1980). Feedback provides information to accomplish two key performance improvement goals: (1) improve the training program and identify necessary corrective actions, and (2) reinforce the positive outcomes and benefits of using the learned skills in the workplace.

Senior Management Support and Commitment

The foundation of any successful organizational training program is senior management support; for example, if an office ergonomic training program is to be successful, senior managers must have the vision and commitment to reduce adverse health effects and increase employee quality of work life, as well as productivity, through the application of office ergonomics. When top decision makers clearly support the mission and purpose of an office ergonomic program, an organizational culture and safety climate change can occur (e.g., see Chapter 8). Without such a commitment, a pervasive organizational change is unlikely.

Training for Supervisors and Middle Managers

Linked to senior management support is training for supervisors and middle managers. These individuals interact daily with the workers who are ultimately responsible for implementing the new strategies. Mid-level managers also need the support of upper-level management in implementing the new skills, knowledge, and practices in the work system. This support can take many forms, but certainly includes the time to attend appropriate training courses. With this commitment, supervisors will have the opportunity to use their own enhanced skills and knowledge. Most importantly, they will know what behaviors to look for to reinforce in their trained employees.

Using an Instructional Systems Design Approach

Designing an effective training program should include several processes and activities including: (1) conducting a needs or front-end analysis, (2) designing the training materials, (3) developing the training materials, (4) implementing and delivering the training, and (5) evaluating and measuring the effectiveness of the training. There are several ISD models, and each one incorporates these basic processes or phases to create effective training programs (e.g., Gagne, Briggs, & Wagner, 1988; Gordon, 1994; Goldstein, 1993; Mager, 1984). For each instructional system development stage, different types of information are conveyed and different techniques are used to analyze the collected information. See Fig. 12.1 for the ISD model and phase activities.

Conduct a Front-End Analysis—Phase 1

A front-end analysis is conducted to determine strategic training needs and to assess the company's return on investment in training. A training design team first identifies the organizational and trainee needs and constraints before it begins designing the training program. Analysis is the foundation for all later work related to implementing a training program. A thorough needs assessment helps reduce the risk of funding inappropriate or unnecessary training. Training should be developed by a team that includes, among others, training professionals, a human factors/ergonomics practitioner, and, as applicable, various maintenance or operations supervisors and workers.

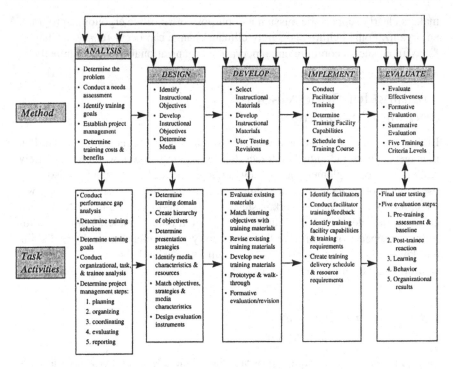

FIG. 12.1. Instructional system design phases, activities, and feedback loops.

Needs Analysis. In the needs analysis phase, an organizational, task (job), and person analysis is conducted. Two questions need to be answered: "What is the *current* performance of the organization and the workers?" and, "What is the *desired* performance for the organization and the workers?" The derivation of the training objectives and their linkage to the corporate goals are accomplished at this stage. If the performance problem is identified as a lack of knowledge, skill, or ability of the workforce, then a training program should be designed as a necessary intervention strategy. However, if the needs analysis uncovers that the performance problem is due to a poorly designed work system, then training is not the only intervention solution. Also accomplished in this phase is developing the criteria for evaluating and measuring training effectiveness. These criteria are linked to the training objectives and are established at the trainee, departmental (strategic business unit), and organizational performance levels.

Organizational Analysis. Organizational analysis consists of (1) an evaluation of the organization and/or industry in which the trainee performs his or her job and (2) an evaluation of the organization or unit expected to provide the training. There are a number of methods that can be used to obtain information and collect data for an organizational analysis. Probably the most efficient approach is a combination of document analysis, interviews, and questionnaires; that is, the training designer would start by obtaining and analyzing organizational documents. Once this has been completed, a few key interviews will yield additional relevant information. This information is verified on a wider population sample via questionnaires. The training design and development team is selected during this phase of the process. It is critical to the success of the training program to include representatives from all of the essential areas of the organization. Likewise, senior management must allow team members to devote the necessary time to the project. Organizational analysis determines the probable cause(s) of performance gaps and includes a distinction between needs that can be solved by training and needs that must be addressed by a change in organizational procedures or policies. Of course, if an incompatible work system design is suspected, a macroergonomics analysis of the work system's structure and processes and any resultant work system changes should be completed first (see Chapters 3 and 4).

Tasks Analysis. Task analysis determines the tasks required in a job, the subtasks that costitute each task, and the knowledge and skills required to successfully complete these. Task analysis helps the instructional designer determine exactly what the learner needs to be able to do. It also allows the designer to develop objectives from the task elements. A task analysis may be accomplished by: (1) observing a skilled and knowledgeable employee, (2) reviewing documents/manuals on the tasks, (3) interviewing employees who perform the tasks, and (4) performing the tasks. It is critical to remember that the best sequence for instruction might be quite different from the sequence of tasks in the workplace. An instructional designer may be able to generate the instructional objectives directly from the task analysis results.

Trainee Analysis. Trainee analysis is performed to identify the relevant characteristics of the learners who will be participating in the training program. These characteristics and associated learning issues are presented in Goldstein (1993) and Gordon (1994).

Design—Phase 2

In the design phase, results of the needs analysis are used to determine how the training objectives are going to be met. This involves determining the: (1) training prerequisites, (2) trainee population, (3) desired learning outcomes (e.g., knowledge, skills, and abilities in cognitive, affective, and psychomotor domains), (4) training media and techniques, (5) training environment, (6) learning conditions (e.g., individual differences), (7) instructional strategies, and (8) learning principles. Contextual factors involving organizational, environmental, and social issues, such as designing training for newly hired workers versus experienced workers, are determined in this phase. Because training occurs within an organizational culture, how the organization values the training and its integration and link to the corporate strategy must be accounted for in this design phase. Information derived from the needs analysis phase is used to identify these factors.

Specifying the selection of instructional strategies must be determined before developing the training materials themselves. This selection also is needed to outline how the instructional activities will relate to the accomplishment of the training objectives and goals. Gagne, Briggs, and Wagner (1988) propose a series of nine instructional events that must occur for learning to take place that apply learning theories to the development of training. These nine events include: (1) gaining attention of the trainee, (2) informing the trainees of the training objectives, (3) using recall or transferring from the trainee's existing experience, (4) presenting training material to be learned, (5) providing learning guidance or elaboration, (6) eliciting desired performance, (7) providing feedback, (8) assessing performance, and (9) enhancing retention and facilitating transfer of training to actual task performance.

Optional selection of the training delivery system is conducted by matching the training media strengths with the identified training objectives as determined in each instructional event. Different instructional media have different capabilities and strengths for providing the various events of instruction. For example, Fig. 12.2 presents a media decision table in which an instructional designer makes a series of decisions depending on the type of learner reaction(s) required. In developing an instruction strategy, the choice of the training delivery system can be assigned event-by-event and objective-by-objective in order to accomplish the training goals.

Evaluation of the resources available for the development and delivery of the training program also is necessary. This consists of identifying various constraints, such as the availability of equipment, time, money, and instructors. This information is transformed into a set of functional design specifications, a specific list of training goals, and system requirements that will provide the boundaries of the training program.

This phase of the ISD model involves developing the goals and objectives of a training program, selecting the instructional content, specifying the instructional strategies, designing evaluation instruments, and specifying the training media. The design process consists of four levels: (1) program, (2) curriculum, (3) course, and (4) lesson (Hannum & Hansen, 1992). Program and curriculum are associated with a *macro* (general) type of organizational analysis (Hannum & Hansen, 1992). Training is linked with the strategic plans of the organization and a series of course needs is identified for different groups of trainees. The course and lesson levels are defined as the *micro* type of planning, such as developing instructional objectives, learning task hierarchies, evaluating and testing

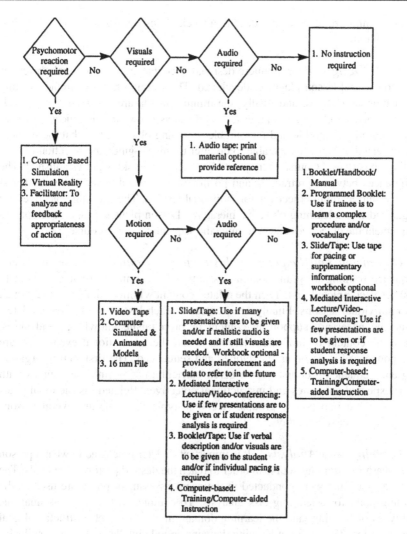

FIG. 12.2. Media selection model.

procedures, and selecting media (Hannum & Hansen, 1992). At this point, decisions are based on instructional theory and research. Thus the designers are concerned with the learners' ability to understand, remember, and transfer the training concepts to the work site. A training program can include several curricula. Each curriculum consists of a series of courses and each course will usually include a number of different lessons. The time required to design a curriculum containing several courses varies from a few weeks to a few months, depending on the level of effort expended and the complexity of the curriculum. Designing a course might require several days to a week. A lesson usually can be designed in 1 to 3 hours.

Micro Design.   During micro design, the course and instructional objectives and course and lesson plans are developed. The learning task hierarchies, testing evaluation procedures, and finally, the training media are all created and developed. Course-level design requires a careful analysis of goals and objectives to ensure they are observable and measurable. Learning objectives—what the trainees are expected to know and be able to do after training—guide the selection of "enabling objectives." Course prerequisite knowledge and skills are established. The differences between pretraining and posttraining knowledge and skills are transformed into learning objectives for individual lessons. The learning objectives are organized into a learning objective hierarchy. Lesson plans are then developed to implement the objectives and determine the sequencing of the course.

Organizing and Sequencing Instructional Content.   To organize and sequence training program content, assigning large units of related content to courses is completed first. Then the related content within each course is grouped into individual lessons. Finally, the content of each lesson is analyzed to determine the necessary supporting content or prerequisite knowledge and skills. Once the content of individual lessons is set, the instructional events that are components of a lesson are developed and sequenced. The first step in organizing and sequencing courses for a given curriculum is to link the courses with their expected outcomes and instructional objectives. Related instructional goals are organized into course groups. A course typically contains several lessons completed over several days.

Learning Task Analysis.   It is important to determine exactly what a person must learn in a training course in order to reach the desired performance goals. The task analyses that were conducted in the needs assessment phase are used as the starting point for a learning task analysis. To complete a learning task analysis, the types of learning, and the learning domain implied by each instructional goal are identified. The different learning domains are: (1) intellectual skills—skills including understanding and articulating concepts, rules, and procedures, (2) information—ability of individuals to verbalize declarative information, (3) cognitive strategies—acquiring strategic knowledge, (4) attitudes—emotions and values adopted by trainees, and (5) psychomotor skills—skills involving muscle development and coordination (Gagne, Briggs & Wagner, 1988). Because the learning outcomes are different in each domain, analyzing these learning outcomes requires different techniques. Once the particular domain is identified, the appropriate analytical technique can be applied. As human performance is organized by these categories of learning outcomes, understanding and identifying these learning outcomes help determine the appropriate instructional conditions for the trainees.

Sequencing Instructional Content. There are two aspects to sequencing lessons: (1) instructional content and (2) instructional events (Hannum & Hansen, 1992). Instructional content includes those facts, ideas, concepts, skills, etc., that are defined in the instructional objectives. This is the material instructional designers expect the learners to master by going through the lesson. Instructional events are those features of a lesson that, when present, facilitate learning. These include informing the learner of the objectives of the lesson, providing examples, and giving feedback. The content in different learning domains has a different natural organization; for example, motor skills have procedural structures, intellectual skills have learning prerequisite structures, and attitudes have a behavioral structure. These different structures imply different instructional sequences. There is not one instructional sequence that is effective for all types of instructional outcomes. Learning is more complex than that. Different types of instructional content, i.e., different domains of learning, require different instructional sequences.

Learning Objectives and Hierarchy. The learning objectives related to each learning outcome are ordered so that an instructional designer can specify the level or depth to which the potential learners must be brought in order to know the information. The use of learning categories ensures that instruction is properly focused. Identifying and understanding the learning outcomes can ensure that it is implemented in a logical and sequential manner.

Performance Objectives. Developing and writing performance or instructional objectives should follow the "ABCD" format as described by Knirk and Gustafson (1986). It is more important to include all four components than to follow the exact sequence implied by the format. Clearly stated and written performance objectives establish the basis for evaluating the training. Basically, instructional designers are trying to determine whether the trainees have successfully accomplished the stated instructional objectives. Learning objectives can serve as an organizer for trainees. These objectives explicitly state what trainees are expected to learn and demonstrate.

Specifying Training Media. After instructional objectives have been developed, designers must select instructional strategies and media. These decisions tend to be interrelated and should be made concurrently. Many instructional strategies use a combination of methods and media to deliver training. The instructional medium should present instructional stimuli in an efficient, easily understood manner. Complex media, which tend to be costly and time consuming, are often inefficient and unreliable. Use the least expensive medium that will result in trainees' attaining the desired objectives within a reasonable amount of time. Choose as the primary medium of instruction one that is appropriate for the majority of objectives—one that can be used throughout the instructional program. Additional media, such as simulations or animated visualization, can be used for emphasis or motivation. It is desirable to have a mix of instructional methods that actively involve the learners. Frequent media changes are often confusing, time consuming, and expensive. Interactive exercises, role playing, and games or simulations that involve the trainee, promote the sense that the training is relevant and useful.

A method of making media selection decisions is to relate the general domain of each objective to student grouping requirements (Goldstein, 1993; Salas, Dickinson, Converse & Tannanbaum, 1992); for example, if the course objectives are at the lower end of the cognitive taxonomy (e.g., knowledge of specifics and comprehension), then certain types of teaching activities or media are more appropriate for individualized instruction. Others may be more appropriate for group instruction.

Training Plan: Course Outline and Lessons. A training plan provides a blueprint for training development. This includes providing a description of the training objectives, content, media, training aids, and other elements required for actual instruction, and estimated time required for each training topic.

Design of Evaluation Instruments. Evaluating the training program involves measuring the degree to which the learning objectives were met. Using the information from the task analyses and the learning objectives hierarchy, one can establish performance criteria that can be subsequently measured and evaluated. These training evaluation instruments may include: (1) questionnaires, (2) observations, (3) interviews, (4) verbal protocols, (5) task performance measures, and (6) work unit and organizational performance measures. Certain instruments collect essential training-related performance data before and after the training takes place. These assessment tools are used at different times during the training evaluation process (Cannon-Bowers et al., 1993; Goldstein, 1993; Gordon, 1994; Kirkpatrick, 1979). Table 12.1 presents a typical evaluation assessment process (Cannon-Bowers et al., 1993; Gordon, 1994; Hannum & Hansen, 1992; Kirkpatrick, 1979). Creating an evaluation plan during the design phase will ensure that the process is more efficient and provides useful information. The evaluation process establishes the link between the goals and objectives of the training program and its results. Information collected from the course evaluation creates an important feedback loop—demonstrating the overall effectiveness of the training. This information is also useful for course revision.

Development—Phase 3

Developing the training materials is the next phase in the instructional system design processes that includes developing all training materials, in whichever media are selected during the design process. It is important to note that the

TABLE 12.1

Evaluation Process—Five-Level Process and Evaluation Measures

Five-Level Process	*Examples of Evaluation Measures*
1. Baseline Assessment	Measurements taken prior to implementing the alternative (intervention). Measures could include:
	• Health and safety performance indicators (e.g., lost work days, absenteeism, frequency, and severity rates)
	• Productivity data (e.g., individual performance, strategic business units, organizational performance, market share, customer satisfaction, balance scorecard)
	• Users' and managers' current skills, knowledge, abilities, attitudes, and opinions
	• Time series and trend data is preferred, several months to a year of data continuously collected before implementation of alternative
2. Reaction	Users' and managers' reaction to the intervention; including managerial response. Measures could include:
	• Surveys
	• Semistructured interviews
	• Users' perception of the usefulness and relevancy of the intervention to their job
3. Learning	Users' and managers' degree of learning (knowledge, skills, and abilities). Measures could include:
	• Surveys (same as preknowledge)
	• Semistructured interviews
	• Observations; attitude change, opinions
4. Performance	Users', Managers', and business unit performance. Measures could include:
	• Surveys; self-reported performance, perceptions, intent of behavior changes
	• Semistructured interviews: attitude, behavior, and productivity changes
	• Observations: behavior changes; team performance changes (collaboration and communication)
	• Unit or departmental performance measures (e.g., products, project completion; quantity and quality of service)
5. Organizational Results	Performance and productivity measures. Measures could include, similar to baseline tracking measures:
	• Safety and health performance measures (e.g., lost work days, frequency, and severity rates)
	• Strategic organizational performance measures (first to market, product innovation, customer satisfaction)

instructional design process itself determines the effectiveness of a training program, not the training media and technology. If sound instructional design principles are used, the instructional designer will choose instructional or other technologies that meet the functional and training requirements. Because various media interact with certain instructional methods, the development phase involves piloting and walkthroughs of all training modules. During this phase, instructional strategies are applied sequentially to each training activity, then the most effective media delivery techniques are selected. Additionally, the format of delivering the training is determined, that is, it may be instructor/facilitator controlled, performance controlled, or trainee self-paced style.

Training Plan Development. The primary activities in the development phase are developing training materials and media, as well as developing and testing prototypes. Training materials are modified during this phase, based on the results of prototype and user testing. An outline of typical activities for selecting and developing training materials is shown below (Gordon, 1994; Hannum & Hansen, 1992; Knirk & Gustafson, 1986):

- Create a development plan
- Search for existing content related training materials
- Evaluate existing instructional materials
- Match objectives with training content and materials
- Make trade-offs of objectives and training materials (economic and effectiveness)
- Examine copyright requirements (obtain copyright permissions)
- Revise/modify existing training materials
- Develop and produce new training materials
- Develop facilitator and participant handbooks
- Prototype and walk through new training materials
- Revise/modify new training materials and handbooks
- Final user testing of training materials and handbooks
- Final development of training materials and handbooks.

A training development project should specify the following elements (Gordon, 1994):

- Personnel—training and human factor specialists, content specialists (maintenance)
- Budget—money to develop the training materials and handbooks; personnel cost of developing the course; travel time and expenses; evaluation costs
- Equipment—technical equipment, audio/video facility, video cameras, studio equipment, editing equipment, audio equipment
- Outside services and consultants—scriptwriters, actors, graphic designers, videographers, computer programmers.
- Tasks and activities to be completed and by whom
- Training tasks and activities timeline.

Developing Training Materials. The media selection model(s) chosen in the design phase will serve as the framework for developing training media.

Before beginning production, instructional objectives should be reviewed to confirm their sequential order. Conceptual sketches and outlines of the audio-visual aides are developed and then reviewed by other members of the design team and relevant subject matter experts.

Storyboard Scripting. Developing a script requires the instructional designers and other team members to think visually. Sound and the written word are often not as reliable as a visual presentation for enhancing the trainees' retention. Visuals can carry the message and narration can be used to clarify and reinforce the visuals. The development team should be certain that the graphics, written material, and audio support each other. Scripting the training materials requires the instructional designer to: (1) clarify difficult points through visual illustrations, (2) simultaneously present to two different human senses, i.e., seeing and hearing, (3) determine the best approach to convey the message quickly and clearly, and (4) isolate and focus the trainees' attention on the central points specified by the performance objectives (Gagne et al., 1988; Knirk & Gustafson, 1986).

Walkthrough and Formative Evaluation. It is much easier to modify the training materials during the design and development cycle, rather than after the training program has been implemented. Conducting a formative evaluation of the training program while the training materials are in a draft form allows essential and meaningful feedback to be collected from the learners. A simple formative evaluation consists of having a pilot trainee group read and review a storyboard. It should also measure the usefulness of the materials. User teaming will give the development team a good idea whether the general training approach is a sound one. Table 12.1 lists various formative evaluation methods, primarily at the second and third evaluation levels (trainees' reaction and learning) (Cannon-Bowers et al., 1989; Gordon, 1994; Kirkpatrick, 1979; Knirk & Gustafson, 1986). The advantage of user testing is that it allows the design team to solicit meaningful feedback from the users. Early user testing can reduce program costs and increase the probability that the product will perform as required.

Final Development and User Testing. After the prototyping and walk-throughs have been completed, the training materials are put through the final development and production steps. After the production training materials are available, but before they are actually implemented, they should be subjected to one more stage of user (learner) testing.

Facilitator and Trainee Handbooks. After the final user testing is completed and the training program is ready to move into full development and production, facilitator and trainee handbooks should be developed. Typically the facilitator handbook contains the following elements: (1) a detailed outline of the instructional sequence, (2) a description of the course and lesson goals and objectives, (3) a narration of the visuals to be presented, (4) the training time frame, (5) administration issues and guidelines, (6) description of the group exercises, (7) a list of the participant and facilitator materials, (8) a list of reference materials, (9) a description of how to set up the training classroom, and (10) a list and description of the evaluation instruments (Dupont, 1997; Robertson, 1998; Robertson & Endsley, 1997).

Implementation—Phase 4

Implementing the training is the next phase, consisting of scheduling how and when the training is to be delivered. In this phase, the overall training implementation plan is developed and the training is actually conducted. If the training is delivered in stages, it is possible to conduct further formative evaluations and revise the training program and materials before full production and implementation. A spreadsheet can be created that includes the schedule for delivering the training. It also is possible to purchase various computer-based management programs that can provide a framework for structuring the training schedule. An important program element that must be addressed during this phase is to gain (or reconfirm) management's commitment to deliver the training program and provide the necessary resources to successfully implement the course.

Facilitator Training. During the implementation phase, the facilitators must be formally trained. It is possible that the facilitators are part of the design and development team and require only a minimal level of training. However, in many instances the facilitators or instructors will not have taken an active part in designing the training program and will need assistance. If they have been identified previously, facilitators can practice delivering the training lessons during the formative evaluation and user testing stages. In this way, immediate feedback can be provided to the instructors as part of the overall evaluation process.

Evaluation—Phase 5

Planning for evaluation should take place during the design phase. Evaluation is important in order to: (1) determine if the training meets the objectives, (2) determine if the entire training program meets its goals, (3) provide feedback to the facilitators, (4) provide feedback to the trainees, (5) provide feedback to top managers and the organization as a whole, and (6) review and improve the training program. The evaluation process should measure the effects of training on the variables that have been identified as being important. Evaluation "criteria" are those variables that represent the specific factors course designers have targeted during the development process. These criteria are based on the training objectives and goals and are established in the needs assessment and design phases.

Evaluation Process. There are two types of evaluation: formative and summative (Gagne et al., 1988; Goldstein, 1993; Knirk & Gustafson, 1986). Formative evaluation was discussed in the design and development phases as part of the user prototyping activities. Summative evaluation is conducted after the training course has been developed, implemented, and delivered. Summative evaluation typically determines the extent to which the training program has been successful in meeting its stated training, behavioral, and organizational objectives. It also determines the value of the training program, and what modifications need to be made to make it more effective. A summative evaluation should be conducted using the following general principles: (1) conduct the evaluation in an environment that is as similar to the ultimate job environment as possible, (2) conduct the evaluation after a realistic period of time (preferably 2, 6, and 12 months following training), and (3) conduct the evaluation based on the targeted job tasks and conditions (Gordon, 1994; Knirk & Gustafson, 1988).

Evaluation Model. Table 12.1 lists various methods for evaluating training courses (Cannon-Bowers et al., 1993; Goldstein, 1993, Gordon, 1994; Hannum & Hansen, 1992; Knirk & Gustafson, 1986; Kirkpatrick, 1979). These levels of evaluation and the types of data that can be collected provide for a solid framework in evaluating a training program. Evaluating the effectiveness of the training program and providing feedback to the organization and trainees is the last phase in the instructional system design. When the training results match the training goals and objectives, the training program can then be concluded to be effective.

A systematic, multiple measures, five-level training evaluation model can be used for evaluating training effectiveness (Gordon, 1994; Kirkpatrick, 1979). This five-level evaluation model process includes: level I, training baseline assessment; level II, reaction to the training program; level III, learning of principles, facts, techniques, and attitudes; level IV, behavior relevant to job performance; and level V, results of the training program related to organizational objectives. Measurements that may be taken at each of these training evaluation levels are as follows: level I: baseline measures of performance and organizational measures prior to training; level II, posttraining questionnaire asking the trainee to evaluate the value and usefulness of the training; level III, pre- and postquestionnaires/tests assessing how well the trainee learned the information taught as well as observations/interviews with the trainee; level IV, assessment of the trainees' behavior on the job—how well the trainee was able to transfer the knowledge and skills to the job—this may be completed by observations and/or interviews; and level V, results and impacts of the training program on organizational performance measures, which requires benchmarking and tracking of organizational performance measures.

CASE STUDY: HUMAN FACTORS
TRAINING IN AVIATION MAINTENANCE

Background of Program

Aviation maintenance operations are complex, demanding, and dependent on good communication and teamwork for their success. Success in aviation maintenance is measured by the safety and quality of a maintenance operation. Aviation maintenance operations are most successful when crews function as integrated, communicating teams, rather than as a collection of individuals engaged in independent actions. Over the past decade, the importance of teamwork has become widely recognized (Maurino, Reason, Johnston & Lee, 1995; Robertson, 1998; Rogers, 1991; Taylor & Robertson, 1995). This has resulted in the emergence of maintenance resource management (MRM) training programs and other safety-related programs within the aviation community.

MRM training is a human factors intervention designed to improve communication, effectiveness, and safety in airline maintenance operations. Effectiveness is measured through the reduction in maintenance errors and the increase in individual and unit coordination and performance. MRM training also is used to change the "safety culture" of the work system by establishing a positive attitude toward safety among the maintenance personnel. Attitudes, if positively reinforced, can lead to changed behaviors and performance. Safety typically is measured by occupational injuries, ground damage incidents, reliability, and aircraft airworthiness. MRM improves the reliability of the technical operations processes by increasing the coordination and exchange of information among team members, and among teams of airline maintenance crews.

Within their programs, airlines may use MRM principles differently. Based on human factors principles (e.g., engineering, cognitive psychology, work physiology) and the research literature in the behavioral sciences (e.g., industrial psychology, organizational behavior), MRM programs link and integrate traditional individual aviation human factors topics, such as, equipment design, human anthropology and biomechanics, cognitive work load, and workplace safety and health. MRM principles are best understood through training programs, however, the goal of any MRM training program is to improve work performance and safety, and reduce maintenance errors through improved coordination and communication.

One of the early activities when starting a MRM program (or, as previously noted, any training program) is to gain the understanding, commitment, and visible support of the senior management in the company. Management must actively support and value MRM. The relevance of the MRM program to business objectives must be clear or management will question the investment in time and costs associated with such a program. It may also be necessary to develop some simple return on investment models to justify implementing the program; for example, the cost of one ground damage incident, inflight shutdown, or turnback, versus the cost of a MRM training course and related reduction in maintenance errors and increased safety results is a positive return on investment.

Once support has been established, the first step toward designing a MRM training program is to take a macroergonomic perspective and step back and view the entire maintenance operation as a larger "system." This system is composed of numerous subsystems, including: aviation maintenance technicians (AMTs), engineering, quality control, planners, document support, inspectors, maintenance control, materials and stores, management, and administrative support. When viewed as a system, it is apparent that the overall success of the maintenance operation is dependent on the quality of information exchanged among the team members making up each function, and among the functions themselves.

Once the functions involved are identified, and their roles understood, a MRM training program can be designed. However, as with senior management, it is important to establish a clear rationale for all employees in maintenance operations about the relevance of the MRM program to the airline business. For instance, if an objective of the airline is to reduce errors and increase safety, then the training program should include examples of how the principles and concepts being taught in the MRM training are directly related to these goals. It is important for employees to understand the relevance of any changes they make in their work, and the effort they must put into that change, to the broader business objectives of the airline.

An MRM training program has many facets, but they are all focused on improving communication, coordination, and safety. A typical MRM training program addresses each of the following components (Dupont, 1997; Robertson, 1998):

- Understanding the maintenance operation as a system
- Identifying and understanding the basics of human factors issues
- Recognizing contributing causes to human errors
- Situation awareness
- Decision-making skills; leadership
- Assertiveness (how to effectively speak up during critical times)
- Peer-to-peer work performance feedback techniques
- Stress management and fatigue
- Coordination and planning
- Teamwork skills and conflict resolution
- Communication (written and verbal)
- Norms

Training Concepts for MRM Programs

The intent of this section is to present the training concepts that are most directly applicable to establishing an MRM training program. Each core concept is briefly described and, in the next section, specific methods to use in the development of a MRM training program are provided.

Systems Approach

Maintenance resource management, as with other human factors–oriented processes, is based on a systems approach. It incorporates a variety of human factors methods, such as job and work design, and considers the overall sociotechnical maintenance system. For example, the SHELL model (S = Software, H = Hardware, E = Environment, L = Liveware) shows how we define human factors as a system and illustrates the various interactions that occur between subsystems and the human operator (Hawkins, 1993; Robertson, 1998).

The interactions in this model can affect both individual and team performance. MRM training typically focuses on the interaction between the individual AMT and other team/crew members; liveware/liveware interactions in SHELL terminology. This person-to-person interaction can be considered the micro level of communication and team building, while the interactions among teams and departments is at the macro level. There also are external forces that can affect individual and team performance. These include political and regulatory considerations (e.g., FAA, OSHA, NTSB) and economic factors (e.g., global competition). Achieving the goals of MRM requires improving interactions at both the micro and macro levels. These improvements must occur within the context of external factors, and require an understanding of their effects. To this end, the SHELL model (Fig. 12.3) depicts a macroergonomic or systems approach to integrating human factors methods and principles to design an MRM program (Robertson, 1998).

Instructional Systems Design (ISD). The concept of the systems approach is exemplified by a hierarchical, top-down and bottom-up structured approach to instructional design and development. Identifying organizational

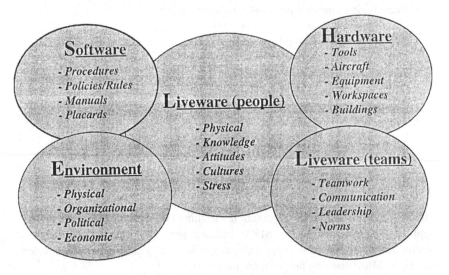

FIG. 12.3.   SHELL model: human factors.

needs and performance gaps focused on the macro level provides the program structure. Involving end-users and subject matter experts in designing and developing training programs incorporates the bottom-up approach. Using a systems approach to design, implement, and evaluate an MRM training program ensures that it will meet the needs of both the learners and the organization.

Continuous Learning and Improvement. The idea of continuous change and adaptation is fundamental to making any system responsive to the needs of its users. Continuous improvement is not a short-term activity. Rather, it requires a long-term management commitment of financial and organizational resources to continuously adapt and improve the program.

Measurement and Evaluation

Once an MRM program is in place, we must determine how well (or whether) it is working. This measurement and evaluation process is typically the weak link in the systems approach. However, it is necessary to measure the effects of the MRM program over time. It also is necessary to use multiple measures in order to gauge the effectiveness of the program. It is important to keep in mind that we are measuring not only the effects of the MRM training program, but also the effects of on-the-job MRM practice.

Evaluation tools can include assessment instruments and questionnaires, behavioral observation and analysis, and unit and organizational performance measures (Cannon-Bowers, Salas & Converse, 1993; Kirkpatrick, 1979; Robertson, 1988). These tools can be used to assess the cost of designing and delivering an MRM training course. In addition, performance measures establish the basis for calculating the company's return on investment regarding the effects of the training program.

Feedback

Feedback provides information to accomplish two performance improvement goals: (1) identifying necessary corrective actions to improve the program, and (2) reinforcing the positive outcomes of using MRM skills. Internal newsletters can be used to describe specific MRM outcomes; for example, one organization reports how an MRM group exercise led to initiating a positive change in a specific maintenance operation. A more active feedback method is to have AMTs write their own "MRM story" describing their experiences using MRM principles and skills (Driscoll, 1996).

Participation

Participating in the creation, development, and implementation of an MRM program promotes a feeling of individual ownership and a sense of commitment to supporting the MRM program goals. Being a member of a team that is developing and implementing a MRM program is motivating, rewarding, and beneficial to the individual and organization. Working together on a cross-functional, interdisciplinary team also provides a unique strength in designing and developing such an MRM program (Robertson, 1998).

Active Learning

As noted earlier, effective instructional methods, sometimes called "inquiry" or "discover" learning, emphasize the involvement of learners (Gordon, 1994). Active learning involves students by having them participate in problem-solving activities and group discussions. The strength of this approach for the maintenance environment is that the use of group exercises and maintenance-related case studies promotes an active and motivating learning environment, because since students are doing more than just passively receiving information—they are actively applying and using the various concepts and skills. To further strengthen this approach, training courses are cofacilitated by subject matter experts in maintenance operations (e.g., AMTs, inspectors, QA). They encourage students to participate by bringing "real workload" experiences into the classroom.

Transfer of Training

For training to be effective, AMTs must be able to apply their newly acquired skills in their real work environment—a positive transfer of training. Transfer of training is enhanced by reinforcement from coworkers and supervisors. A practice period occurs when AMTs return to their workplace after a training course. The reactions of others during the practice period either reinforces newly learned MRM skills or discourages their use. Therefore, it is important that managers receive MRM training in advance of the workers. Transfer of training also is enhanced when classroom exercises are similar to actual workplace experiences.

MRM Evaluation: One Company's Experience

Evaluation results of an MRM training program at a major airline company demonstrate positive and significant effects of the MRM training program (Robertson & Taylor, 1996; Robertson, Taylor, Stelly & Wagner, 1994; Taylor & Robertson, 1995). A systematic evaluation of the effects of the MRM team training program on maintenance personnel attitudes and behaviors was used based on the five-level evaluation model described earlier (Robertson & Taylor, 1996; Taylor & Robertson, 1995). Multiple measures and assessments of the managers' attitudes and self-perceptions of behaviors, as well as maintenance performance results, were used spanning a 4-year period. This provided a unique opportunity to measure and track the long-term training effects. Analyses of the association between attitudes and organizational performance over time were conducted. Data were gathered through the use of the "Maintenance Resource Management/ Technical Operations Questionnaire" (MRM/TOQ), on-site interviews and observations, trends of maintenance performance measures, and attitude-performance analysis (Robertson & Taylor, 1996; Taylor & Robertson, 1995).

Overall results of the evaluation demonstrated a positive and significant effect of the MRM training on attitudes, behavior, and organizational performance. The significant and positive improvements of the maintenance personnel attitudes reflected the expected and intended training effects on the participants' attitudes and their stability over time. Results of each of the evaluation steps are presented below.

Step I Evaluation—Baseline Assessment. Two baseline measurements were taken before the training intervention occurred to measure any changes in the trainees' MRM attitudes and knowledge before the training commenced. With these two measurements prior to the training a stronger quasi-experimental field research design was created. There were no significant changes found in the AMTs' attitudes and behaviors as measured by the baseline and pretraining MRM/TOQs.

Step II Evaluation—Reaction. This level of evaluation involved the participants' written reactions to the value and usefulness of the team training program as measured by the MRM/TOQ. Several questions were developed to assess the trainees' reactions to the training course materials, objectives, organization, training climate, and instructor skills. This level of evaluation also served as a formative evaluation of training materials and delivery methods in the initial phases of the training program. Level II evaluations showed that the participants' immediate responses to training were positive, as over 90% rated the training as "very useful" or "extremely useful," and over 96% felt that it was one of the best training courses they had attended. Another positive aspect of the course was having a mix of participants in the class. This was beneficial as the managers were able to gain an appreciation of other managers' job functions, what their constraints and problems were, and how the outputs of their jobs affected others in the work system.

Step III Evaluation—Learning. The knowledge gained, the immediate changes in the participants' attitudes, and the stability of these changes in time were measured by the pre- and posttraining MRM/TOQ questionnaires. Changes in relevant attitudes measured immediately before and after training were significant, with positive changes following training for three of the four attitude indices measured ("command responsibility," "communication and coordination," and "recognizing stressor effects"). The fourth attitude measure, "assertiveness," rose significantly between the postmeasure and the 2-month follow-up survey. Follow-up results indicated that all four scales remained high and stable over the 2-, 6- and 12-month surveys following training.

Step IV Performance—Behavior. Step III evaluation results, derived and content coded from the open responses on the follow-up surveys, indicated how the trainees actually used the training on the job. The trainees' self-perception of their behavior on the job significantly shifted from "passive" responses (e.g., "be a better listener" and "being more aware of others") to improvement of more "active" responses (e.g., "having more daily meetings to solve problems," "gathering more opinions," and "getting more feedback from others"). Field interviews and observations over a 1-year period were conducted to validate the contents of the self-reported behaviors.

Step V Organizational Results. Step IV evaluation examined trends in maintenance performance before and after the onset of the MRM team training program. One of these performance trends for occupational safety was the rate of lost time injuries per 1,000 hours worked, for 55 work units. Overall, the injury rate remained at a low level for the year and a half after training was introduced. To correlate attitude changes with performance, the individual maintenance personnel data were combined into averages for the units to which they belong. The organizational performance measures included were aircraft safety (ground damage), personal safety (occupational injury), dependability based on departures within 5 minutes, and on-time maintenance. Results from this analysis of the follow-up surveys showed a significant number of positive correlations between maintenance unit performance and attitudes.

Future Directions. Using a systematic training evaluation process provides a framework to demonstrate the effects of an MRM training program. This company's MRM training program is still being conducted, with several new courses being developed and implemented (Endsley & Robertson, 2000; Robertson & Endsley, 1997). Evaluations of these MRM courses are being conducted using the same process as described above (Taylor, Robertson & Choi, 1997). Other companies are currently evaluating their MRM training programs and are showing positive and significant effects of the training on maintenance personnel attitudes and behavior (e.g., Driscoll, Kleiser & Ballough, 1997; Dupont, 1997).

CONCLUSION

The above application of a macroergonomic approach to instructional system development, implementation, and evaluation of an MRM program illustrates its value. Similar results should be achievable in applying this macroergonomic ISD approach to any work system to ensure the effectiveness of its personnel in achieving both organizational and personal goals.

13

Macroergonomics in Large-Scale Organizational Change

Brian M. Kleiner

Virginia Polytechnic Institute and State University

INTRODUCTION

Traditionally, ergonomics has focused on the improvement of productivity, quality, health and safety, and quality of work life. With macroergonomics in particular, there has been a greater emphasis on ergonomists achieving large-scale organizational improvement (Kleiner, 1996). Macroergonomics can change an organization's culture and can achieve 60% to 90% performance improvements (Hendrick & Kleiner, 2001). Large-scale organizational change has also been the laudable goal of many consultants and managers. However, this elusive goal has however been difficult to achieve. Furthermore, it is one of those terms that is interpreted differently by people.

The orientation of this chapter is to define large-scale organizational change in terms of organizational performance. Once performance criteria and metrics are developed and understood for an organization, significant rates of change and/or amount of change in performance constitute large-scale change. The traditional ergonomic metrics—productivity, health and safety, and quality of work life—are necessary to individuals and organizations, but may not be sufficient, as emphasized by the recent focus on community ergonomics and other "macro system" concerns. Nickerson (1992) stressed the need for ergonomists to play a larger role in society. Ergonomists have begun to apply their technologies to go beyond traditional measures to impact regional economic development as well. Community ergonomics has been proposed as a way for ergonomics to have a positive impact in the local community. This approach has been shown to have a positive impact on individuals in the local community (e.g., Smith, Conway & Smith, 1996). One very important component of a society is its economy. Many economics scholars have proposed that the very essence of economic success is the community's ability to manufacture goods. As Deming (1986) suggested, if manufacturers prosper, a chain reaction will cause jobs to be created and society to benefit, as was demonstrated in postwar Japan.

Many researchers and consultants have proposed methodologies for large-scale organizational change. It is contended that macroergonomics offers an ideal framework for organizing these various tools and methodologies. In fact, macroergonomics offers its own sociotechnical approaches to change. Thus, rather than detail the specific tools and methods of organizational change, this chapter focuses on the organization of the toolbox itself and presents the key characteristics and components of the toolbox necessary to achieve large-scale change.

WHAT IS LARGE-SCALE ORGANIZATIONAL CHANGE?

Macroergonomics presents a sociotechnical framework for studying both the macro and micro issues associated with large-scale organizational change. Within this perspective, performance is viewed as multidimensional, characterized by multidimensional criteria and measures related to various checkpoints in the work system (Kleiner, 1996). In this context, large-scale change has been operationally defined as significant improvement (i.e., greater than 50%) of one or more of these performance variables or less than 50% improvement but in a relatively short time span (Kleiner, 1996). According to Hendrick (1995b), an important outcome of macroergonomic intervention is also a culture change, in which organizational culture is primarily defined by the organization's core values.

It is also important to note at the outset a simple but profound assumption regarding large-scale organizational change: that the change is valid change. That is, the targeted change should be supportive of and aligned with the organization's purpose. Normally, structural changes support strategic changes and all strategic changes should be aligned with the organization's purpose. Given this, we can assume that most change will also result in some type of performance improvement and/or culture change as discussed earlier.

However, it should be recognized that many organizations pursue change that does not necessarily meet these criteria. In some organizations, the "program of

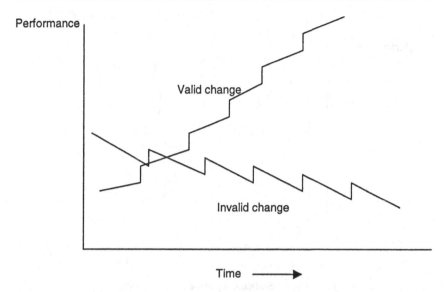

FIG. 13.1. Valid versus invalid change.

the month" characterizes the culture. Over time, employees learn to meet each new program with skepticism or resistance. One of the ironies in an organization undergoing invalid change is that to the change managers and facilitators, the change can feel valid. That is, in attempting to get employees to change behaviors and attitudes, the resistance encountered can be physically stimulating. Adrenaline flows, stress levels increase, and at the end of the day, the leader of change feels exhausted and thus fulfilled. Periodic (e.g., monthly) spikes of improvement reinforce the positive feelings. As illustrated in Fig. 13.1, in reality, preceding each improvement spike may have been a period of performance decline.

Figure 13.2 illustrates the performance curves typically experienced in large-scale change. Small-scale improvement as seen in continuous improvement intervention and efforts can be portrayed as an "S-shaped" learning curve. However, for "breakthrough" or large-scale change, performance shifts to a new S-shaped curve (Sink & Morris, 1995).

Performance Improvement

Sink and Tuttle (1989) proposed that organizational performance can be measured or assessed using seven criteria or clusters of measures: efficiency, effectiveness, productivity, quality, quality of work life, innovation, and profitability or budgetability. Specific measures can be derived related to performance criteria. These can be subjective, as in the case of attitude surveys, or they can be

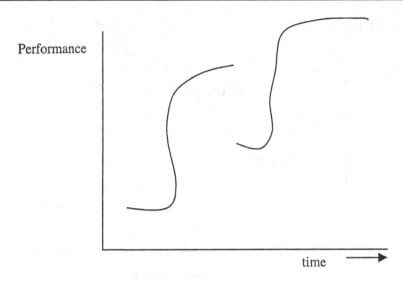

FIG. 13.2. S-shaped performance curve.

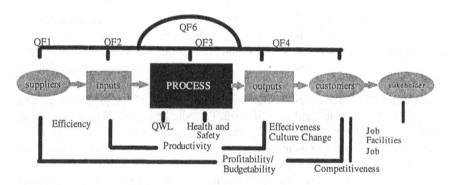

FIG. 13.3. Organizational performance criteria.

based on objective data. The seven performance criteria relate to specific parts of the organization as represented by an input-output model. This approach introduced the issue of measurement and specifically suggested that a management system has an effect on organizational performance as characterized by multidimensional criteria. As illustrated in Fig. 13.3, Kleiner (in-press) later added "flexibility" to the checkpoints previously allotted to quality. Especially in manufacturing system design, flexibility is an important criterion at different points in the system. In addition, Kleiner (2001) emphasized for ergonomists the criterion of health and safety.

PROCESS OF ORGANIZATIONAL CHANGE

Phases of Change

Passmore (1988) proposed a framework for managing sociotechnical change in an organization. Each phase includes several analyses that should be performed in order to validate and manage the change process.

Define Scope of System to Be Redesigned. During this phase, the need for change is defined. The scope of change is defined by defining the organizational unit or units targeted for change. Within this context, stakeholders are identified and their expectations are understood. Finally, an assessment of the potential for success is performed.

Determine Environmental Demands. External stakeholders are identified and their current and future demands are understood. An appropriate response to these demands is developed. Finally, organizational goals are derived.

Create Vision Statement and Charter. During this phase, a philosophy and values for the change effort are defined. Desired outcomes for the change effort are clarified and a vision statement is drafted. This "charter" is reviewed with change sponsors, usually key managers in the organization.

Educate Organizational Members. Passmore (1988) recommended that personnel receive sociotechnical systems education before beginning the change effort. In addition, particular skills training can be offered to supplement the education.

Create the Change Structure. A design team is then created. Intergroup meetings are held between the design team and a steering committee. A communication strategy is developed to support the change process. The design team and steering committee receive sociotechnical systems education as well. Then, several strategies are developed; one for participation, one for resourcing, and one for change.

Conduct Sociotechnical Systems Analysis. During the Sociotechnical Systems (STS) analysis phase, Passmore (1988) suggested identifying resources for the change effort. Analytical training should be provided to those on the design team as well. Then the system can be analyzed and results reviewed with the steering committee. Following this communication, information can be shared with the remainder of the organization.

Formulate Redesign Proposals. During this phase, design inputs are reviewed and expectations for desired outputs are clarified. Then, specific proposals are formulated. The impacts of the proposals are systematically examined through a cost-benefit analysis. Specific proposals are then selected and reviewed with the steering committee.

Implement Recommended Changes. The selected proposals are then communicated with the organization and are reviewed with upper management. Once upper management has provided their approval, an implementation plan is created. To support the plan, employees and supervisors are trained. Finally, the plan is executed.

Evaluate Changes and Redesign as Necessary. For the purpose of continuous improvement, an evaluation methodology is developed. Data is then collected and reviewed against established goals. Results are communicated to the organization and management. As necessary, redesign takes place.

Macroergonomist as "Change Master"

Sink and Morris (1995) focused on the "change master" as a manager of change. It is contended that increasingly, macroergonomists will be placed in such a position. The authors noted that the following change process states can overlap in some cases and that these states describe individual, group, and organizational change.

Situation Appraisal. During this state or phase, the organization is focused on a particular problem area. The change master seeks to reach an agreement at this phase about the problem areas and process approaches. Strategies include identifying key stakeholders. Sink and Morris (1995) also recommended a participatory approach in the diagnosis through interviews and structured group processes to get consensus about current problems. A walk-through assessment is usually the framework for this phase.

Catharsis. During this phase, the organization "lets go" of past blame, problems, etc., to focus on the current change effort. Before letting go, there is typically some emotional distress in the organization as personnel attempt to understand what caused the current problem(s). The goal of this phase is to expose the stresses, put them aside, and move onto a rational view of change (Sink & Morris, 1995). Again, various participatory methods are used to confront the stresses and move to a more healthy state. For the change master, much can be learned during this phase about the organization's history, power distribution, motivational state, culture, etc.

Self-Awareness. During this phase, the organization uses various tools and techniques to better assess the current system. The goal is to achieve a shared, explicit, open understanding of the present situation (Sink & Morris, 1995). This phase establishes the motivational starting point for change. The change master uses data to create a shared awareness of the need to improve and how to improve. He or she uses historical data and uses objectivity to persuade the organizational members to see the facts as they are, putting aside past biases and beliefs.

Self-Evaluation. Here the organization comes to some conclusions about its current state. Specifically, Sink and Morris (1995) indicate the organization reaches some conclusions about the effectiveness, efficiency, or cost-benefit relationship relative to the present methods and systems in use. The goal is to achieve a clear conceptual model of goals, standards, expected levels of performance, criteria, and measures of performance. A measurement system may be developed during this phase. Understanding statistical variation in data is also important during this phase.

Self-Diagnosed Change. During this phase, the organization plans or accepts plans for new behaviors, methods, and systems. It is very important to have organizational members participate in the redesign process. The new process attempts to remain sensitive to past history and to take into account current individual differences. The change master during this phase adopts one or more of the following roles: stimulator, technical expert, honest broker, and/or support person (Sink & Morris, 1995).

Try Out New Methods and Behaviors. Here the organization experiments with new systems and/or methods. The change master supports, encourages, and facilitates the organization's testing of the newly designed systems and methods. A primary goal during this phase is for the organization to experience the proposed changes and redesign them to be sensitive to the current realities of actual use. Change master skills in experimental design, decision analysis, statistical analysis, etc., are helpful during this phase.

Reinforce the New Behavior. This is where the new behaviors are reinforced, typically with a new reward system. During this phase, the new methods and systems are often demonstrated to others in the organization. Here again, the change master's objectivity can be helpful in cultivating support for the required changes.

Culture Change

As Hendrick (1995b) has stated, culture change is often the outcome of macroergonomic intervention. Organizational culture is related to the norms, beliefs, unwritten rules, and practices in an organization. At its core, culture is created by underlying values that are unseen. On the surface of an organization, the rites and rituals, heroes, villains, symbols, and behaviors of members are visible.

Culture can be differentiated from organizational climate. The former is much more permanent and pervasive, while the latter is the temporal reaction to critical incidents and events.

Culture can be changed in several ways. First, major policy changes can affect the culture of an organization; that is, change can, in a sense, be mandated. Again, we are reminded that not all change is valid change. Sometimes, policy and strategic change are invalid reactions to turbulence in the organizational environment or, at worst, are the personal whims of an executive who is "shooting from the hip." Second, changing the behaviors of organizational leaders can create a culture change. By seeing their leaders "walk the talk," organizational members are far more likely to modify their own behaviors and attitudes, prompting a culture change. Thirdly, selection and training can help change cultures. Regarding selection, many organizations now screen employees for having an attitude consistent with the organization's culture. While some can question the ethics and/or legalities of the practices, personality tests, role playing, and other mechanisms are commonplace in today's interview environment. In addition to selecting employees with value systems aligned with that of the organization, many human resource departments attempt to use training once employees are hired to adjust and align attitudes and behavior. Finally, when appropriate, a comprehensive work system design change can and often will result in a change in culture. The latter, using methods as described in Chapter 3 and supported by valid changes in policy, leadership, and training, is the best approach to achieving desired culture change and performance improvement.

A SOCIOTECHNICAL
SYSTEMS UNDERSTANDING
OF LARGE-SCALE CHANGE

Macroergonomics is concerned with the optimization of work systems through consideration of relevant social, technical, and environmental variables and their interactions. The work system is comprised of personnel interacting with technology, within internal environments, external environments, and an organizational structure (Hendrick & Kleiner, in press). Building on the sociotechnical theoretical foundation established by the Tavistock Institute, and most directly, Emery and Trist (1965), macroergonomics takes a top-down and bottom-up (e.g., participatory ergonomics) sociotechnical systems approach to the design of organizations, work systems, jobs, and related interfaces between humans and machines, users and system, and humans and environment (Hendrick, 1995a). This results in a fully harmonized work system at both the macro- and micro-ergonomic level (Hendrick, 1995a). Macroergonomic intervention begins with an assessment of relevant sociotechnical variables and their implications for the design of the structure of the work system and processes. Once the overall work system has been evaluated, microergonomic interventions, such as how to optimally allocate functions and tasks to humans and machines or computers, can be accomplished (Hendrick, 1986).

Macroergonomics can be used to perform an important role in society by helping to retain and create jobs, and can thereby impact industrial expansion which, in turn, positively influences regional economic development. However, as one moves from micro- to macroergonomics, the nature of success measures changes. By definition, macroergonomics applies to functioning organizations, constantly adapting to changing environments. Thus, increasingly aggregated measures, such as job retention or facilities expansion, cannot be uniquely causally related to interventions. Also, the interventions are not uniquely under the ergonomists' control. As Drury (1991) noted, when operating in a wider arena, ergonomists no longer operate alone.

In the author's experience, the macroergonomic approach taken led to major industrial turnarounds in the northeastern United States, and was credited with retaining and creating thousands of jobs. As reported in Kleiner and Drury (1999), in one company, joint optimization was accomplished by analyzing the personnel and technological subsystems in a Malcolm Baldrige—winning, nonunion plant. The company, an electronics operation, required workplace redesign, managerial intervention, structural realignment, and a move to a group-based quality culture for work system optimization to occur. Specifically, in a clean room environment, technical performance was assessed to determine the technical capability of the area. In addition, the physical demands on the operators in this relatively stable environment were assessed. Using participatory ergonomics, operators were interviewed to determine their subjective feelings about working in the clean room environment. Based on these analyses, ergonomic changes were given to architects and engineers for incorporation into the new plant's design. Changes were based on operator preferences and demands for improved task performance.

At a second company, operator–machine interface design required training support systems, managerial intervention, and decentralization to improve the work system. In this unionized company, in addition to similar analysis of the technological and personnel subsystems, corporate takeover required the researchers to develop quickly a working understanding of, and appreciation for, national cultural issues and their interaction with corporate culture. As objective and neutral third parties, the researchers could transfer these new insights and understanding to company managers and operators and incorporate cross-cultural understanding and sensitivity into the program.

Sociotechnical systems theory can be used to better understand the mechanisms and likely causes of these and other macroergonomic successes. As illustrated in Fig. 13.4, there are several important work subsystems operating and interacting: the personnel subsystem, technological subsystem, internal environment, external environment, task, and organizational design.

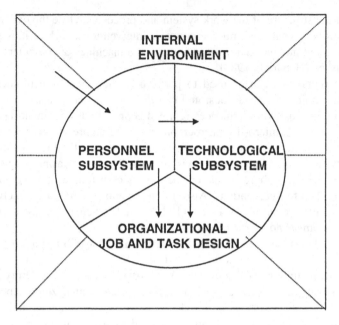

FIG. 13.4.   Work system subsystems.

According to sociotechnical systems theory, the personnel and technological subsystems are jointly subjected to environmental influences and, therefore, an organization's structure should be designed as a function of the joint design of the personnel and technological subsystems. In many cases, including those reported by Kleiner and Drury (1999), major changes in the environmental subsystem occur during interventions. A strictly microergonomic approach might have ended intervention at the onset of such events. It is hypothesized that the dramatic results achieved through macroergonomics are achieved because macroergonomists facilitate the interface between stakeholders in the environmental subsystems (i.e., subenvironments) and other subsystems. In the cases reported by Kleiner and Drury (1999), the environmental subsystem consists of components such as the regional, political, legal, and educational resources. By working with state officials and institutions, organized labor, and corporate headquarters, interventions with center personnel focused on the political interface with local organizations. Capability data had to be collected and reported in usable form to support the subenvironment interfaces. By working with management/union arbitrators and credit institutions, the legal interface was facilitated. By transferring knowledge acquired during environmental system analyses, educational programs supported most interventions. Joint optimization of personnel and technological subsystems was thus pursued, as was organizational, job, and task redesign, but these interventions began and supported the environmental interface.

UNDERSTANDING INDIVIDUAL DIFFERENCES

In managing large-scale change, it is necessary to understand the fundamental definition of an organization: two or more people working together to achieve a common purpose; that is, an organization is comprised of individuals, and these individuals will differ with respect to their motivational reaction to change. Large-scale change occurs as the result of several small changes. Attitudinal and behavioral adjustments need to be made by the individuals in an organization in order for systemic change to occur. It is therefore most useful to predict the kinds of reactions one will observe when managing or facilitating change.

It is intuitive that not all change is perceived as positive. Experientially, some change has been perceived by individuals as positive and some negative, and thus new change is met with some skepticism. Scholtes (1990) reported that when all else is constant, organizational members will exhibit a normal distribution when reacting to change, as illustrated in Fig. 13.5. Some individuals, based on their experiences, personalities, etc., will be exceedingly positive when confronted with change. On the other hand, some will be quite negative when confronted with the prospects of change. However, it is interesting that most organizational members will be neutral. From a change master's perspective this is important, because they will not remain neutral for long. Their attitudes can turn

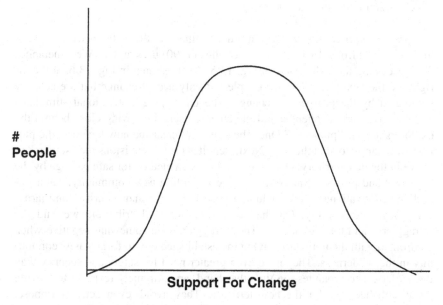

FIG. 13.5.   People's reaction to change.

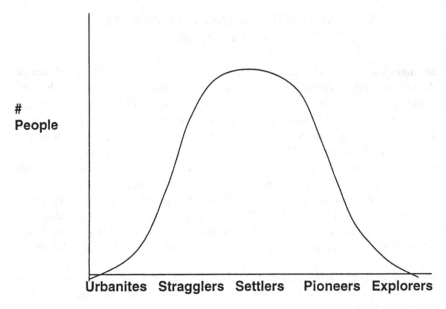

FIG. 13.6. People's reaction to change (adapted from Scholtes, 1990).

either positive or negative, and the manager of change can be a factor in terms of which direction the individual turns.

Motivational Differences. In order to illustrate the differences people exhibit when confronted with change, Scholtes (1990) uses an Old West metaphor for typecasting the individuals in Fig. 13.5. As illustrated in Fig. 13.6, at the far right are the "explorers." These people not only accept change but are actually stimulated by the prospect of change. The underlying motivational stimulants include adventure, challenge, and the change itself. Following close behind the explorers are the "pioneers." Once the explorers clear the new territory, the pioneers are happy to lead the way. Next, Scholtes (1990) envisions the "settlers" to arrive in the new territory. Once the way has been cleared for safe passage by the explorers and pioneers, the settlers arrive to build a new community. Their underlying motivational needs include a need for safety and security, familiarity, constancy, and stability. On the other side of the distribution, we find the "stragglers" and the "urbanites." The "stragglers" are somewhat negative when confronted with the notion of leaving the established city to forge a new community in the wilderness. They have even greater need for safety and security than the settlers. The urbanites, on the other hand, are extremely resistant to leaving the established city for the new territory. They avoid, even actively oppose, change.

A Strategy for Managing Change. What is most instructive about this model is to understand that individuals have different needs and wants in the context of change. An effective change master will learn how to manage these differences. Scholtes (1990) provides a useful strategy for managing this normal distribution of people through a change process.

The first step is to understand the authority or power distribution in the organization. All organizations have people with *formal* and *informal* authority. Formal authority or authority of position (Barnard, 1938) is that power bestowed on an individual by virtue of his or her job title. Formal authority is contrasted with informal authority or authority of leadership (Barnard, 1938), which is based on charisma or other nonorganizational attributes related to the informal organization. As illustrated in Fig. 13.7, a table can be constructed mentally to organize thoughts about the power or authority infrastructure in the organization.

Next, the individuals can be understood in terms of their reaction to change as illustrated in Fig. 13.5 through Fig. 13.7. As a result, Fig. 13.8 can be developed. Note it is not good practice to categorize and certainly not a good idea to communicate overtly where individuals are perceived to reside in the distribution. However, it is useful for the change master to understand mentally the individual differences of organizational members with respect to authority and reaction to change and adjust the mental model over time.

Scholtes (1990) provides a simple strategy to correspond to Fig. 13.8. In leading a change effort, he says that it is most important to get the help of the "1"s and "4"s. These members have both authority in the organization and support the change effort. The reason, according to Scholtes (1990), for securing the commitment of these individuals is the vulnerability of the "2"s and "5"s. Recall that most of the people will initially reside in this "undecided" portion of the distribution. If they are not persuaded to support the impending change, they could be persuaded by those who are negative, such as the "3"s and "6"s. The so-called "3"s are a particular concern because they have informal or formal authority and thus can persuade more neutral members to align with them. Scholtes (1990)

	No influence	Moderate	Powerful
Formal Organization			
Informal Organization			

FIG. 13.7. Power/authority in an organization.

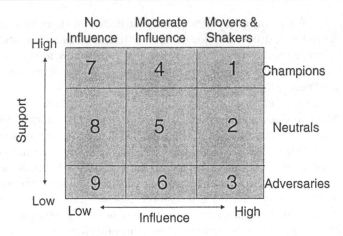

FIG. 13.8.   Strategy for managing change (adapted from Scholtes, 1990).

recommends that "3"s be invited to participate, even help facilitate the change effort, but as with everyone else, not forced to change. In the end, if they are unwilling or unable to support the change effort, Scholtes (1990) says it is best to isolate them in positive ways. This can be accomplished through special assignments and through relocation. Again, it is assumed that these alternatives are preferred by the person as well as the organization, and thus is as close to a "win-win" solution as possible.

Probably the most effective skill needed during this process is active listening. It is important to understand why the various members position themselves at different points in the support distribution. Why is a "3" so resistant? What need must be addressed in order for them to become supportive, or at least not so destructive." Why is a "1" so supportive? Can this attitude be replicated? These are some of the questions that can be answered as the change master attempts to statistically think about his or her distribution of organizational followers.

LEARNING AND ADOPTING
APPROPRIATE CHANGE ROLES

According to Sink and Morris (1995), the change master must assume various roles during various points in the change process. These roles help move the distribution of people previously discussed through the desired threshold of change.

Inquirer.   Active listening is the skill utilized in the inquiry mode (Sink & Morris, 1995). The key in this mode is not to be judgmental and to encourage organizational members to clarify issues constructively and thus relieve anxiety. This role can be performed both actively (i.e., with inquiry) and passively (e.g., listening).

Data/Fact Gatherer. At various points in the change process as exemplified by the methods discussed in Chapter 3, the macroergonomist as change master must collect data and convert that data to information. When going beyond data collection or design of the system of evaluation, it is useful to have organizational members participate in the process.

Collaborator. The macroergonomist as change master will become part of the organizational design and/or implementation team. In this role, Sink and Morris (1995) point out that the change master is a team member and should take care not to dominate the team.

Structure Provider. The change master will also need to be skilled at organizing and facilitating group processes (Sink & Morris, 1995). The most challenging aspect of this role is to maintain a focus on process and not get involved in the content of the group's business, unless called on as an expert.

Expert. The change master may be called on for his or her technical knowledge. In this capacity, he or she is acting as an expert. This role could involve providing recommendations, converting data into information, and/or problem solving.

Teacher and Skill Developer. In this role, the change master assumes the role of educator. He or she may be required to transfer knowledge, skills, or both and may do so in a classroom environment or individually.

Challenger. A fact of large-scale change is that it is not easy and certainly not always pleasant. One role the change master from time-to-time finds himself or herself in is that of challenger. In this capacity, the change master uses his or her objectivity and neutrality to question what an individual, group, or the organization itself has decided or implemented. When used judiciously, this role can be instrumental to achieving needed change.

In summary, the macroergonomist as change master will have to be flexible and adopt various roles as different situations or phases arise. Not all roles will be easy, and some will come more naturally than others. For example, most will have difficulty with the need to be a challenger. However, it is advisable for the change master to receive training in those areas in which he or she is less confident. In the case of the challenger mode, for example, conflict resolution and negotiating courses may be applicable and useful.

CONCLUSION

The perspective and methods of macroergonomics hold promise for large-scale change in organizations and communities. Specifically, in a work system, which is comprised of personnel interacting with technology in internal and external environments and organizational structures, it is necessary but not sufficient to be concerned about human–machine interface design—the traditional purview of ergonomics. The interfaces among the human–machine work unit and the other subsystems require attention as well. Of particular importance is attention to the various subenvironments in the external environment and the need to couple a macro- and micro-ergonomic approach to interface design at all levels. As a manager and facilitator of large-scale organizational change, including culture change, the macroergonomist becomes a "change master."

14

Community Ergonomics

John Henry Smith, William J. Cohen,
Frank T. Conway, Pascale Carayon,
Antoinette Derjani Bayeh, Michael J. Smith

University of Wisconsin—Madison

INTRODUCTION

Community Ergonomics (CE) was developed from two parallel directions. On the one hand, CE emerged from existing theories and principles in human factors and ergonomics and behavioral cybernetics that could be used for situational assessment and solution-finding purposes at the societal level. On the other hand, CE also emerged from the careful and thorough documentation of specific applications in communities, which in turn led to additional principles and theories more contextually relevant to the societal level of analysis. As such, CE is a robust approach in its principles and a flexible practice in its applications; this combination is one key to the optimization of complex systems.

Community ergonomics can be defined as a design approach to the interfaces between people and system design in societal contexts. CE theory is built on the foundations of human factors and ergonomics, sociotechnical systems theory, quality improvement, behavioral cybernetics, and the problem-solving technique of Breakthrough Thinking™ (see Smith et al., 1994). The CE approach focuses on distressed community settings characterized by poverty, social isolation, dependency, and low levels of self-regulation (and control). Inner cities are examples of such communities. The practice of CE seeks to identify and implement a community–environment interface to bridge the gap between the disadvantaged residents of a community and the resources defining the social environment within which they function; hence, achieving a "fit," through interface design, among people, environment, and community.

CE SOCIOTECHNICAL SYSTEM
(STS) MODEL

The CE model (shown in Fig. 14.1) is built on the foundations of sociotechnical systems theory (Emery & Trist, 1965; Pasmore, 1988; von Bertalanffy, 1968) and then applied to complex societal systems. Other theories contributing to the CE STS model shown in Fig. 14.1 are the fit theory (adapted from the P:E fit theory of Caplan, Cabb, Franch, von Harmon, and Pinneau, 1975; van Harrison, 1978), the CE interface design and implementation process (adapted from the Breakthrough Thinking™ and planning and design approaches to solution finding of Nadler (1981) and Nadler and Hibino (1994)), and community self-regulation principles adapted from behavioral cybernetics (K.U. Smith, 1966, among others). The CE STS is comprised of the social subsystem (defined by community residents), the technical subsystem (defined by institutions, services, policies, etc.), and the community ergonomic process (linking the social and technical subsystems and the environment; e.g., economy, employment, education). These are the three circles shown in Fig. 14.1.

One of the principles of CE proposes that community residents need to be able to track their environments and other community members within it. In addition,

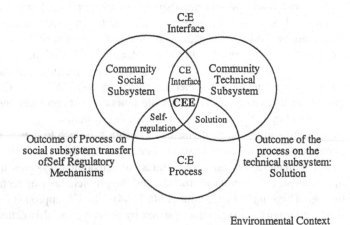

FIG. 14.1.   The community ergonomic system model.

it is important that people exert control over their lives within those environments, while increasing awareness of the impact of their actions on others and the environment as well. The process referred to in this framework as the community environment (C:E) interface design process results from the complementary effects of two parallel mechanisms. One such mechanism is intended to identify, design, and implement purposeful solutions (C:E interface). The other mechanism enables community leaders and residents to participate actively in building self-regulating mechanisms of learning, social tracking, and feedback control. The key solution sought through the C:E process is the implementation of an effective C:E interface intended to bridge the gap between the community residents and the environment or community structure. An effective self-regulating interface is one that bridges the identified gap between community residents and their environment, as well as allowing for the generation and implementation of subsequent solutions to address emerging challenges in a continuously changing and turbulent environment. Finally, the common area in which the three circles in Fig. 14.1 meet (labeled CEE) brings into the C:E design process, the community ergonomic engineer (CEE) or professional who plays the role of an interface professional. As such, the CEE leads and manages the C:E design process such that a solution is identified and implemented, while providing community members and participants with self-regulating mechanisms to track and control their environments better. The model in Fig. 14.1, while simple in appearance, conveys the complexity of community ergonomic systems design. Ultimately, the model suggests that communities act as closed loops, subject to changes in a turbulent environment setting. Consequently, these communities may exhibit seemingly intractable conditions, which CE theory does not consider as impossibly hopeless to improve, but instead a challenge to overcome. In the foreword to *Separate Societies: Poverty and Inequality in U.S. Cities* (Goldsmith & Blakely, 1992), Harvey Gantt said simply, "There is no permanent poverty, permanent joblessness, permanent hopelessness . . . ."

The Context in the United States

Residents in the inner cities of the United States are isolated from basic human needs and resources, therefore leading to what CE theory defines as "cumulative social trauma" (CST). CST results from long-term exposure to extreme detrimental societal conditions leading to a vicious cycle of dependency, social isolation, and learned helplessness. However, as a proactive solution-finding approach, CE asserts that CST is not a condition without preventive measures or solutions. Instead, the CE approach insists on the fact that CE interfaces with built-in self-regulatory mechanisms may be implemented to bridge existing gaps between the community environment and its residents such that social isolation may be prevented or remedied.

Historically, welfare, public housing, public health, and public safety have not provided sufficient services and products to match the needs, desires, or capabilities of those living in poverty in declining areas of inner cities (Cohen & Smith, 1994; Sainfort & Smith, 1996; Smith & Smith, 1994). Often, poor people cannot follow the bureaucracies of public institutions,[1] and neither can these institutions understand the behavior of poor people sufficiently to provide services effectively. The inability of people and social institutions to follow or accurately track each other results in severely impaired systems performance. The necessary conditions and obstacles encountered on the path to progress in the declining areas of inner cities can be defined in terms of human errors and a lack of sociotechnical systems reliability. Human errors are decisions, behaviors, and actions that do not result in desired and expected outcomes (i.e., keeping a job, getting off of public assistance, paying bills, and not being involved in illegal activities). Social and economic system reliability is the ability of public and private institutions to achieve specific and general intended objectives for positive community performance outcomes (Smith & Smith, 1994). Economic achievement and growth of competency for individual residents are determined by the design and arrangement between the characteristics of the community residents and those of the community environment. In the case of substantial misfits between people and environment, poor urban residents are likely to make repeated "errors" in life behaviors, which result in poor economic and societal outcomes. Institutions with low sociotechnical systems reliability do not provide adequate feedback (performance information or direction) or services for their constituencies to correct their errors. The fundamental purpose of community ergonomics is to improve the fit between environmental conditions and personal behaviors in poor inner cities to reduce resident errors and achieve optimal social system performance.

Numerous previously implemented public programs have repeatedly failed to improve the living conditions in inner city. This is in part due to the fact that simply forcing inner city residents to change their behavior in order to fit some predetermined desired state cannot ameliorate such problems. Nor will focusing on only one aspect of the sociotechnical system, such as housing or employment, provide a complete solution to a very complex problem. This focusing of interests and resources is like putting a patch on a hole in the dike. If the water pressure on the dike remains high, then a leak will just occur somewhere else in the dike and another patch will have to be applied. This will repeat itself again and again. Inner cities are like a dike with huge water pressure ready to cause numerous holes simultaneously. In order to be effective, the solution has to account for the total system and hence decrease the water pressure. In the inner city context, both the environment and people need to be dealt with to bring about the most desirable

---

[1] A case study showing bank lending practices to low-income residents is presented later in this chapter.

fit. The theory and practice of community ergonomics is based on the assumptions that individuals or groups must attain and maintain some self-regulation, and that individuals need to have control over their lives and their environments in order to achieve a personal ability to thrive. We believe this can be achieved by having the residents actively participating in their own self-improvement, and in the improvement of the inner city environment in which they live.

THE COMMUNITY ERGONOMIC APPROACH

Lodge and Glass (1982) recognized the need for a systems approach for managing issues at a societal level. They advocated a systematic approach that involves cooperation from multiple parties in order for improvement to take place. They point out that various systems need to be aligned in order to achieve overall system improvement. For instance, providing jobs is insufficient if training, day care, and other support systems are not in place. These reinforcing links represent the elements of a human factors system.

The community–environment system is one with multilateral and continuous interactions among populations, conditions, institutions, situations, and objects. These are linked through feedback, forming a unified whole leading to the vital functions of lifespan development. An interacting population of people, often with common social, political, economic interest, and characteristics, represents the "community." The community is surrounded by other communities and institutions, which may have very different beliefs, values, and modes of behavior. Likewise, the community is surrounded by aggregate communication systems, architecture, transport, energy systems, and other technology that influence and act on the community, and ultimately determine the survival and life quality of individuals and groups in the community. This community–environment system may include institutions for education, financial transactions, government and politics, commerce and business, law enforcement, transportation, and housing. The "organizational" complexity of the system defines how the environment can influence behavior, and the ways through which individuals and groups can try to control the environment through their behavior. The community–environment system can be characterized as the closed-loop feedback interaction of two or more social or economic systems to achieve a desired output, state of quality, and/or purpose. This is based on the assumption that the design of the human–environment interface for reciprocal guidance and control influences performance, which is a recurrent theme in psychology and engineering (McCormick, 1970; Smith & Smith, 1996; Smith and Smith, 1966).

The community environment management system is focused on a human-centered concept of community design, which seeks to achieve better community investment and higher community quality outcomes. A community ergonomic process is employed to bring about management integration of the community and the environment as a total system enterprise.[2] The aim is to build compliance between the community and the environment. This framework provides the foundation for the community ergonomic practitioner to design community environment systems in order to augment community residents' perception, action, control, feedback, production, consumption, and quality of living within an urban municipal setting. The management system seeks to establish positive social tracking patterns between the community and the external environment so that interpersonal behavior, group, and societal interactions are efficient. Within this framework, the practitioner determines the task to be performed, human resources needed, financial requirements and resources, and the ergonomic "fit" design requirements to achieve the desired results.

Community environment economic systems contain inputs and outputs among households, financial institutions, businesses, municipal government, and educational institutions, organized and linked by means of feedback to achieve and maintain the monetary cycles among various community enterprises. Community productivity, the measure of input to a community enterprise used to create a specified output, can be quantified in community environment systems. Community productivity values for community enterprises, combined with the input and output values of other community enterprises, provides a measure of total community productivity value. Community productivity values can be measured in terms of increasing, decreasing, or consistent trends over time. Total community productivity is a function of the closed-loop feedback relations between community enterprises in achieving and maintaining the monetary cycle requirements for community quality. This perspective is a measure of growth or stagnation as a function of the rate, efficiency, and effectiveness of community productivity.

Community ergonomics operates from the perspective of self-control and self-governance of vital human functions such as development, perception, motivation, and learning through self-regulation of individual physiological processes, the external social and cultural environment, and bureaucratic institutions. The individual is not viewed as a helpless and passive prey of direct external stimulation of the environment to which he or she responds in reflex fashion. Nor is it thought that cognition, information processing, physiological homeostasis, or internal drive regulates behavior. The individual is viewed as an organized system within a community in which the guidance and course of behavior is determined by the ability of the person to develop reciprocal feedback control over the economic, social, and cultural aspects of the community environment. The feedback

---

[2] A comparative case study is presented later on in this chapter to illustrate different approaches to enterprise network building and resulting outcomes.

concept is a significant aspect in the development of a community ergonomic perspective of individual behavior. It specifies the reciprocal manner in which self-generated activity and self-controlled activity determine the course, rate, and degree of individual learning and behavior in relation to the community system.

The combined self-regulatory behaviors of community residents determine the acceptable norms of behavior, which can also be defined as habit patterns for the community (Smith, 1972). According to Smith (1972), habit is defined as self-governed learning in the context of control of the social environment as well as one's own physiological processes. The self-governance of learning becomes patterned through sustained and persistent performance, which is critically dependent on time schedules. As these timed patterns become habituated, the individual can predict and anticipate social events, which is a critical characteristic in the management of social environments.

Self-control and guidance in social situations further enhances the ability of the individual to adjust to various settings by following or tracking the activities of other individuals or groups. Social tracking patterns during the habit cycles determine critically significant behavior. Individuals, groups, and communities develop and maintain their identity through the establishment and organization of these social habits. In addition, the maintenance of group patterns, and the adherence to this process, is achieved through social yoking (or mutual tracking), so that individuals sense each other's social patterns (Smith, 1972). Community environment systems are dependent on critical time tracking factors such as work and schooling as a developmental aspect of daily life. The difficulties faced by many communities stem from the lack of or disruption of critical time factors. This results from faulty design of the timing processes, and from conditions that negatively distort time and the self-governing capacity of individuals and organizations.

Community ergonomic designs develop a tracking system to aid in the sensing of time and the development of a record and memory vital in determining future courses of action and expectations. In other words, these tracking systems are aimed at building good "habits." Absence of such critical timing factors results in delays in functioning, social disruption (trauma), organizational adjustment problems, personal emotional difficulty, and community quality failures. We believe that for effective individual and community functioning, the design of the community environment system and its operations must comply with the critical temporal processes of the people living in the community. According to Smith (1972), the optimal system design will be achieved when the development of individuals and communities are promoted and the design fits with the built-in or learned behavior of individuals. On the other hand, system design can adversely affect behavior, learning, and development of individuals and communities when individual needs are not considered. Cumulative social trauma was defined earlier as a condition resulting from repeated exposure to poorly designed environments and/or long-term social isolation that leads to ineffective individual performance or coping abilities. CST is the repetition of an activity or combinations of activities, environmental interactions, and daily life routines which develop gradually over a period of time and which adversely affect motivation, skill, and emotions leading to social and behavioral disruption, psychosomatic disorders, and mental distress. We believe it is possible to identify the "operational hazards" (i.e., the hazards faced by members of the community in their daily interaction with various system elements) that lead to the development of CST in communities. Such hazards need to be detected, monitored, and controlled to prevent CST from occurring in order to enhance community quality.

COMMUNITY ERGONOMIC PRINCIPLES

This section will describe the principles that constitute the philosophy driving the CE approach to societal solution finding and CE interface design. These principles are listed below and will be described in detail in this section:

1. Action-Oriented Approach
2. Participation by Everyone
3. Diversity and Conflict Management
4. Encouraging Learning
5. Building Self-Regulation
6. Feedback Triad
7. Continuous Improvement and Innovation

1. Action-Oriented Approach

The community ergonomic approach is action-oriented in nature. The assumption is that rather than trying to change people in order to "cure them of unproductive behavior," it is more effective to address the environmental factors that lead to situational dysfunction in the first place. The objective of the CE approach is to reach collective aims and perspectives on issues of concern and to meet specific goals and aspirations through specified actions. This process requires a purposeful formulation of plans to reach a state of solution implementation. We believe this approach is more likely to achieve a satisfying result because it is based on purposes, goals, and aspirations rather than on existing community skills, or needs assessments, or the qualification for external funding sources. Other approaches have tended to become bogged down fulfilling the institutional requirements of some sponsoring organization and pursuing the right solution to the wrong problem (Nadler & Hibino, 1994). Our approach

(CE) prevents these problems by focusing on "selected aims and goals" rather than on the current problems.

2. Participation by Everyone

Inner city improvements often fail because the residents are not substantially involved in the process of selecting the aims, objectives, and goals. It is essential to have resident participation from start to finish. Such participation is a source of good ideas, a means of motivation for the residents, and a way to educate residents to new ideas and modes of behavior. There are many mechanisms for participation including individual involvement, action groups, and committees, and often a combination of all of these. In addition, the early involvement of strategic persons and organizations in a process allows the resulting design plan to incorporate the necessary concepts, technical expertise, and capital in the process and the solutions. It is risky to spend the time developing solutions only to have to sell the idea to someone whose understanding of and commitment to the project is minimal or nonexistent. While participation by every member of the community in a community ergonomic project may not be possible at first, the goal is to get everyone involved at some point. It is expected that reluctant and passive involvement will be minimized and will continually decrease throughout a project or activity. Effective information and action transfer among individuals, organizations, and institutions are made possible through the community ergonomic process. The CE interface can be viewed as a continuously changing medium that allows links to be made across gaps or barriers in the community, leading to more permanent pathways of exchange. The interface is most often a person who coordinates information among individuals and groups. Even though today's information technology can be used to achieve this function, this is usually a luxury not commonly found in inner cities.

3. Diversity and Conflict Management

In the CE development process, the creation of a CE professional to serve as the interface is valuable mainly in that it may provide for easier and more effective access to resources and services. Problems in the inner city are never neat and compartmentalized, and a community system is complex in nature, making it difficult to comprehend. That is why the role of the CE professional is essential in helping to formulate a process for managing diversity, conflict, and confusion that occurs in urban communities. We must realize that urban communities typically have low levels of self-regulatory capabilities and high levels of diversity and dissention. In other words, community ergonomics becomes a method to reduce the instability that occurs in community environment tracking and interactions bringing together diverse community elements. It is important to spend the necessary time designing solutions in concert with community leaders and residents so that the residents feel comfortable with the approach. This is in contrast to the more conventional method of studying problems and precisely quantifying just how bad things are in inner city communities, or generating community-wide plans or solutions with only residents or only government agencies taking part in the process. The CE professional can facilitate interaction and coordinate the interactive processes as well as communication. One instance of such facilitation techniques may be the integration of diverse information into a coherent statement.

4. Encouraging Learning

This principle suggests that a well-designed process will allow individuals and organizations to interact positively and effectively even under the conditions of a highly turbulent environment. It is expected that the community residents and planners will learn from each other and from participating in the process. In addition, these learning effects will be transferred to subsequent related endeavors, with or without the presence of the community ergonomic professional to facilitate the process. Thus, there is a transfer of "technology" (transfer of control, transfer of knowledge, transfer of skills) to the community in the form of the process and learning experience. Furthermore, system participants are expected to enhance their abilities in leadership roles learned while involved in the CE process. We believe that formally documenting the system management process within the group setting as it occurs provides for a better understanding and management of the community ergonomic process in future endeavors of a similar nature.

5. Building Self-Regulation

One aspiration of the community ergonomic process is to provide participants in the CE process with an increased ability and capacity for self-regulation. Self-regulation, defined as the ability of an individual or group to exert influence over the environmental context, is enhanced by creating specific tasks, actions, and learning opportunities that lead directly to the successful development of a plan for action. When a project based on purposes achieves specific goals (i.e., funding, improvements in the environment) as part of its implementation strategy, then new abilities and skills to self-regulate are developed. This serves as motivation toward more community improvement activity.

6. Feedback Triad

Feedback is an important premise of the CE process. Reactive, instrumental, and operational feedback mechanisms are integrated into a feedback triad allowing for closed-loop control, social tracking, and self-regulation principles to take effect in the process of designing and implementing community environment improvements. It is necessary to provide these sources of feedback (as described by Smith & Kao in 1971) within the planning and design process for true self-regulation to develop in the participants. Furthermore, in the absence of feedback, the desired outcomes for either the participants or environment may not occur. Without indication or means of tracking the cause(s) for system failure, how can improvements be made?

Reactive feedback is the personal sense that one's actions (or the actions of a group of individuals) result in a perceived outcome on the environment. Instrumental feedback is sensed from the subsequent "movement" of an organization or other entity in the form of milestones achieved and output produced (e.g., reports, timelines, and future action to be completed). Operational feedback comes from the results of activities such as planning, designing, installing, and managing group intentions, as well as the new policies, laws, buildings, and institutions resulting from such activities. These are examples of persisting results that can be directly sensed by community environment social tracking systems.

Without the feedback triad, self-regulation by group participants would not be very effective and the system could quickly degenerate. Participants must sense that their personal actions, words, and participation have effects on themselves, on others, and on the environment. People must sense that the results of their actions, words, and participation are perceptible as outcomes on the environment. These are perceived in the form of persisting, tangible evidence of actions taken.

7. Continuous Improvement and Innovation

The community ergonomic approach recognizes the need for continuous improvement, which can be achieved by continuous planning and monitoring of results of improvements. Private industry can be encouraged to provide guidance and feedback on the aims, goals, and successes of improvement initiatives. Benefits from private industry and governmental programs can be utilized to promote an entrepreneurial spirit that promotes effective community habits. These can be benchmarked against other communities and other programs. Valuable information can be elicited by studying the effects on an implemented solution over a period of many years in order to document the redevelopment of a dysfunctional system and give members of troubled communities opportunities for better lives. This implies the need for ongoing monitoring and research studying the operational requirements for implementation, measuring effectiveness, and use of feedback to alter the programs already in existence. Consistent monitoring of citizen needs, desires, and values must be established in order to verify that programs and products are accessible, usable, useful, and helpful to community residents.

The community ergonomic approach is comprised of seven activities that lead to the identification, design, and implementation of effective community

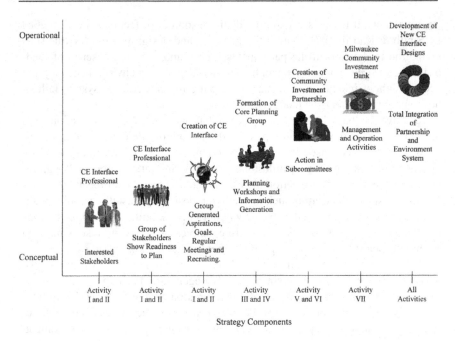

FIG. 14.2. The CE process and general activity timeline.

environment interfaces; with learning and self-regulatory mechanisms built-in at each stage. These activities are shown in Fig. 14.2. Each activity has specified purposes and expected outcomes that should occur in a general sequence that allows for parallel, iterative, and continuous achievement of activity expectations. Thus, the purposes, solutions needed, and expected outcomes are specified for each activity. Information needed, consumed, and transformed at each stage is also specified. This figure also illustrates how the process moves from a conceptual stage of problem definition to the operational stage of solution implementation.

An actual implementation of the process shown in Fig. 14.2 will be presented in the following section on case studies.[3] This particular application took place in the City of Milwaukee (Wisconsin, United States). Throughout the duration of the implementation of a target solution for a given environment, participants took an active part in structured activities designed to develop and increase self-regulatory abilities in the participants and the group. The purposes of the process were three-fold: (1) participants purposefully learned to self-regulate and control an interface between two or more aspects of their subenvironment (i.e., part of the city of Milwaukee), (2) the community ergonomic process produced a tangible and useful tool for community improvement applications, and (3) the process was monitored and measured so that it could be improved.

---

[3] The case study referred to here is the Milwaukee Community Investment Partnership (MCIP).

CASE STUDIES OF COMMUNITY
ERGONOMICS IN ACTION

The following case studies are presented to illustrate some of the principles and theory of CE. The first case study presented here describes the process and outcomes of a community-based partnership in an American inner city environment intended to equip community residents with the tools and skills to exert control over their lives and environment through self-regulation and continuous solution-finding techniques. Principles of social tracking, feedback, learning, self-regulation, and continuous improvement are brought up in this case study. The second study presented is a survey exploratory study, which brings to light several factors that seem to influence bank loan approval to community residents in financial need. Some of these factors include home ownership and active bank accounts. The last study will compare top–down and bottom–up approaches to network building among local enterprises in different regions of France. Principles of participation, action-oriented approaches, learning, and feedback control will be examined in the context of community quality management.

The Milwaukee Community Investment
Partnership (MCIP)

For illustration purposes, statistics describing the community environment where this partnership was born will be presented. This will provide a clear picture of the conditions characteristic of a community considered to present a CST profile based on the CST definition provided earlier. The physical or geographic area is where the CE interface design process is carried out and will be referred to as the investment zone. The resulting interface from the implementation of the CE design process will be referred to as the MCIP for "Milwaukee Community Investment Partnership." The MCIP will be described in detail in terms of how it came about and the subsequent outcomes.

A Picture of the Investment Zone. The *MCIP* is representative of the type of communities that the community ergonomic approach attempts to redress. Freedman's 1993 *From Cradle to Grave: The Human Face of Poverty in America* describes the "plight of America's poor" growing "more severe each year." Indeed, he noted that the 1992 Bureau of the Census "reported that 37.5 million people had fallen below the poverty line—the highest level in 27 years, since the start of President Lyndon Johnson's War on Poverty in 1964." Although the zone definition and boundaries are outcomes of the Partnership activities and represent an aspect of the results, they provide a clear context for understanding the severity of concerns for similarly dysfunctional communities in Milwaukee. A problem with providing laundry lists of horror story statistics is that they are "released to the public without sufficient explanation, [and] racial stereotypes are reinforced" (Wilson, 1987). The zone statistics are based on 1990 census data and encompass a total of 13,745 households (including those that are not considered to be families) and represents a total of 46,870 people. The demographic breakdown by ethnicity/racial background in the zone is 88.2% African-American, 5.2% Caucasian, 3.6% Hispanic, and 3% American-Indian, Asian-American, and "others." In comparison, African-American residents comprise 30.2% of the population for the City of Milwaukee and 4.9% for the State of Wisconsin. This lends support to the claim that Milwaukee is a highly segregated city within a highly segregated state.

Of the entire population in this area, 57.7% live below the poverty level. Compared to 22.2% for Milwaukee in its entirety and 10.6% for the State of Wisconsin, these families represent a very impoverished population indeed. Forty-nine percent of the households (which consist of one or more persons) in the zone subsist on less than $10,000 per year. Of these, 22.3% are 0 to 5 years old, while 35.4% are between the ages of 6 and 17. Thus, nearly 58% of the impoverished people living in the target zone are children. With regard to educational attainment, 57.2% of the target zone residents (25 years or older) do not have a high school diploma, compared with 28.5% for the city proper and 14% for the state. Unemployment statistics reported for all Milwaukee and state census tracts were 8.9% compared to 5.2% for the state. In similar inner city neighborhoods, it is typical for unemployment rates to surpass 40% (Committee For Economic Development, 1995). For the target zone, 29.1% were unemployed, a far different picture than for the rest of the state. Married couples represent 19.2% of the households compared to percentages of 38.6% for the city and 57.5% for the state. Forty-eight percent of the households are headed by a female compared to 19.8% for the city and 9.6% for the state. A disturbing statistic is that on December 31, 1991, 72.6% of Milwaukee county inmates held in Wisconsin institutions were African-American. The opinion of the Chicago Tribune's (1986) *The American Millstone: An Examination of the Nation's Permanent Underclass* was that, "Society already appears to have made some choices and apparently is relying on prisons to handle social problems." These percentages, based on census tract information, portray a community unlike those seen in most other areas of Milwaukee or anywhere else in the state. This is a community in which nearly 81% of the population subsists on incomes of less than $25,000. Evidently, the target zone needs help. The target population to be served is comprised of residents at or below 50% of median income for the City of Milwaukee.

An assumption made about the population of the inner city communities of Milwaukee was that a large majority of residents in these areas were in great need of effective, functional, and accessible interface designs for, at least, economic reasons. A systems perspective considers these grim statistics as part of and relating to a variety of social factors. The City of Milwaukee currently spends tens of millions of dollars in communities such as the ones represented above. However, the problems remain starkly evident, further demonstrating the difficulty of applying money as a sole remedy. Harrington (1962) "begged" the reader to forget the numbers game. Whatever the precise calibrations, it is obvious that these statistics represent an enormous, unconscionable amount of human suffering in this land. Similarly, the statistics cited above should evoke a similar sentiment, but there are no solutions in such sentiment; only the implicit need for the method to problem solve.

The CE Interface: The MCIP.   Planning the MCIP was the framework to pursue solutions for problems evidenced by the census information describing the target zone (as discussed above). This process involved key individuals and used information and knowledge necessary to design for continuing change and improvement. MCIP was designed and created to decide what was needed, why it was needed, who was needed, when things were needed to drive the solution, and where resources would be needed in order to implement solutions. Once these decisions were formulated, the framework for the plan was established and the planning and implementation process to create new community systems was initiated as an overall Partnership activity process. The planning and design process created a framework that established purposes to be achieved, organized a strategy to be pursued, specified and presented solutions, involved people critical to planning and implementing the solution, developed and used information and knowledge appropriately, and arranged for changes and improvements to the solution and the solution-finding process. The need for a partnership, carefully crafted from the talents and skills of community residents, leaders, and interested others, represented a strategy in the solution-finding process. It was this framework of people and ideas interacting over time and sharing in decisions that was crucial to the successful outcome.

The Partnership was comprised of members with varying experiences in community improvement. The partners represented community-based organizations (CBOs), community development corporations (CDCs), industry, small business, the political structure, financial institutions and policy makers, large insurance agencies, entrepreneurs, and religious organizations. Each partner had demonstrated long-term commitment to the communities and the communities of others. The partners' willingness (as well as skepticism) to work on the Partnership resulted from their expressed frustrations with failures of past efforts to improve inner city problems. An underlying frustration with the results of decades of policy making and programs aimed at the inner city pervades the attitudes of the partners. Each partner could speak at length about why inner city communities are in such terrible shape. Each could also articulate the hopes for improvement in global as well as specific terms and ideas, none of which represent systematic solutions.

The members of the Partnership possessed great strength in their quests to improve their blighted neighborhoods. Often, the help came from concerned citizens from outside the immediate area trying to "give something back to society." Whatever their reasons, the Partners represented passionate commitment for positive change, a passion not typically evident. They possessed great insight into the politics of community planning and reasons for the multitude of past failures. Their greatest asset was the willingness to play a believing game after deciding that the CE team was comprised of capable individuals. Even though past successes might not have changed the face of Milwaukee poverty, the combined experience of the partners represented a formidable power of knowledge. Ultimately, their commitment to the community residents in these impoverished areas was beyond reproach. The CE approach to CE interface design process simply guaranteed a greater likelihood of something positive happening in the community. In fact, the idea of a bank (which was an outcome of applying the solution finding process by members of the MCIP) was only a moniker given to a new entity comprised of solutions that the partners perceived as important to improving quality of life in the community. Fig. 14.3 illustrates how the CE interface design process went from the creation of the MCIP to the subsequent installation of the Milwaukee Community Investment Bank.

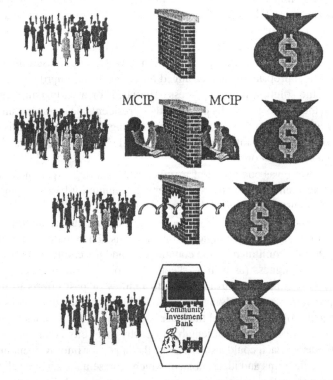

FIG. 14.3. Evolution of the community ergonomic process in the Milwaukee project.

The Milwaukee Community Investment Bank was designed to fit the needs of its potential users and as such was not very "bank-like" in the traditional sense. Its programs and opportunities were also intended to educate those who it was designed to help. The programs to increase home buying used innovative methods to allow people of very low income to afford their own homes. In addition, teaching fiscal responsibility as well as providing a means of attaining "economic emancipation" were provided. These are components that the members of the MCIP built into their banking system. The Milwaukee Community Investment Bank was neither required nor expected to resemble other banking facilities of a more traditional nature. The real importance of the Partnership was not that it created a useful design solution for a small impoverished area, but that it could do so at all. That a group of highly committed community leaders and residents could stand on equal ground with the political leaders of the city and with each other was new. The Partnership was the real interface; it was the first design solution of the CE team and the place where people and ideas interacted repeatedly, but within a framework of specific work to be achieved, and the place that allowed people to enhance their personal abilities for community improvement. Members reported that they felt able to influence their work or personal lives more effectively because of their participation in the effort. In fact, this type of self-regulation was clearly demonstrated through tangible projects or subtle effects reported by the partners. Some of these effects included a greater capacity to initiate, develop, and manage projects on their own; an increased capacity to provide services to area residents throughout work affiliates; an increased capacity to network with like-minded individuals in the city as well as rally and encourage other people. Thus, it appears that involvement in the C:E interface design process increased the capacity of people (who were probably already capable) to work effectively toward their desired goals.

Evaluating Bank Lending Practices to Low-Income Residents. A preliminary evaluation of the community ergonomic framework was conducted by examining the factors related to loan acceptance in two communities. The study was undertaken to improve the investment within these communities to help them prosper. The community ergonomic framework was used as a template to identify the areas in the community where disagreement was identified between the perceptions of how money was loaned and the requirements necessary to receive a loan. Community leaders and the banking institutions would then engage in a dialogue addressing these issues, and determine ways of improving the investment tendencies within the community.

A questionnaire instrument was designed to identify critical issues related to banking practices and perceptions of lending. Residents of the two communities who were 18 years of age and older were eligible to participate in this questionnaire. The planning departments, with assistance from neighborhood association leaders, provided the clusters of residents for each neighborhood in each city.

Participation in the completion of the survey was voluntary. The neighborhood association enumerators delivered a copy of the questionnaire to residents and explained the study to them. Residents who agreed to participate in the study received a clean copy of the questionnaire from the enumerator. The enumerator asked the questions verbally from a separate questionnaire, and allowed the respondents to follow and record their answers. The enumerator could not see what the participants' responses were. Each enumerator was responsible for conducting 10 interviews over a 4-day period and returning the completed results on the fifth day to the community investment leader for each city. Each interview took approximately 1 hour to complete. Of the 600 questionnaires given to survey enumerators to distribute to residents, 558 were returned for a 93% participation rate.

This analysis will focus on the 386 participants who responded that they had applied for a loan with a financial institution. This group represents 69% of the participants. The study design for this analysis included demographic factors, credit activity factors, economic resource items, banking relationship questions, and a question on credit knowledge (knowing how to obtain a credit report). The demographic factors included gender, marital status, age, ethnicity, and city of residence. The credit activity factors included items inquiring about people meeting with a financial institution regarding credit and having seen their credit report in the last 2 years. Economic resources items asked about income sources, annual income, home ownership, real estate ownership, and number of motor vehicles owned. Banking relationship questions included questions about having a checking account, having a savings account, having outstanding loans, owning a credit card, or having no banking relationships. The average age of the participants was just under 40 years. Sixty-two percent of the participants were female. The participants were evenly distributed across the two cities. Fifty-six percent of the loan applicants had their requests accepted. Most of the participants received their income through employment, although 36% had an annual income under $10,000, with a total of 70% of them earning less than $20,000 a year. Seventy-six percent of the participants were African-American. Multivariate logistic regression analysis of loan acceptance rates was run on all factors to determine the relative influence of each factor while controlling for the effects of other factors.

The results showed that participants who owned a home were approximately twice as likely to have a loan request accepted as those with no homeownership (odds ratio = 2.14, 95% C.I. = 1.00, 4.58). Participants with a savings account were nearly four times more likely to have a loan request approved than those with no savings account (odds ratio = 3.95, 95% C.I. = 2.02, 7.71). These results seem to indicate that factors such as homeownership and an established banking record in the community are related to loan acceptance. This implies that some individual economic resources are expected on the individual's behalf for loan approval to be granted. This study serves to illustrate how the community ergonomic framework can help to evaluate the many facets related to access to economic resources and prosperity in a community.

A Case Study of Enterprise Networks in France

Community ergonomics and quality management methods can be used as a tool for developing interrelationships among enterprises in a given geographical zone. Four key objectives of these enterprise networks are (1) to share experiences and expertise on a given topic (such as quality management and training), (2) to improve the quality of the environment, (3) to set up a common structure to achieve goals (for instance, government organizations), and (4) to share resources for a given goal. A key issue in developing such a network of enterprises is to achieve adequate participation of as many enterprises as possible. Without a high level of participation among the targeted enterprises, such a network may not be effective and, therefore, will not be able to achieve its objective. Using the model of participation proposed by Haims and Carayon (1998), we can assume that networks of enterprises are effective and develop over time if the development process using community quality management methods fosters action, feedback, and feedback control. Two different models of enterprise networks are currently being developed in the eastern region of France called Lorraine. These networks use community ergonomics and quality management methods as a tool to achieve their stated objective. The first network is located in the industrial zone of Ludres-Fleville (south of Nancy), while the second network is located in Moselle-est (north of Lorraine). These two networks are used as case studies to provide a better understanding of how community quality management can (and cannot) be used to develop and improve interrelationships among enterprises in a given geographical zone. In order to study the first network, we used a participant-observer method, while in the study of the second network, interviews and analysis of documents were used. Data was collected and analyzed to determine the degree to which action, feedback, and feedback control were built into the development process of these networks.

Network in the Industrial Zone of Ludres-Fleville. The enterprise network in the industrial zone of Ludres-Fleville (ZLF) uses quality management as a methodology, and also as an objective. The ZLF is comprised of about 200 small- and medium-sized enterprises. It is located in the cities of Ludres and Fleville in the eastern part of France. In 1995, a group of political and economic actors in the LF region created an association (Association for Industrial Excellence in Ludres-Fleville—AEILF). The purpose of AEILF was to achieve two main objectives: (1) quality in the enterprises of the ZLF (e.g., implementation of ISO 9000 programs), and (2) quality of life and work in the ZLF itself. About half of the enterprises of the ZLF joined the AEILF. Based on a survey of enterprises of the ZLF, seven working groups were formed to examine problems common to the enterprises of the ZLF using quality management problem-solving tools and principles. A project manager hired by the AEILF led these groups.

The groups examined the following issues: (1) safety, (2) traffic transportation, (3) mail multimedia, (4) communication on the ZLF family issues, (5) training relationship between enterprises and education/training institutions, (6) environment, and (7) shared services (e.g., energy, human resources management). A flow diagram was built by the steering committee of the AEILF to define the overall functioning of the working group. In addition, problem-solving guidelines were given to each group.

Network in Moselle-est.   In contrast to the network in the ZLF, which used more of a top–down approach, the network of enterprises in Moselle-est used a bottom–up approach. In response to demands from local small- and medium-sized companies located in Moselle-est, in the eastern part of France, an association was created in 1991 (European Space-Initiative-Competitivity—EEIC). The EEIC includes more than 40 enterprises located in Moselle-est. Problems and difficulties experienced by the enterprises comprising this association or network were used as the starting point to create a work group or subnetwork. The creation of a new group or subnetwork was carefully studied before proceeding. Managers and employees from the member enterprises led each subnetwork. The EEIC's role focused on logistics as well as providing administrative support for the various subnetworks. There were 6 subnetworks addressing the following topics of interest to enterprises: (1) quality assurance, (2) continuing education and training, (3) quality measurement, (4) the environment, (5) exports, and (6) quality management. Recently, the EEIC has expanded its domain by creating a subnetwork on quality in the public sector. Analyses of the network in the ZLF show the difficulty in maintaining a high level of participation among enterprises of the industrial zone in activities related to the development of the network. This second network seemed to be more successful in sustaining a high level of participation of enterprises. The difference in participation levels may be due to the history of the networks (e.g., how the network was started, age of the network), but also to the characteristics of the networks in terms of action, feedback, and feedback control.

CONCLUSION

This chapter introduces the theory and practice of community ergonomics to the framework of macroergonomics. CE is a long overdue answer to the application of human factors engineering principles to address complex societal problems. The robustness of the foundations defining CE lies on the theories on which it was built. The CE principles presented in this chapter are the canons driving the effective practice of the CE approach. The CE interface design process is flexible enough in nature to accommodate numerous situations and overcome unforeseeable challenges, yet structured enough to ensure successful completion of the process with solutions having a positive long-term impact in the community in question.

Community ergonomics is a philosophy, a theory, a practice, a solution-finding approach, a process, all in one. That is why the role of the community ergonomics professional or CEE (community ergonomics engineer) is essential in the successful implementation of this approach to improving complex societal systems showing signs of CST (cumulative social trauma). CST is not to be taken lightly; the costs are immense in every respect; financial, human, social, health, and developmental. When any one group of people, a community, or region is isolated, alienated, and blocked from access to resources, the consequences are detrimental. CST is indeed a social ailment affecting the core of society, the human element, the intellectual resource, the social fabric that inhibits the evolution, development, and growth of human kind as contributing members of society. CE provides the tools, the techniques, and the ideology necessary to break through the barriers that destine communities to experience CST and bridge the gap between people and the community environment in which they function.

It is no coincidence that the case studies presented here show that as we enter the 21st century we still encounter situations in which people do not have equal access to economic resources (as shown in the bank lending practices study), or segregation (as shown by the shocking statistics describing the MCIP), or that different approaches in enterprise network building lead to different outcomes in terms of participation, learning, and feedback control (as shown in the comparative study of two regions in France). However, it is important to emphasize that as valuable as CE is in unveiling and overcoming these societal problems, the role of a CEE is essential in ensuring that desirable outcomes are achieved and social conditions are improved in the communities in question. Every effort should be made not to cause further damage in areas showing signs of CST such as isolation, alienation, and learned helplessness. CE is indeed a powerful tool and promising in many respects and at many levels. As shown in the case of the MCIP, the benefits at the individual as well as group and community levels are priceless and long-lasting. Instilling in people a sense of self-esteem and self-efficacy and providing them with tools to exert control over their environment and actions through social tracking, self-regulation, learning, and feedback control gets to the true essence of CE. It is community residents in these now disadvantaged areas who will break the cycle of detrimental effects of long-term exposure to CST conditions and build new healthy, strong generations of people able to self-regulate and make valuable contributions to their communities and the world.

15

Macroergonomic Considerations in Technology Transfer

Houshang Shahnavaz
Lulea University of Technology

INTRODUCTION

Technological development has contributed enormously to economic growth and social progress in the industrialized world. It is beyond doubt that wealth in today's world is mainly of applied technological origin and is primarily the product of applied science and technological knowledge. Technological advancement has elevated mankind standard of living and contributed greatly to the reduction of many sources of occupational accidents, injuries, and stresses. However, the influence of technology on the quality of working life is two-sided: along with wealth, advanced technology has brought new sources of work stress and injuries. The complexity of modern technology, the changing nature of work, the work organization, and the production system has placed extra demands on the workforce. As a result the total work demand (mental and physical) has sometimes increased.

Further, increased competition in global markets imposes higher challenges on innovation, productivity, and product quality. This, in turn, requires a knowledgeable workforce with broader skills who are flexible and motivated with access to advanced technology and information. To achieve this an organizational structure and management system is required in which operators' knowledge and skill are continuously improved through education and training and the total company resources are fully utilized for optimum performance and for dealing with rapidly changing problems at work. In this endeavor, progressive firms are shifting from a technology-centered to a human-centered approach, regarding people as central to all development initiatives. This notion highlights the importance of ergonomics in the development and application of technology to ensure that the technology is accepted and fitted to its users and to its operating environment for safe and effective use.

In the industrialized countries (ICs), technological advancement has also contributed to the growing awareness and recognition of the importance of ergonomics. The sphere of ergonomics (micro and macro) has thus been expanded. Development of new organizational structures, management systems, and work methods, as well as recommendations, directives, and standards concerning these issues have become an important part of ergonomic activities. Issues that deal with people and how they perform are considered as important as the technology per se and are linked to the overall company policies. At the same time, cooperation between managers and employees in a participatory problem-solving process has become a successful practice at many workplaces.

It could be stated that IC are trying to keep pace with technological advancement through development of know-how, legal, and administrative procedures for protecting their workforce, keeping their competitive advantage and making optimum use of technology.

SITUATION IN THE INDUSTRIALLY DEVELOPING COUNTRIES

Industrially developing countries (IDCs) strive for an overall improvement in the quality of life to be achieved through economic growth. Industrialization, mostly through importation of advanced technology, often is considered as the quickest way of achieving economic development and social progress. However, because of several complex sociocultural, economic, human, and technological factors, this policy is not always successful in terms of leading to any significant improvement in the IDCs' economies or people's quality of life.

It is generally accepted that the characteristics of a technology are mostly determined by the prevailing conditions of the technology-producer country. It usually reflects the specific requirements and availability of both human and material resources, income level, resource costs, organization and political systems, infrastructure, and sociocultural conditions of the country for which the technology is designed. When the technology is transferred from an IC to an IDC that has different requirements and characteristics, some adaptation is needed to fit the transferred technology to the recipient country. However, due to ignorance and lack of demand from technology recipient countries, very seldom do companies take the initiative to adapt their technology to the conditions of the recipient countries.

The developing world, with about 80% of the global workforce, can have access (depending on their accessibility[1]) to advanced technology without having developed sufficiently their technology-absorptive capacity or their legal or administrative infrastructure to successfully utilize the imported technology and control its adverse consequences.

Industrial development through technology transfer requires, firstly, conscious and rational selection of the technology from the wide range of technology available in the world market, and then adapting it to local requirements. Secondly, it requires preparing local conditions for optimum absorption of the technology and adapting the society and social services to the changes required by the new technology. Many social functions and management practices in IDCs are incompatible with the demand of industrial progress. Lack of a safety-conscious culture, low educational standards, inadequate training in handling the new technology, lack of a proper environment for innovation, and inappropriate organizational structure for optimum utilization and maintenance are some of the contributing factors to technology transfer failure.

Industrially developing countries share some common problems, such as poverty, low productivity, and low product quality. The importation of inappropriate technology, which could not be fitted to local conditions (in every sense of the word), is considered as one of the reasons for the current problems in most IDCs. This has created a vicious circle of unemployment, chronic poor health, high rate of accidents, low motivation, increased physical and mental stress, and low productivity (Shahnavaz, 1987). The aim of using ergonomics in the technology transfer process is to break this vicious circle of failure by creating productive, safe, and satisfactory condition; for technology users.

APPROPRIATE TECHNOLOGY

Technology—defined in terms of physical products, techniques, know-how, information, skill, labor, and organization—is an integrated part of a country's structure. It is also a part of the dynamic element of culture. Consistent with sociotechnical systems theory (see Chapter 1), any changes in technology have an impact on the social, political, and economic systems.

An appropriate technology is a technology that is (a) suited to its user population and to the environment in which it is used, (b) appropriate to the needs and means of the local population in a long-term perspective, and (c) considers local

---

[1] The World technology is narrowed down to the technology in use in IDCs depending on their communication ability and selection mechanism.

educational, social, cultural, infrastructure, economic, and political aspects. Technology, in a broader sense, is a means by which a country undertakes to change its circumstances to better satisfy its needs. It therefore is not the objective of development, but rather the principal means for attaining the required development.

The broad range of circumstances that exist within the various countries of the world (with regard to economy, climate, demographics, culture, customs, political systems, and social conditions, etc.) highlights the need for different development strategies and the selection of alternative technology. With regard to IDCs, which are mostly dependent on technology-producing countries for their industrialization, the right selection and proper application of transferred technology is of vital importance.

Appropriate technology is not a general concept and there is no uniform prescription or single system that one can regard as the appropriate technology for all IDCs at all times. It differs according to place and stage of development.

For IDCs trying to achieve the goal of strengthening their autonomous capacity for using and creating technology for meeting their needs through their own efforts, it is important to identify which technology to acquire, how to transfer it, how to implement it, how to adapt it to its users and its operating environment, how to maintain it, and how to build on it. Ergonomics is of significant help in this process.

PROCESS OF TECHNOLOGY TRANSFER

The aim of technology transfer (TT) is to ensure that imported technology makes the maximum contribution to the technical progress and economic and social development of a country. However, the often-used technology selection strategy that has only an economic objective in mind has many shortcomings and cannot succeed in the long run. Further, in the technology transfer process, emphasis usually is placed on the engineering aspect of technology (technoware). As a result of this narrow approach, many IDCs are struggling not only with a low rate of technology utilization, and low product quality, but also with the ill health of their workforce due to poor working conditions. In order to avoid past mistakes, a more holistic and systematic approach, considering all of the interacting technology components, should be adapted with regard to technology transfer.

According to the Technology Atlas Team (Sharif, 1988; TAT, 1987), technology comprises four interrelated components:

1. Technoware (machines, equipment, tools, physical processes, capital and intermediary goods, etc.)
2. Human (human labor, human capacity, capacity for systematic application of knowledge and problem solving, skill, know-how, ideas, etc.)

3. Information (scientific and other forms of organized knowledge, technical information, data, computer software, recommendations, standards, etc.)
4. Organization (organization of products, processes, tools, and services, social arrangements, means for using and controlling factors of production, etc.)

All four components of technology are complementary to one another and are required for the production of goods and services. Depending on the nature of the activity, the relative importance of each of the four components may differ. Furthermore, technology does not operate in a vacuum. Its use takes place within an operating environment (technology climate). Therefore, not only the four components but also their interactions within the operating environment must be fully considered. The technology climate includes both external and internal factors. The external environment includes factors such as physical infrastructure, support facilities, R&D institutions, and political and legal systems as well as country culture and administrative institutions. The internal factors include the firm's culture, physical environment, economic conditions, and social and political conditions.

MAKING TECHNOLOGY TRANSFER A SUCCESS

The transfer of a technology from a firm in an IC to a firm in an IDC depends on the decisions made both by the technology supplier (IC) and the technology receiver (IDC), as well as the dynamic interaction and communication between the two partners. The environment surrounding the technology transfer is also another major component in this process. The environment influences the development of a particular relationship between the partners. The success of the process is determined by communication and cooperation between partners— something that can be developed only during long-term contacts. Furthermore, it is generally agreed that three characteristics of the technology recipient's environment are decisive for successful technology transfer: government policy, technical absorptive capacity, and cultural distances (Robinson, 1988). Figure 15.1 illustrates the model of technology transfer between two partners.

ERGONOMIC CONSIDERATION IN THE PROCESS OF TECHNOLOGY TRANSFER

The level of ergonomic awareness by the technology supplier and receiver firms, as well as their commitment to ergonomic issues, will greatly influence their decision regarding how appropriate transferred technology will be put into effect.

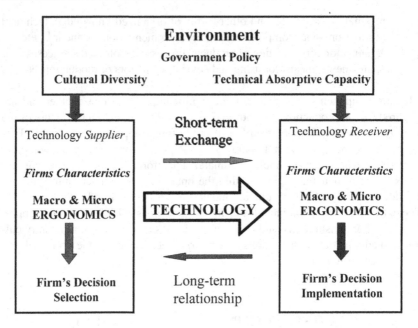

FIG. 15.1. Technology supplier and receiver decision making and action.

These can be analyzed by closely examining the firm's characteristics and attitudes because they generally reflect the firm's own micro- and macroergonomic conditions. The better the ergonomic conditions of a firm and the firmer its commitment to ergonomics (both at micro and macro levels), then the better the choice and utilization of technology, leading to a more appropriate technology transfer. However, it should be remembered that even an "ergonomically designed product" (or system) made for a certain population cannot be used efficiently and safely by a different population in a diverse environment (Shahnavaz, 1989).

Ergonomic considerations, from both the classical human–machine aspects of micro-ergonomics and the organizational and sociocultural issues of macroergonomics, ensure successful transfer of technology. Ergonomics contributes to a "good fit" among technology, its users, and the operating environment, creating the needed compatibility within system components for a harmonic and efficient interaction. It also is a useful tool for evaluating the choice of technology and its implementation. As technology becomes increasingly more complex, ergonomic considerations are more important in order to eliminate possible negative effects.

Ergonomics at the Micro Level. At the micro level, ergonomics is concerned with the design of products or systems with the objective of creating a "usable" user–machine–environment interface. A mismatch during humans, technology, and/or technology climate can be physical (anthrophysiologic), mental (cognitive or psychological), or both. Issues of user–interface design concerning human dimensions, physical and mental capacity, abilities, and preferences (psychophysical) are of micro-ergonomic interest.

Issues that need to be considered in the process of technology transfer at the micro-ergonomic level are: anthropometry, physical work capacity, functional abilities, physical working environment, and cognitive and cultural differences that exist between technology producers and users.

MACROERGONOMIC ISSUES

Macroergonomics refers to the proper design of organization and management systems, including their software aspects. There is evidence that, because of differences in culture, sociopolitical conditions, and educational level, many successful managerial and organizational methods developed in industrialized countries could not be successfully applied in third world countries (Negandhi, 1975). On the other hand, technology transfer is not just a hardware transfer. The physical infrastructure and human resources, as well as the organizational structure and management system, must be compatible for a technology to work.

An organization and its structure, values, and behavior pattern are usually a cultural product, which is more strongly emphasized in the traditional societies of IDCs than in ICs. Compatibility must exist at all level of culture—societal, organizational, and professional. Kedia and Bhagat (1988) consider social and organizational compatibility as the most important determinants of a successful technology transfer.

In most IDCs, an organizational hierarchy and the downflow of authority within the organization is a common practice. Western values, such as democracy, empowerment, or power sharing in decision making, which are regarded as key issues in modern management for proper utilization of human resources (with regard to intelligence, creativity, problem-solving potential, and ingenuity), do not agree with the cultural sense of hierarchical power. The feudal system of social hierarchy and its value system are widely practiced in most industrial workplaces. This makes, for example, the participatory management approach (which is regarded as essential for the new production mode of flexible specialization and a motivated workforce) a difficult endeavor. However, there are reports that confirm the desirability of introducing autonomous work systems in these cultures (e.g., Ketchum, 1984). Research evidence also shows that if a proper approach is adapted, many firms in IDCs also are eager to make use of these macroergonomics finding (Helali & Shahnavaz, 1996, 1998).

A major problem of many transferred technologies is the lack of macroergonomic consideration. Macroergonomic issues are more complex than micro-ergonomic issues because they are widely influenced by the firm's culture which, in turn, is much affected by the sociocultural structure of the country.

A problem pertaining to the implementation of macroergonomics is the lack of awareness regarding its benefits. The concept is regarded as both new and a Western product. Therefore, macroergonomics often is resisted by IDC managers. A useful strategy for ergonomists dealing with these issues is to find familiarity between ergonomic principals and the sociocultural values of the IDC organization and try to aim for "small wins'." Because these issues are both great and complex to deal with, it is better to find linkages between the ergonomic principals and the accepted norms and values in the society for easier acceptance and implementation. For example, in Islam, consultation in problem solving is recommended (*Koran*, Sora, 42:37). This statement can be used as a fundamental argument for participatory ergonomics and as a basis for organizational reform in traditional Islamic countries.

Macroergonomics is better accepted if it is justified within the national culture, such as in religion, folk tales, stories, phrases, and expressions. According to McWhinney (1990),

> . . . conflict is avoided by making changes in accordance with the path of the myth and stories that guide our lives. Changes are most effectively accomplished when we have uncovered the core stories of the relevant culture and use our skills and courage to advance along the paths that are natural to the person, the organization, and the culture.

A clever strategy would be to make the best use of the positive features of a society's culture for promoting ergonomic ideas and principles. A good example of cultural appreciation with regards to designing management strategy is the implementation of the "seven tools" technique for quality assurance in Japan. Seven tools are the minimum weapons a samurai warrior carried with him when he went out to fight. The quality Control Circle pioneers named the basic tools that are required for quality improvement the "seven tools"—a familiar term to everyone in Japan—in order to involve all employees in the quality work (Lillrank & Kano, 1989).

Organizational Design and Management

An organization is a social structure wherein employees play a decisive role in improving its performance. Decision making and action taking should be concentrated in the heart of the operation to reduce the risk and duration of system failure and to better utilize resources as well as increase system reliability and availability. However, organizational change is a difficult, time consuming, and expensive process. Cultural factors, including the way people interact with each other in an organization and commit themselves to organizational goals, are complex matters that have significant bearing on the success of an organizational change (Shahnavaz, 1998). It, therefore, is necessary to match management methods and techniques to the local conditions.

Societal and organizational culture-based differences should be considered when designing or introducing change in an organization. With respect to information technology, De Lisi (1990) indicates that networking capabilities will not be realized unless the networks fit the existing organizational culture.

In some cultures the family has a strong role to play in work organization. Among some communities in India, for example, a job is generally regarded as a family responsibility and is collectively performed by all family members (Chapanis, 1975).

Zhang and Tyler (1990), in a case study related to the successful establishment of a modern telephone cable production facility in China, supplied by a U.S. firm, stated:

> Both parties, realize, however, that the direct application of American or Essex management practices was not always practical nor desirable due to cultural, philosophical, and political differences. Thus the information and instructions provided by Essex were often modified by the Chinese partner to be compatible with the conditions existing in China.

Design of many macroergonomics features are closely influenced by culture. Further examples are the design of a maintenance program (both preventive and regular) and housekeeping and work schedules, which are a common problem in many IDC workplaces.

The traditional culture of agricultural societies, predominant in many IDCs, is not compatible with the requirements of industrial work and how activities are planned. For example, traditional agricultural activity does not require vigorous formal preventive and corrective maintenance programming or precision work, and is generally not under time pressure. In the field, it is usually left to nature to take care of the maintenance and house-keeping work. Designing maintenance program and housekeeping manuals for industrial activities should consider these cultural constraints and provide for adequate training and supervision.

With regard to work schedules, in the Islamic culture, for example, people are required to break work at least once within a work shift for praying, and to fast for 1 month each year from sunrise to sunset. These cultural constraints require special work organization considerations.

Unfortunately, in many IDC companies, narrow, tightly controlled, fragmented jobs and hierarchical organization are still common practice. They usually are the cause of low motivation among workers, having adverse effects on both individual and organizational performance. On the other hand, advanced technology requires a human-centered organization for its effective operation (see Chapter 1). Introducing organizational changes and new management systems could easily be achieved in IDCs through a process of training and education in macroergonomics for both management and employees. This training and education will bring about awareness and action, aiding in the introduction of necessary changes. It has been shown that this is quite possible and that the synergistic effect of a macroer-gonomic approach is significant (Helali & Shahnavaz, 1996, 1998).

The organization should regard employees as its future problem-solving agents, and must thus allow for enhancement of their skills and innovative capaci-ties. To facilitate employees' participation in solving technical and psychosocial problems at work, and to create a safe and productive environment, management must institute a vigorous program of education, training, and self-improvement.

Combining bottom–up and top–down approaches, macroergonomics takes full advantage of utilizing broad participation within the organization. This will create a shared vision and a program for change acceptable to both employees and management. This is especially essential to develop more complex and sophisticated systems.

Workers' Participation

Participatory ergonomics is a useful macroergonomic approach for solving various work-related problems (see Chapter 2). Because the introduction of new technol-ogy needs to be legitimized by the workers as well as by management, a number of studies have emphasized the significance of worker participation in the introduc-tion process (Noro & Imada, 1991; Wilson, 1991). Worker participation can help improve the quality of the decision and implement the technology to its fullest.

In this competitive world, a company cannot survive without the total involve-ment of its people in technology utilization. Although participation is a culture-free concept, political as well as sociocultural factors must be considered for technology's proper applications in IDC workplaces. According to Liker et al. (1988), ". . . effective participatory ergonomics program can take many forms. The best program for any plant in any culture may depend on its own unique history, structure and culture."

Participation between the technology supplier and receiver is needed at all stages of the technology transfer life cycle. It should start at the design concep-tion stage and continue throughout the technology transfer to the actual utiliza-tion of the technology.

Communication

Culture is the key to effective and clear communication and dialogue, which is the means of information flow between the technology supplier (S) and technology receiver (R). Because technology utilization is a continuous process, proper communication and interaction of the two partners (S&R) are essential. Bridging cultural differences and learning to understand each other is a key for successful technology transfer and building a long-term relationship. Clear and effective communication plays a decisive role in all five phases of technology transfer (i.e., initiation, analysis, selection, implementation, and utilization). In each phase, the information necessary for making mutual decisions is exchanged between the two partners. Information is invaluable in order to assess the environmental characteristics of the technology receiver (e.g., the capacity to absorb a new technology). It is needed to predict the cultural differences as well as to provide knowledge regarding a firm's characteristics, including the current state of its macro- and micro-ergonomics, incorporation of ergonomics in the technology transfer, and the necessary actions for a successful transfer.

The present globalization of information and communication technology is the fastest growing technology transfer in the world. However, in order to make full use of the technology, variations in culture, legal aspects, infrastructure, human resources, and organizational issues need to be considered. Walsham (1993) noted that when the design and development process of an information system strictly reflects a common top–down structure of an IDC, the system is most likely to fail. Meshkati (1986) described the systematic integration of economic and ergonomic analyses that needs to be performed in each phase of TT through communication and joint action by the technology supplier and receiver.

Environmental Factors

Government Policy

The macroergonomic parameters that should be considered in government policy include (a) proper assessment of the country's needs with regard to the nature and type of technology, (b) establishment of relevant research institutes and support organizations for promotion, implementation, and control of safety and ergonomics at work, (c) development of national statistics and data banks, (d) recommendations, standards, and directives for ergonomic implementation, and (e) monitoring of work environment practices, collection of health and safety statistics, and establishing an effective information network.

Because the many factors that influence the nature, extent, and diversity of technology transfer problems are specific to each IDC, it is necessary for each country to make research, education, and training an indispensable part of its industrialization/development policy. Likewise, proper legislation and inspection procedures have significant roles in the encouragement of ergonomic issues. These procedures also can have a crucial impact on companies with respect to the maintenance of their ergonomic programs. Further, it is an important way of ensuring a healthy, safe, and ergonomically sound workplace, thus contributing toward the national aims of prosperity and higher quality of working life.

The macro-level actors, such as government, business organizations, and organized labor, are the major determinants of macroergonomics at work. The higher their knowledge and commitment to these issues, the better will be the firm's organization from a macroergonomic viewpoint.

Technology Absorption Capacity

In the context of technology absorption capacity, one of the main problems for IDCs is the lack of scientific and technological infrastructures and training facilities for improving the workforce's level of education, skill, and understanding of safe and effective operation, maintenance, and development of the imported technology. These are important because technology will be absorbed better at the local environment if it is in harmony with its users and its operating environment. Certain concrete steps can be taken to meet this problem; for example, in most cases the necessary ergonomic knowledge for making the right choice with regard to purchasing a new technology or its proper utilization does not exist among companies in IDCs. In other cases, adaptation of the imported technology to local conditions, or modification and correction of technology, may be needed to make the technology more efficient and to minimize its negative effects. In such cases assistance can be sought from an ergonomic consultancy or research organization. A well-established support and information system for providing ergonomic information at the country level ensures proper choice and utilization of the imported technology.

Cultural Diversity

Vis-à-vis cultural distances, the local culture of the technology recipient firm and its similarities with the technology producing firm greatly influence the success of the transferred technology. Cultural values and behavioral patterns have a direct bearing on people's willingness and ability to adapt and absorb technology. A successful transfer thus requires that the cultural barriers be overcome and cultural issues considered in the life cycle of the technology transfer process. However, it should be remembered that the ultimate responsibility for successful technology transfer rests with the technology receivers—the recipient country's policymakers involved in development planning as well as the firms actually importing the technology.

At the national level, effective institutions are needed for the formulation and implementation of sound policies. It is vital that they support the local firms, conducting inspections and control to ensure appropriate technology transfer and utilization. At the firm's level, serious consideration of macro- and micro-ergonomics, including proper interaction between technology supplier and technology receivers, will greatly help not only in selecting the right technology, but also in utilizing it efficiently in the long run.

16

Macroergonomics and Aviation Safety: The Importance of Cultural Factors in Technology Transfer

Najmedin Meshkati

University of Southern California

Facts are not pure and unsullied bits of information; culture also influences what we see and how we see it. Theories, moreover, are not inexorable inductions from facts. The most creative theories are often imaginative visions imposed upon facts; the source of imagination is also strongly cultural.
—Dr. Stephen Jay Gould, renowned Harvard University professor of geology, biology, and the history of science (Gould, 1981, p. 22)

INTRODUCTION

Whenever you fly an aircraft, either sitting in the cockpit or back in the passenger cabin, land at a "foreign" airport and your flight safety is likely to be at the mercy of macroergonomic considerations. These considerations include systematic incorporation of cultural factors in work system design and operation. The above succinct epigraph, which provides the first compelling reason, is also a testament to this contention. It may be anathema to many scholars and practitioners of "hard" sciences and other engineering-dominated fields, but recent rigorous research has proven that even *scientific* theories, facts, and

323

practices—that determine and govern aviation systems' operations—are strongly culturally based.

For instance, according to an investigative article in *The New York Times* (May 23, 2000), an attempt to ban communication in French between pilots and air traffic controllers at Charles de Gaulle airport near Paris lasted only 15 days, primarily due to cultural factors—more specifically, because of "stubborn Gallic pride that persists among French pilots and air controllers." As all non-French speaking seasoned pilots who have approached this airport suggest, this decision could have serious long-term safety implications.

Macroergonomics is focused on the overall people–technology system level and is concerned with the impacts of technological systems on organizational, managerial, and personnel subsystems. It includes areas such as training, management, the planning process, information systems, internal review/inspection programs, performance measurement systems, reward structure, initial employee qualification assessments, and personnel selection criteria (Hendrick, 1986). In addition to micro-ergonomic factors, the aviation industry could greatly benefit from systematic incorporation of macroergonomic considerations in aviation systems design, operation, and maintenance.

Moreover, understanding the impact of cultural factors, which constitute the dormant and, probably, the most subtle subset of macroergonomic considerations, is of paramount importance in aviation safety because the aviation industry will become even more "international" in the near future, and because international traffic is expected to continue providing a greater overall share of world air transport. Boeing expects it to account for 54% of all air travel in the next 20 years, with domestic traffic making up the other 46% (*Financial Times*, May 21, 1994). This increasing trend heightens the importance of aviation technology transfer from primarily Western aircraft and systems manufacturers to some 600 airlines and air traffic control centers in countries around the world. Thus, not only should cultural factors be considered, but they should be given even greater attention in global aviation safety.

The objectives of this chapter are three-fold: to define the dimensions of national cultures; to provide factual justifications for, and examples of, the critical impact of cultural factors in aviation safety; and to identify important culturally based parameters affecting aviation safety.

NATIONAL CULTURAL FACTORS

National culture, according to anthropologists, is the way of life of a people—the sum of their learned behavior patterns, attitudes, customs, and material goods. According to Azimi, the culture of a society consists of a set of ideas and beliefs (Azimi, 1991). These ideas and beliefs should have two principal characteristics or conditions: first, they should be accepted and admitted by the majority of the population; and second, the acceptance of these beliefs and ideas should not necessarily depend on a scientific analysis, discussion, or convincing argument. Also, national culture, in the context of technology transfer and utilization, could operationally be defined as the "collective mental programming of people's minds" (Hofstede, 1980a).

National culture affects not only the safety, but also the success and survival of any technology. National cultures differ on at least four basic dimensions: power distance, uncertainty avoidance, individualism-collectivism, and masculinity-femininity (Hofstede, 1980b).

Power distance refers to the extent to which a society accepts the fact that power in institutions and organizations is distributed unequally. It is an indication of the interpersonal power or influence between two entities, as perceived by the least powerful of the two [Boeing Commercial Aircraft Group (BCAG) (1993)].

Uncertainty avoidance concerns the extent to which a society feels threatened by uncertain and ambiguous situations. It also refers to attempts to avoid these situations by providing greater career stability, establishing more formal rules, not tolerating deviant ideas and behaviors, and believing in absolute truths and the attainment of expertise.

Individualism refers to a "loosely knit" social framework in which people are supposed to take care of themselves and their immediate families only, while *collectivism* is characterized by a tight social framework in which people distinguish between in-group and out-group; they expect their in-group members (e.g., relatives, clan, organization) to look after them, and in exchange, they owe absolute loyalty to the group.

The *masculinity* dimension expresses the extent to which the dominant values in a society are "masculine," as evidenced by decisiveness, interpersonal directness, and machismo (Johnston, 1993). Other characteristics of masculine cultures include assertiveness, the acquisition of money and material goods, and a relative lack of empathy and reduced perceived importance for quality-of-life issues. This dimension can also be described as a measure of the need for ostentatious manliness in the society (Boeing Commercial Aircraft Group, 1993). *Femininity*, the opposite pole of this continuum, represents relatively lower assertiveness and greater empathy and concern for issues regarding the quality of life.

The four cultural dimensions discussed above also have significant implications for the performance, reliability, and safety of most complex technological systems. For instance, according to Helmreich (1994) and Sherman, Helmreich, and Merritt (1997), there is evidence that operators with high power distance and high uncertainty avoidance prefer and place a "very high importance" on automation. Furthermore, it is known that the primary purpose of regulations is to standardize, systematize, and impersonalize operations. This is done, to a large extent, by ensuring adherence to standard and emergency operating procedures. On many occasions it requires replacing operators' habits with desirable intentions that are prescribed in procedures or enforced by regulations. However, according to several studies, an operator's culturally driven habit is a more potent predictor of behavior than his or her intentions; and there could be occasions on which intentions cease to have an effect on operators' behavior (Landis, Triandis & Adamopoulos, 1978). This fact questions the effectiveness of those regulations and procedures that are incompatible with operators' culturally driven habits.

CULTURAL FACTORS
AND AVIATION SAFETY

The International Civil Aviation Organization (ICAO), which is a specialized agency of the United Nations, has acknowledged the critical importance of cross-cultural issues in the aviation safety. According to Maurino (1994), ICAO experience supports the conclusion that the effectiveness of human factors training may be diminished—or even denied altogether—by the context within which such endeavors take place (Smith, 1992). The lessons from the ICAO Human Factors Programme have also shown that "safety deficiencies that could be addressed by human factors training in North America may not be effectively addressed *at all* by training in other regions of the world" [emphasis added] (Maurino, 1994, p. 174). Johnston (1993) and Merritt (1994) have suggested that North American approaches to crew resource management training may not be applicable in many cultures. According to Helmreich (1994a, p. 17), "this raises the important research question of how to measure significant cultural differences and how to adapt training to reflect them."

Cultural factors contributed significantly to the crash of an Avianca flight 052 (the airline of Columbia) Boeing 707 in Cove Neck, New York, on January 25, 1990. As a result of this accident, 73 of the 158 persons aboard were fatally injured. According to the National Transportation Safety Board (NTSB):

> The NTSB determines that the probable cause of this accident was the failure of the flight crew to adequately manage the airplane's fuel load, *and their failure to communicate an emergency fuel situation to air traffic control* before fuel exhaustion occurred. (emphasis added; NTSB, 1991, p. 76)

> The world *"priority"* was used in procedures' manuals provided by the Boeing Company to the airlines. A captain from Avianca Airlines testified that the use by the first officer of the word "priority," rather than *"emergency,"* may have resulted from training at Boeing. He stated that these personnel received the *impression* from the training that the words priority and emergency *conveyed the same meaning* to air traffic control. The controllers stated that, although they would do their utmost to assist a flight that requested "priority," *the word would not require a specific response* and that *if a pilot is in a low fuel emergency and needs emergency handling, he should use the word "emergency."* (emphasis added, p. 63)

The NTSB concluded:

> The first officer, who made all recorded radio transmissions in English, *never used the word "Emergency,"* even when he radioed that two engines had flamed out, and he did not use the appropriate phraseology published in United States aeronautical publications to communicate to air traffic control the flight's minimum fuel status. (emphasis added, p. 75)

Helmreich's (1994b) comprehensive analysis of the AV052 accident has thoroughly addressed the role of cultural factors. He contended that:

> Had air traffic controllers been aware of cultural norms that may influence crews from other cultures, they might have communicated more options and queried the crew more fully regarding the flight status . . . The possibility that behavior on this (flight) was dictated in part by norms of national culture cannot be dismissed. It seems likely that national culture may have contributed to (the crew's behavior and decision making) . . . Finally, mistaken cultural assumptions arising from the interaction of two vastly different national cultures (i.e., crew and ATC) may have prevented effective use of the air traffic control system. (parenthetical statements added; p. 282).

In Helmreich's (1994b) judgment, the important role the national culture of Avianca's cockpit crew played in that accident also should be considered:

> In a culture where group harmony is valued above individual needs, there was probably a tendency to remain silent while hoping that the captain would "save the day." There have been reported instances in other collectivist, high power distance cultures where *crews have chosen to die in a crash rather than disrupt group harmony and authority* and bring accompanying shame upon their family and in-group. (emphasis added, p. 282)

High uncertainty avoidance may have played a role (in this accident) by locking the crew into a course of action and preventing discussion of alternatives and review of the implications of the current course of action. High uncertainty avoidance is associated with a tendency to be inflexible once a decision has been made as a means of avoiding the discomfort associated with uncertainty.

The importance of cultural factors for automation in the aviation industry has been further highlighted by several important studies. Helmreich and Merritt (1998), in their study of national culture and flightdeck automation, surveyed 5705 pilots across 11 nations. Based on their results, these researchers noted that "the lack of consensus in automation attitudes, both within and between nations, is disturbing" (p. 14). They concluded that there is a need for clear explication of the philosophy governing the design of automation. Most recently, the U.S. Federal Aviation Administration Human Factors Study Team (1996) issued a report entitled *The Interfaces Between Flightcrews and Modern Flight Deck Systems.*

This team identified several "vulnerabilities" in flightcrew management of automation and situation awareness that are caused by a number of interrelated deficiencies in the current aviation system, such as "insufficient understanding and consideration of cultural differences in design, training, operations, and evaluation" (p. 4). The study team recommended a host of further studies, under the title of Cultural and Language Differences. The proposed studies included pilots' understanding of automation capabilities and limitations, differences in pilot decision regarding when and whether to use different automation capabilities, the effects of training, and the influence of organizational and national cultural background on decisions to use automation.

ETHNIC VARIABLES, CROSS-CULTURAL ISSUES, AND THE AVIATION COMMUNITY

Traditionally, more than in any other industry, the aviation community around the world has been sensitive to ethnic variables. [For a general review of "ethnic variables" in human factors, refer to Chapanis (1974, 1975).] However, until recently, attention was limited to considering physical, anthropometric, and physiological characteristics from among all of the ethnic variables of different user populations. One of the pioneering and noteworthy studies of ethnic factors and their effects on the safety and performance of different aviators was conducted by Pierce (1963). According to this work, Japanese pilots faced problems with fit when they had to use an American full-pressure suit—even the smallest-sized suits were too long for many of the Japanese pilots. This triggered an anthropometric survey of Japanese pilots involving measurements similar to those taken for the U.S. Air Force Survey. Differences in the amount of oxygen consumption of the pilots from different ethnic backgrounds were also studied and taken into account in cockpit design. It has traditionally been a requirement for American pilots to take supplemental oxygen when flying in an unpresserized plane above 10,000 feet. Peruvian pilots, on the other hand, who are natives to high lands, do not need the additional oxygen due to special development of their lungs and cardiovascular systems. They could fly in comfort at 20,000 feet without the use of the oxygen equipment (Pierce, 1963).

The Airplane Safety Engineering department of Boeing Commercial Airplane Group (BCAG) conducted an exhaustive analysis of hull loss accidents for Western-built commercial jet transport over 60,000 lb. (27,216 kg.) worldwide, for the period 1959 through 1992. According to this work (BCAG, 1993, p. 58), there was "strong correlation" between accident rate and two cultural indices (i.e., *individualism* and *power distance*). This fact led the BCAG to recommend that "this is an area that needs further analysis. The industry must continue its efforts to identify the cultural impacts on aviation safety" (p. 61).

Airline executives and airplane manufacturers have also acknowledged the importance of cultural variables (Graeber, 1994). Robert A. Davis (1993), Vice President of Engineering and Product Development of Boeing Commercial Airplane Group, in his keynote address at the 1993 Annual Meeting of the Human Factors and Ergonomics Society, entitled "Human Factors in the Global Marketplace," referred to "diverse cultures" as one of six "human factors challenges" facing his industry. He stated, "Successful penetration of future growth markets will require us to do better at taking into account the culture that our customers' employees and passengers bring to our airplanes. Furthermore, our designs must span a wide range of human abilities across these cultures" (p. 9).

IMPORTANT CULTURALLY BASED PARAMETERS AFFECTING ORGANIZATIONAL FUNCTIONING, TECHNOLOGY UTILIZATION, AND AVIATION SAFETY

The following issues and examples attempt to demonstrate some important culturally based behaviors affecting organizational functioning, technology utilization, and particularly, aviation safety (adapted from Kofler & Meshkati, 1987; Meshkati 1989, 1994, 1996):

- risk perception
- attitude toward work
- work group dynamics
- attitude toward technology
- attitude toward organization, hierarchy, procedure, and working habits
- attitude toward time and time of day
- religious duties and their effects on work
- achievement motivation and orientation
- population stereotype (e.g., color association)
- the "If it ain't broke, don't fix it" attitude

CONCLUSION

In summary, behavior analysis models and theories that are developed in one country reflect the cultural characteristics of that country and may not be fully applicable to other countries. Therefore, an organization that transfers technology to a country with a different culture should attempt to be fully adaptive to the host country's cultural dimensions (Demel & Meshkati, 1989).

Research has demonstrated that technology utilization, without the incorporation of the necessary human factor and cultural considerations, is doomed to failure (Meshkati & Robertson, 1986). It is incumbent on the world's airlines and aviation industry (aircraft and equipment manufacturers, air traffic controllers, and civil aviation authorities) to systematically take into account the physical and psychological factors, as well as the cultural attributes, of their user populations in the design and operation of passenger aircraft and aviation systems. Cultural factors have a significant effect on the realities of operating a complex technology such as modern aviation, and the nature of these effects must be understood and accommodated if aviation systems are to operate safely. As noted by the late Nobel physicist Richard Feynman (1986), in reference to the space shuttle Challenger explosion: "For a successful technology, reality must take precedence over public relations, for nature cannot be fooled."

17

Macroergonomic Root Causes of Large-Scale Accidents: Three Mile Island, Bhopal, Chernobyl

Najmedin Meshkati
University of Southern California

There is no such thing as an accident. What we call by that name is the effect of some cause which we do not see.

—Voltaire (1694–1778), in *Letters de Memmius*, III

INTRODUCTION

A common characteristic of large-scale technological facilities such as chemical processing plants, refineries, energy conversion and generation systems (e.g., nuclear, fossil fuel, thermoelectric power plants, gas processing facilities), off-shore rigs, and high-capacity compressor and pumping stations is that large amounts of potentially hazardous, flammable, combustible, or pressurized materials are concentrated and processed in single sites under the centralized control of a few operators. The effects of human error in these facilities are often neither observable nor reversible. Consequently, error recovery is either too late or impossible. Catastrophic breakdowns of these systems, created by manmade and natural causes, pose serious threats and long-lasting health and environmental consequences for workers in the facility, for the local public, and possibly for the neighboring region and the whole country. The accident at the Chernobyl nuclear power plant in the Soviet Union in 1986 attests to this. Chernobyl demonstrated, for the first time, that the effects of any such nuclear accident would not be localized, but rather would spill over into neighboring countries and have global consequences. The radioactive fallout resulting from Chernobyl was detected all over the world, from Finland to South Africa. Specifically, the Europeans, in addition to serious health concerns, have had to deal with significant economic losses and serious, long-lasting environmental consequences. This phenomenon has been described most succinctly as *a nuclear accident anywhere is a nuclear accident everywhere.*

For the foreseeable future, despite increasing levels of computerization and automation, human operators will remain in charge of the day-to-day control and monitoring of these systems. Thus, the safe and efficient operation of these sociotechnical systems is a function of the interactions among their *human* (i.e., personnel and organizational) and *engineered* subsystems.

It is essential to realize that the Bhopal accident should not be considered an isolated event that is unique to and could only happen in developing countries. The findings of an extensive comparative analysis presented at a UN-organized international conference demonstrated that with the present safety precautions, this accident could happen at any comparable plant in any developed country, e.g., the Federal Republic of Germany (Uth, 1988). It could also happen, as easily, in the United States—"there is no question that Bhopal would happen in the U.S. . . . You are dealing with such terribly dangerous chemicals that human failures or mechanical failures can be catastrophic. The potential is here and it could happen, maybe today, maybe 50 years from now" (cited in Weir, 1987, p. 123). Moreover, according to a 1985 report by the Congressional Research Service, about 75% of the U.S. population lives "in proximity to a chemical plant" (cited in Weir, 1987, p. 116). In fact, according to an expert with the Environmental Policy Institute, there have been "17 accidents in the U.S., where each of which released the Bhopal equivalent toxic gases . . . Only because the wind was blowing in the right direction, Bhopal did not happen here" (*USAToday*, August 2, 1989).

The underlying rationale and major objective of this chapter is to highlight and demonstrate the critical effects of micro- and macroergonomic factors in the safety of hazardous, large-scale sociotechnical systems. This is done by analyzing the three well-known accidents at such systems: Three Mile Island (TMI), Bhopal, and Chernobyl. Moreover, by integrating the common causes of these three accidents, a policy framework and/or guideline facilitating adherence to those identified, safety-ensuring factors is suggested.

GRAVE CONSEQUENCES OF HAZARDOUS, LARGE-SCALE SYSTEM ACCIDENTS

The aftermath of most hazardous, large-scale sociotechnical system accidents has serious and long-lasting health and environmental consequences. The physical and epidemiological consequences of the accident at the Three Mile Island nuclear power plant in March 28, 1979, are not fully known yet. However, according to many studies, residents in the vicinity of TMI exhibited elevated symptoms of stress (as measured by self-report, performance, and catecholamine levels) more than 1 year after the accident in 1979 (Baum, Gatchel & Schaeffer, 1983). Baum's (1988) study of TMI and other technological calamities, especially with the involvement of toxic substances, suggested that they cause more severe or longer-lasting mental and emotional problems than do natural disasters of similarly tragic magnitude. The technological accident-induced chronic stress and the associated emotional, hormonal (i.e., epinephrine, norepinephrine, and cortisol), and immunological changes may cause or exacerbate illness. Moreover, based on Davidson and Baum's (1986) findings, symptoms of post-traumatic stress syndrome persisted among the residents in the vicinity of TMI even as long as 5 years after the accident. The investigative work of Prince-Embury and Rooney (1988) following the restart of the TMI nuclear plant in 1985 also found that psychological symptoms of stress in the residents in the vicinity of the TMI "remained chronically elevated" (p. 779). Furthermore, it took only $700 million to build the plant, but to clean it up after the accident, 400 workers had to work for 4.5 years with a cost of $970 million (*New York Times*, April 24, 1990).

The leak of methyl isocyanate (MIC) at the Union Carbide pesticide plant in Bhopal, India, on December 4, 1984, resulted in the deaths of approximately 3,800 people and the injury of more than 200,000 (*New York Times*, September 12, 1990). And, on the average, 2 of the 200,000 people who were injured at the onset of this disaster die every day. Those who initially survived the gas "are continuing to suffer not only deterioration of their lungs, eyes and skin but also additional disorders that include ulcers, colitis, hysteria, neurosis and memory loss" (*Los Angeles Times*, March 13, 1989). Moreover, the MIC exposure has even affected the second generation; "mortality and abnormalities among children conceived and born long after the disaster to exposed mothers and fathers continue to be higher than among a selected control group of unexposed parents" (*Los Angeles Times*, March 13, 1989). Also, based on a recent report issued by the National Toxic Campaign Fund, on the basis of lab tests, several toxic substances that were found in the Bhopal environment are raising further questions about the additional long-term effects of the disaster (Jenkins, 1990).

Among these, three accidents, Chernobyl has had the most widespread effects. According to recent charges, "the Chernobyl accident released at least 20 times more radiation than the Soviet government has admitted" (*Time*, November 13, 1989, p. 62). The immediate (short-run) aftermath of the 1986 Chernobyl nuclear power plant accident in the Soviet Union included: 300 deaths (*New York Times*, April 27, 1990); a $12.8 billion cost of disruption to the Soviet economy (*New York Times*, October 13, 1989); twice the normal rate of birth defects among those living in the vicinity of the plant (*Los Angeles Times*, March 27, 1989); thyroid glands of more that 150,000 people were "seriously affected" by doses of radioactive iodine; rates of thyroid cancer are 5 to 10 times higher for the 1.5 million people living in the affected areas; leukemia rates among children in some areas of the Ukraine are two to four times normal level, one or two children a week in a Minsk Hospital are dying of leukemia compared to one or two a year before the accident; the death rate for people who have been working at the Chernobyl plant since the accident is 10 times what it was before the accident (*New York Times*, April 28, 1990); one-fifth (1/5) of the republic of Byelorussia's more than 10 million people have had to be moved from areas contaminated by radiation, including 27 cities and more than 2,600 villages (*Los Angeles Times*, June 20, 1990); $26 billion is allotted for the resettlement of the 200,000 people still living in the irradiated areas (*New York Times*, April 26, 1990); evacuation of people in all the contaminated areas is estimated to cost $70 billion (*Los Angeles Times*, June 20, 1990); and it will take up to 200 years to "totally wipe out" the effects of the accident in the affected areas (*Los Angeles Times*, June 20, 1990).

The other grave, long-term environmental consequences of this accident are yet to be realized. For instance, due to Chernobyl's contamination, about 2 million acres of land in Byelorussia and the Ukraine cannot be exploited normally and Byelorussia has lost 20% of its farmland (*Insight*, January 15, 1990). In March 1990, these facts led the Ukrainian Supreme Soviet, the Ukraine republic's legislature, to order the government to close the remaining reactors at Chernobyl permanently and to "decide on the stoppage of further development of additional nuclear power stations in Ukraine" (*Los Angeles Times*, March 3, 1990). The Ukrainian decision follows a similar move last year in the southern republic of Armenia to close an atomic energy plant.

Moreover, most of the injuries inflicted on the people in such accidents, such as exposure to radiation and toxic gases, are very difficult and sometimes even impossible to cure. Most of the people exposed to high doses of radioactive material at Chernobyl have died of cancer. Even exotic medical techniques such as bone marrow transplants have not been effective in saving their lives (Gale & Hauser, 1988).

MAJOR MICRO- AND MACROERGONOMIC CAUSES OF THE THREE MILE ISLAND ACCIDENT

The Three Mile Island nuclear power plant accident is the most investigated accident in the history of the commercial nuclear industry. The title of the Comptroller General Report to the Congress is "Three Mile Island: The Most Studied Nuclear Accident in History." The following, however, is a summary account of only the most critical human factor causes of this accident [for further information and detailed analysis, see Kemeny (1980), Perrow (1981, 1984), and Rogovin (1980)].

The lack of human factor considerations at the design stage was most evident in TMI's control room. It was poorly designed with problems including: controls located far from instrument displays that showed the condition of the system; cumbersome and inconsistent instruments that often looked identical and were placed side-by-side but controlled widely differing functions; instrument readings that were difficult to read, obscured by glare or poor lighting or actually hidden from the operators (many key indicators were located on the back wall of the control room and many of these indicators were faulty or misleading); contradictory systems of lights, levers, or knobs—lever up may have closed a valve, while pulling another lever down may have closed another. For instance, in the case of the pilot-operated-relief-valve (PORV; a pressure relief valve to release water from the core), when it stuck open, its indicator did not show whether the valve was actually open or closed. The indicator only showed the position of the operating switch that was supposed to open or close it. Moreover, there was no direct way or any designated indicator to read the exact water level in the reactor core. This was partly responsible for one of the most significant errors by the operators: they cut back and failed to maintain the high-pressure injection (HPI) system. Furthermore, the HPI throttle valves were operated from a front panel, while the HPI flow indicator was on a back panel and could not be read from the throttle valve operating position.

In the control room of TMI, there were three audible alarms sounding and more than 1,600 lights blinking at the time of accident. The TMI operators had to literally turn off alarms and shut down the warning lights (Perrow, 1981).

The lack of proper operator training in general, and "stress training" in particular, was a major contributor to the TMI accident. It was a critical human factor consideration that should have been paid attention to at the design stage of TMI. The Comptroller General's (1980) report to the U.S. Congress concluded: "Most training, including simulator training, was geared toward preparing the operator to run the plant during routine situations, instead of understanding or coping with the unexpected [emergencies]" (p. 25). According to Rogovin (1980), "other than being required to memorize a few emergency procedures, [TMI] reactor operators are not extensively trained to diagnose and cope with the unexpected equipment malfunction, serious transients (temporary electrical oscillations), events that cannot be easily understood." In other words, the TMI operators were only trained to handle the discrete events and not to deal with "multiple-failure" accidents. These accidents were not simulated in the training (Perrow, 1981).

Another lack of human factor consideration at the design stage, the organizational factors, played a major role in TMI accident. Some typical (generic) problems were due to the hierarchical organizational structure, such as problems of mismatches in the response times at the different levels in the hierarchy, and of information overload (cf., Meshkati, 1991). In fact, according to Perrow (1984), "the dangerous accidents lie in the system, not in the components, and the inherent system accident potential can *increase* in a poorly-designed and managed organization" (p. 351). One of the recommendations of the President's Commission (1979) investigating TMI accident was that "to prevent nuclear accidents as

serious as TMI, fundamental changes will be necessary in the organization, procedure and practice." I would add that a macroergonomic analysis of nuclear work
systems' structures and processes, such as described in Chapters 3 and 4, should
be conducted to help determine what these fundamental changes should be.

Due to enormous human factor problems of TMI at the design stage, the operating stage's problems may seem insignificant. However, there were numerous
instances of misjudgment by the operators. Finally, as also referred to by Senders
(1980), the Kemeny report (1980) made it plain that the causes of the TMI accident went far beyond the errors made by a few front-line operators:

> while the major factor . . . was inappropriate operator action, many factors con-
> tributed to the action of the operators, such as deficiencies in their training, lack of
> clarity in their operating procedures, failure of organizations to learn proper lessons
> from previous incidents, and deficiencies in the design of the control room. These
> shortcomings are attributed to the utility [the Metropolitan Edison Company], to
> suppliers of equipment, and to the federal commission that regulates nuclear power.

MAJOR MICRO- AND MACROERGONOMIC
CAUSES OF THE BHOPAL ACCIDENT

The lack of human factor considerations, both at the design and operating stages,
played a very significant role in the accident at the Bhopal plant. The overall
design and safety of the plant's control room had many inherent human factor
problems. The room was not, in fact, designed plausibly at all, based on a
thorough operators' task analysis. A highly critical pressure gauge that should
have indicated buildup of MIC pressure around the relief valve was missing from
the control room. It was located close to the valve's assembly, somewhere in the
plant's site, and was supposed to be monitored manually with no link to the control room or warning system (*Chemical and Engineering News*, February 11,
1985). Furthermore, according to *Time* magazine's special report on the Bhopal
accident (December 17, 1984, p. 25), "an important panel in the control room
had been removed, perhaps for maintenance, thus preventing the leak from
showing up on monitors." Moreover, approximately half an hour after the start of
the gas leak, MIC began to engulf the control room. Many of the operators, not
having oxygen masks, could neither see nor breathe and had to run from their
workstations. This happened at the most critical time of the plant's life, when
they were needed most. More specifically, the problems with the visual displays—"gauges" and "meters"—at the Bhopal plant were addressed by almost
all the studies. The gauges were consistently either broken, malfunctioning, off
the scale, giving wrong data, or considered "totally unreliable."

According to the International Confederation of Free Trade Union (ICFTU)
report (1985), "broken gauges made it hard for the MIC operators to understand

what was happening. In particular, the pressure indicator/control, temperature indicator and the level indicator for the MIC storage tanks had been malfunctioning for more than a year" (p. 9). Having broken gauges "(was) not unusual at the factory," according to the Bhopal plant operators (*New York Times*, January 30, 1985).

Based on the *Chemical and Engineering and News* analysis (February 11, 1985, p. 31), the pressure meters monitoring the leak from the MIC storage tank (number 610) gave abnormally low pressure readings. A pressure of 20 psi was given as 2 psi. About 2 hours before the MIC leak, the pressure gauge had risen from 3 to 10 psi. At first, operators thought that the pressure gauge was faulty, as was often the case at the Bhopal plant. Half an hour later, according to Agarwal, Merrifield, and Tandon (1985), operators started to detect an MIC leak; not through any gauge or sensor, but because their eyes started to tear. Also, the volume indicator for tank number 619, which was supposed to be empty and usable as a spare, "incorrectly read that it was 22% full" (Everest, 1986, p. 28). Furthermore, the temperature gauge of the MIC storage tanks, which sometimes in the summer went off the scale, could not be relied on (*New York Times*, January 30, 1985). Moreover, the increase in temperature (of the MIC tank), according to the Union Carbide investigation (1985), was not signaled by the tank high-temperature alarm, because it had not been reset to a temperature above the storage temperature.

Of the two major safety devices at the Bhopal plant, one was the vent-gas scrubber, a system that was designed to pour caustic soda on the MIC so that it would decompose. The other device was a flare tower that would ignite the gas and burn it off in the air harmlessly. (This system was not operational because of a missing piece of pipe and other maintenance problems.) According to an engineering analysis conducted by Naschi (1987), the scrubber unit was not turned on until after the situation had gone out of control. Furthermore, a flow meter also "failed to indicate" that a flow of caustic soda had started (MacKenzie, 1985). The latter problem could have contributed to the overall failure of the vent-gas scrubber's functional performance because, according to Bowonder, Kasperson, and Kasperson (1985), operators, not having the correct and required information, neglected to augment the flow of caustic soda required to neutralize the MIC.

In the MIC control room of the Bhopal plant at the time of the accident, the operators were extremely overloaded and found it "virtually impossible to look after the 70-odd panels, indicators and console and keep a check on all the relevant parameters" (Agarwal et al., 1985, p. 8). The foregoing catalogue of problems may have led Krishnan (1987) to conclude that "the Bhopal tragedy was a pathetic example of how careless display control design can end up in a catastrophe" (p. 5).

The Bhopal plant's rigid organizational structure, according to Kleindorfer and Kunreuther (1987), was one of the three primary causes of the accidents. Moreover, the Bhopal plant was plagued by labor relations and internal management disputes (ICFTU, 1985). For a period of 15 years prior to the accident, the plant had been run by eight different managers (Shrivastava, 1987). Many of them came from different backgrounds with little or no relevant experience. The last managers, who served the plant at the time of accident, were originally transferred from a UCIL's battery plant. This group was not fully trained about the hazards and appropriate operating procedures for the pesticide plant (ICFTU, 1985).

The discontinuity of the plant management, its authoritative and sometimes manipulative managerial styles, and the nonadaptive and unresponsive organizational system, collectively, contributed to the accident. The latter element, i.e., organizational rigidity, was primarily responsible for not responding and taking the necessary and corrective course of actions to deal with the five reported major accidents occurring at the plant between 1981 and 1984. This leads one to conclude that the catastrophic accident of 1984 was only the inevitable and natural byproduct of this symptomatic behavior. This is in accordance with Mitroff's thesis (1988) that "crises often occur because warning signals were not attended to" (p. 18). Moreover, the Bhopal plant's organizational culture also should be held responsible for not heeding the many operator warnings regarding safety problems, such as the one after the October 1982 combined release of MIC, hydrochloric acid, and chloroform, which spread into the surrounding community. Bhopal's monolithic organizational culture, as the plant's operational milieu, only fostered the centralization of decision making by rules and regulations or by standardization and hierarchy, both of which required high control and surveillance. This was diametrically different from Weick's (1987) contention that characterized organizational culture as the "source of reliability," and suggested that high system reliability could only be achieved by simultaneous centralization and decentralization (which allows and encourages operators' discretion and input).

According to ICFTU (1985), training was a major problem at the Bhopal plant and many operators had been given little or no training about the safety and health hazards of the MIC or other toxic substances in the plant. Language also may have contributed to the lack of understanding about MIC operations and hazards. All signs regarding operating and safety procedures were written in English, even though many of the operators spoke only Hindi.

In addition to emergency training, the lack of human factor considerations at the operating stage at the Bhopal plant, for instance, was reflected in the task-related training. The concern that operators did not have adequate task-related training was also raised by Union Carbide's safety audit of 1982. This, of course, was partly due to a high turnover rate. Many key personnel were being released for independent operation without having gained sufficient understanding of safe operating procedures. There also was concern that the training relied too much on "rote memorization of steps" instead of an "understanding of the reasoning behind procedures" (*Chemical and Engineering News*, February 11, 1985). [This is the so-called carbon copy of the training problems at TMI.]

MAJOR MICRO- AND MACROERGONOMIC
CAUSES OF THE CHERNOBYL ACCIDENT

The critical role of micro- and macroergonomic-related causes in the Chernobyl accident is captured in the following statements and addressed, either directly or indirectly, by a few other studies [Chernousenko (1992), Grimston (1997), Leveson (1995), Medvedev (1991), Meshkati (1991), and Read (1993)]. The late academician, Dr. Valeri A. Legasov, the First Deputy Director of the Kurchatov Institute at the time of the Chernobyl accident, and the head of the former Soviet delegation to the Post-Accident Review Meeting of the IAEA in August, 1986 [quoted in Munipov (1992), p. 340], stated:

> I advocate the respect for human engineering and sound manmachine interaction. This is a lesson that Chernobyl taught us.

The International Atomic Energy Agency's (IAEA) *Nuclear Safety Review for 1987* (p. 43) said:

> The Chernobyl accident illustrated the critical contribution of the human factor in nuclear safety.

IAEA's *Summary Report on the Post-Accident Review Meeting on the Chernobyl Accident* (INSAG-1, 1986, p. 76) noted:

> The root cause of the Chernobyl accident, it is concluded, is to be found in the so-called human element. . . . The lessons drawn from the Chernobyl accident are valuable for all reactor types.

The IAEA's International Nuclear Safety Advisory Group (INSAG) in *The Chernobyl Accident Updating of INSAG-1* (INSAG-7, 1992, p. 24) stated:

> The (Chernobyl) accident can be said to have flowed from deficient safety culture, not only at the Chernobyl plant, but throughout the Soviet design, operating and regulatory organizations for nuclear power that existed at the time . . . Safety culture . . . requires total dedication, which at nuclear power plants is primarily generated by the attitudes of managers of organizations involved in their development and operation.

Mr. Mikhail Gorbachev, General Secretary of the Central Committee of the Communist Party and President of the former Soviet Union, in the wake of the Chernobyl accident [quoted in Illesh (1987), p. 177] said:

> For us, the indisputable lesson of Chernobyl lies in this: the principles regulating the further development of the scientific-technological revolution must be safety, discipline, order, and organization. Everywhere and in all respects, we must operate according to the strictest standards.

Lack of human factor considerations at the design stage is one the primary causes of the Chernobyl accident. Attesting to this is Legasov's statement that one of the "defects of the system was that the designers did not foresee the awkward and silly actions by the operators" (cited in Oberg, 1988, p. 256). He also attributed the accident's cause to "human error and problems with the man–machine interface" (cited in Wilson, 1987, p. 1639). Also, it has been reported that the Chernobyl accident happened because of: (1) faults in the concept of the reactor (inherent safety not built-in); (2) faults in the engineering implementation of that concept (insufficient safeguard systems); and (3) failure to understand the man–machine interface ("a colossal psychological mistake") (in the words of Legasov) [cited in a report by the United Kingdom Atomic Energy Authority, (1988), p. 5.47].

Additional findings of these report indicated that (1) the shutdown system was, in the event of the accident inadequate, and might in fact have exacerbated the accident rather than terminated it; (2) there were no physical controls to prevent the staff from operating the reactor in its unstable regime or with safeguard systems seriously disabled or degraded, (3) there were no fire-drills, and (4) there were no adequate instrumentation and alarms to warn and alert the operators of the danger.

Although a thorough analysis of the design and operations of Chernobyl's control room is not available, according to a published report in the prestigious scientific journal *Nature* entitled "Coping with the Human Factor" (4 September 1986), "the planning of tests (at Chernobyl) seems to have been in tune with the general sloppiness of the operation of the control room at the end of April" [(p. 25); emphasis added].

The lack of proper training, as well as deficiencies in the qualifications of operating personnel, was considered as another contributing factor to this accident by all investigations, including the IAEA's *Nuclear Safety Review for 1987*. The quality of training and retraining personnel was also, however implicitly, acknowledged as a critical factor by the *Soviet Report on the Chernobyl Accident* (1986).

Managerial and organizational factors also contributed heavily to the catastrophic events at Chernobyl. Apart from the design errors, the other cause of the Chernobyl accident in Wilson's (1987) analysis was management error: "there were important admissions of management errors, as distinct from operator error" (p. 1639). Also, it has been reported that there were deficiencies in the plant organization and management (*Nuclear Safety Review for 1989*). The principal "managers" who ran and conducted the test at Chernobyl that caused the accident "were electrical engineers from Moscow. The man in charge, an electrical engineer, was not a specialist in reactor plants" (cited in Reason (1990), p. 144). "Neither the station's managers nor the Ministry of Power and Electrification's leadership had any concept of the necessary actions . . . There was a noticeable confusion even on minor matters" (*Pravda*, May 20, 1986).

According to the IAEA's *Summary Report on the Post-Accident Review Meeting on the Chernobyl Accident (1986)*, one of the main contributing factors to the Chernobyl accident was the potential misunderstanding of the physics characteristics of the reactor by the operators. This fact is corroborated by some comments like, "the staff was insufficiently familiar with the special features of the technological processes in a nuclear reactor. They had also lost any feeling for the hazards involved" (cited in Reason, 1990, p. 144). In response to these shortcomings, the IAEA report recommended that careful attention must be paid to the design of safety and control systems to enable the operators in the control room to "understand" the encountered problems and also to lead them to take the proper course of action(s).

The lack of human factor considerations at the operating stage was highlighted by operator error, which also was identified as one of the major causes of the Chernobyl accident. According to an official report prepared by a team of Soviet investigators, an extraordinary sequence of human errors turned some weaknesses in the reactor's design into deadly flaws. Ramberg (1987), in analyzing the causes and implications of the Chernobyl accident, concluded that it had resulted from "gross operator incompetence—not entirely unlike that which resulted in the accident in 1979 at TMI" (p. 307). Six important safety devices were "deliberately" disconnected on the night of 25 April (Wilson, 1987, p. 1639); the most important of which, the emergency core cooling system (ECCS), was made inoperative. And the reactor was deliberately and improperly run below 20% power.

Finally, a report by the IAEA summarizes the lessons learned (primarily) from the Chernobyl accident as "the root causes of most safety significant events were found to be deficiencies in: plant organization and management; the feedback of operational experience, training and qualification, quality assurance in the maintenance and procedures, and the scope of the corrective actions" (*Nuclear Safety Review for 1989*, p. D61).

COMMONALITIES OF MICRO- AND MACROERGONOMIC PROBLEMS IN LARGE-SCALE SYSTEM ACCIDENTS

The comparison of TMI, Bhopal, and Chernobyl is not unprecedented. In the case of the former two, many authoritative analogies have already been made. In 1984, the President of the World Resources Institute, James Speth (1984), in his statement at the hearing on the "Implications of the Industrial Disaster in Bhopal" before the Subcommittee on Asian and Pacific Affairs of the U.S. House of Representatives, argued that "it is likely that Bhopal will become the chemical industry's Three Mile Island, an international symbol deeply imprinted on public consciousness."

Regardless of the nature of the utilized technology, there are striking similarities and commonalities among the nature and magnitude of the causes of complex, large-scale system failures such as Three Mile Island (TMI), Bhopal, and Chernobyl. Furthermore, it would not be spurious to state that the causes of these accidents are reminiscent of the causes of another past nuclear power plant accident—the accident on January 1961 at the SL1 (Stationary Low Power Reactor No. 1), located at the National Reactor Testing Station, Idaho Falls, Idaho. A quotation from the general conclusions as to the causes of this accident could, as well and almost exactly, be applied to the TMI and Chernobyl cases [as such, one may argue that should it be heeded, these accidents could have been prevented]:

> Most accidents involve design errors, instrumentation errors, and operator or supervisor errors . . . The SL1 accident is an object lesson on all of these . . . There has been much discussion of this accident, its causes, and its lessons, but little attention has been paid to the human aspects of its causes . . . There is a tendency to look only at what happened, to point out deficiencies in the system without understanding why they happened; why certain decisions were made as they were . . . Post-accident reviews should consider the situation and the pressures on personnel which existed before the accident. (Thompson, 1964, p. 681)

CONCLUSIONS AND RECOMMENDATIONS

As demonstrated by Shrivastava, Mitroff, Miller, and Milan (1988), technological system crises (e.g., accidents) are caused by two sets of failures (and their interactions): (1) failure in the system's components (or subsystems) and their interactions; and (2) failure in the system's environmental factors. The former refers to a complex set of human, organizational, and technological (HOT) factors (and their interactions) that lead to the triggering event for the accident. The latter, according the authors, includes regulatory, infrastructure, and preparedness (RIP) failures in the systems' environments. Although RIP is equally important, the emphasis of this work was on HOT factors. As such, the following conclusions and recommendations address only the HOT-related issues.

As also suggested by Shrivastava, Mitroff, Miller, and Milan (1988), technological "organizations are simultaneously systems of production and of destruction" (p. 297). This fact becomes even more critical for the hazardous large-scale systems, such as the ones discussed in this work. These are risky systems, and risky systems are full of failures. Inevitably, these failures will interact in unexpected ways, defeat the system's safety devices, and bring down the system. This is what Perrow (1984) has called a "normal accident." Using Perrow's characterization of these types of industrial accidents, the Bhopal catastrophe, as well as TMI and Chernobyl, could each well be called a "normal accident." Normal in the sense that the accident emerged from the inherent characteristics of the respective system itself and, because of the existing serious micro- and macroergonomic problems at both the design and operating stages, it could have been neither prevented nor avoided.

Many scholars, such as Goldman (1986), Oberg (1988), Wilson (1987), and particularly, Munipov (1992) have implicated the pre-Glasnost Soviet secrecy and the ignorance of TMI's lessons as root causes of the Chernobyl accident. TMI, Bhopal, Chernobyl, previously mentioned SL1, and numerous other accidents will always remind us of George Santayana's dictum that *those who ignore history are forced to relive it.* The continued operation of hazardous systems with secrecy, complacency, or ignorance; not heeding the occasional warnings (incidents); and without change to a proactive, integrated, and total systems approach to design, operations, safety control, and risk management in complex sociotechnical systems, will force us to relive horrors like Chernobyl and tragedies like Bhopal. These accidents were not isolated cases, but were only manifestations—the tip of the iceberg—of the negative effects resulting from the unfortunate and common-practice lack of human factor considerations in the design and operation of major industrial facilities and process plants throughout the world. No matter what the nature or level of the technology, and regardless of the plant's location—in industrialized or developing countries—the human factor-related issues are still universally important. Their absence always causes inefficiencies, problems, accidents, and the loss of property and lives. These and many other past major industrial accidents could have been prevented if the critical issue of complex, large-scale technology utilization was not plagued by sheer political, economic, bureaucratic, and/or technical tunnel vision. In fact, Perrow (1986b) contends that catastrophes are possible when community and regional interests are not mobilized or when they are overridden by national policy, and when "supraorganizational goals," such as the economic health of an industry, are deemed vital.

In the light of the discussion presented above, the following is recommended for the *design stage*: In order to ensure the relative safety of future large-scale sociotechnical systems—such as chemical processing plants, nuclear power plants, and refineries—a holistic, totally integrated, and multidisciplinary approach to system design, construction, staffing, and operation based on sound scientific studies and human factor guidelines is recommended. The total system design (TSD) constitutes such an approach. The TSD, according to Bailey (1989), is a developmental approach that is based on a series of clearly defined development stages. TSD, which has been used extensively for computer-based systems development at AT&T Bell Laboratories, implies that, from the beginning and the inception of the plan, equal and adequate consideration should be given to all major system components (i.e., human, organizational, and technological). [The system development process, therefore, is partitioned into a series of meaningfully related groups of activities called *stages*, each of which contains a set of design and accompanying human factor activities.]

Moreover, as was demonstrated by the TMI, in addition to independent and isolated problems at the workstation (interface), job (task), and organizational (communication) levels, there was a serious lack of cohesive processes of data collection, integration, and coordination. Logically, information is gathered from the interfaces (at the *workstation* site) and analyzed according to the operators' stipulated job descriptions (at the *job* level), and passed through the organizational communication network (according to the *organizational* structure) to the appropriate team members responsible for decision making. Thus, this continuous process in the control room of large-scale technological systems needs: (1) a cohesive and integrated framework for information gathering from the interfaces (at the *workstation* site), (2) analysis according to the operators' stipulated job descriptions (at the *job* level), and (3) its passage through the organizational communication network (according to the *organizational* structure) to the appropriate team members responsible for decision making (Meshkati, 1991).

The early participation of all related and needed disciplines, e.g., human factors, in system design and development also is strongly recommended. This mandates and encourages interdisciplinary dialogue among engineers, managers, human factor and safety specialists. [The need for such a multidisciplinary approach to nuclear safety has also been emphasized in all of the articles in the Special Section of the *Nuclear News*, (The publication of the American Nuclear Society.) on human factors (June 1990).]

At the *operating stage* and in the short-run, efforts should immediately be initiated for the close examination of human operators' physical and psychological needs, capabilities, and limitations in the context of the plant's normal and emergency operation. It should also be coupled with thorough analyses of critical workstations and their design features, job demands, and operators' mental workload (during normal as well as emergency situations), emergency response system, organizational characteristics, training needs, supervisory systems, etc.

The above is a long overdue action, however, and constitutes only a *necessary* step toward ensuring the safety of complex, large-scale systems. To make it *sufficient*, in the long run, we need much more commitment, communication, and cooperation among those who could make these systems safer—the government and regulatory agencies, plant manufacturers and managers, unions, and the human factors and other concerned research communities. We need an overall paradigm shift in dealing with complex technologies' safety and operation. We need more institutionalized interaction among all stakeholders in the public and private sectors. We should learn from lessons of past accidents and should also systematically integrate micro- and macroergonomic considerations into the design and operation stages of our complex, large-scale technological systems.

The lessons of Chernobyl should guide us to devise sound strategies to prevent nuclear accidents in the future. However, the lessons of Chernobyl are reminiscent of another tragic accident—Bhopal. These two accidents have taught humanity enough lessons: Consider the critical role of micro- and macroergonomic factors in the safety of complex, large-scale sociotechnical systems. The late Dr. Valery Legasov has eloquently addressed this fact in the following way [(from Ignatenko, Voznyak, Ya, Kopvaleuko, and Troitskii (1989), cited in Grimston (1997)]:

Naturally, reactor design engineers studied all the accidents which have occurred at nuclear power stations and have, if necessary, adopted additional safety measures. But unfortunately they did not study accidents in other branches of industry. The train of events at Chernobyl NPS, which led to the tragedy, was in no way reminiscent of even one of the emergency situations at other nuclear power stations, but was very, very similar, right down to the last details, to what happened at the chemical works at Bhopal in 1984.

Right down to the last detail. The Chernobyl accident occurred in the night from Friday to Saturday. The accident in India happened early on a Sunday morning. At Chernobyl they switched off the emergency protection; in India they switched off the coolers and absorber which perform a protective function. In India there was a technical fault involving a gate valve, and passage of water resulting in an exothermic reaction, which developed exponentially, with the coolers switched off, whilst here there was an excess of steam and a rise in reactivity. The main thing was that both in India and here, the staff had been able (in spite of this being strictly forbidden) to switch off the protective devices.

If the reactor designers had drawn some conclusions from the Bhopal accident . . . but what use is there in talking about it now? To be fair I would just like to say that it was precisely after Bhopal that chemists knocked on "reactor doors," but such words as "methyl isocyanate," "oxidation," and "chemical reactions" made the problem uninteresting for physicists.

The lesson of Bhopal went unheeded . . .

18

A Vision of the Future of Macroergonomics

Klaus J. Zink

University of Kaiserslautern (Germany)

INTRODUCTION

Macroergonomics is one of the younger "disciplines" of ergonomics. One therefore might ask, why talk about the future? At least two reasons immediately come to mind. First, we are living in a rapidly changing world, and actual definitions or concepts might be obsolete tomorrow. Second, ergonomics as a whole has a lot of challenges to compete with—and macroergonomics might be influenced from this development as well. To learn for the future, it always is worthwhile to analyze the state of the art, based on the past.

RECENT DEFINITIONS OF MACROERGONOMICS— AND SOME REMARKS

In this book—and in another recently published book of the U.S. Human Factors and Ergonomics Society—Hal Hendrick (as "father" of macroergonomics) and Brian Kleiner describe the history and state of the art of macroergonomics.

To create no doubts: All of these ideas brought huge progress to ergonomic thinking—and the chance to deploy theory to practice—and we still have a lot to do to bring these ideas into a much greater number of organizations. There may well be some cultural differences—e.g., in the way microergonomics provides an opportunity for macroergonomic projects [as described by Hendrick and Kleiner (2001) for the United States, or vice versa as stated by Zink (2000) for Germany], but the enrichment for ergonomic approaches is obvious. Though the process of describing a macroergonomic theory has not come to an end, a look to the future should be allowed.

The discipline of human factors/ergonomics can be understood as *human–system interface design*, and the subdiscipline of macroergonomics as *human-organization interface design* (based on a top-down sociotechnical approach). For example, the International Ergonomics Association (IEA) has recently redefined its understanding of ergonomics (or human factors) as *the scientific discipline concerned with the understanding of the interactions among humans and other elements of a system, and the profession that applies theoretical principles, data, and methods to design in order to optimize well being and overall system performance* (IEA, 2000). In developing this revised definition of the discipline, the IEA included "Organizational Ergonomics" as one of the domains of specialization: *Organizational Ergonomics is concerned with the optimization of sociotechnical systems, including their organizational structure, policies, and processes. The relevant topics include communication, crew resource management, work design, design of working times, teamwork, participatory design, community ergonomics, cooperative work, new work paradigms, organizational culture, virtual organizations, telework, and quality management* (IEA, 2000). In addition—but some time before—this same scientific society defined the core competencies for ergonomists with a very broad scope (see IEA, 1999). Therefore, one has first to ask whether the definition of "macroergonomics" is comprehensive enough in this context.

Looking at some other "ergonomic cultures" like Germany, one finds that (macro-)ergonomics is also related to industrial policy or society as a whole (see Luczak, 1993). This also appears to be reflected in some U.S. concepts like "community ergonomics" (e.g., see Chapter 14).

Seeing the sociotechnical systems approach as one of the core elements of macroergonomics, one has to bear in mind that there has been some criticism of this approach in recent years (see Sydow, 1985). Though defining sociotechnical systems as open systems, the primary focus was related to the organization in its social and technical frame conditions as influenced by the environment (e.g., society)—but *influencing* society, or more generally, the environment itself, was not included in this definition. Another aspect, also discussed by Hendrick (1996), is the relevance of economic factors in the design process. Rühl (1975) and later Zink (1984), therefore, started in the 1970s to redefine sociotechnical systems as *sociotechnological* systems. Understanding technology as the way to realize technique, organizational and economical aspects are thus included in such an understanding. But this definition does not include criteria like ecology or other "outside" parties like customers.

Another discussion in recent years focuses on complexity. We all are recognizing that traditional problem-solving approaches based on the knowledge of one discipline are no longer sufficient. Developments like Total Quality Management or so-called Excellence Models and the Balanced Score Card are the result of the necessity of broader, more holistic concepts. Behind most of these concepts we find some sort of stakeholder orientation (see Zink, 1999a).

In recent years, information technology has been dramatically influencing the way we work. Most recently there has been a growing number of teleworkers ("freelancers") and so-called "virtual organizations." More and more people work "any time and any place." How does this refer to the structural dimensions of work systems, like differentiation and integration, formalization, and centralization?

In recent years, we have had many discussions in Western societies concerning the future of work and the work of the future. According to the Club of Rome's recent book (see Giarini & Liedke, 1997), work of the future will be quite different—and perhaps only one third can be compared with our traditional understanding of work in an industrial society. The English philosopher Charles Handy (see Handy, 1994) talks about "patchwork careers." How are these anticipated changes reflected in our picture about the future of ergonomics? And what does it mean for macroergonomics?

Summing up this admittedly incomplete enumeration of possible future changes, one can state that there should be influences from them—both for ergonomics and macroergonomics. Let us first look at ergonomics in general.

THE FUTURE OF WORK
AND THE FUTURE OF ERGONOMICS

If we refer to most of the predictions of changes in work and the working environment, we have many challenges to meet. One of the challenges is included in the vision statements of some human factors and ergonomic societies (e.g., the U.S. and German societies): the improvement of living conditions. The same is true for the International Ergonomics Association.

What does this mean in shareholder-value-oriented companies? Laying off thousands of employees to improve that shareholder value? What does this mean for societies with a decreasing workforce in traditional fields—and patchwork careers? What does this mean for the definition of work at all? How can we also improve living conditions for societies facing, for example, unemployment? This is not the place to answer all these questions—but we have to clarify how ergonomics could help solve some of these problems. As stated above, the complexity of problems does not allow solutions from a single discipline—and the uniqueness of ergonomics is given by its multidisciplinary approach. There is no other science bringing together so many other disciplines to create a holistic problem solving approach (see also the definition of core competencies of the IEA). But not only the integration of different disciplines but also multidimensional target formulation (at least meeting human and economic goals) defines the specific character of this discipline. With these preconditions ergonomics is destined to contribute to problem solving in much more than the "traditional" fields. As president of Gesellschaft für Arbeitswissenschaft (GfA—the German Ergonomics and Human Factors Society), I started and coached a task force dealing with the problems of the future of the "working society." As one result, we wrote a memorandum about topics of research. The following themes were included (see GfA, 2000):

1. Securing employment by
   1.1 designing innovative products and processes by using ergonomics and ecology as factors of competitive advantage
   1.2 helping organizations to deal with change processes (and, therefore, to survive)
   1.3 creating new working time systems (e.g., part-time concepts to employ more people)

2. Saving workability and employability by
   2.1 finding new ways of competence development and preservation
   2.2 developing new concepts for the integration of work and health

3. Rating work in a new way by
   3.1 including unpaid work (at home and in society) and
   3.2 personified services (e.g., including emotional stress and strain)

4. Designing the work of tomorrow by
   4.1 dealing with the possibilities and consequences of information technology
   4.2 considering the changes in individual (work-life) careers and biographies

All of these topics are really just examples for some sort of rethinking in ergonomics—which also has to be reflected in national and international ergonomic societies. As a consequence of a vision discussion, Gesellschaft für Arbeitswissenschaft formulated—among others—the following fundamental principle:

> Gesellschaft für Arbeitswissenschaft *resp.* its members feel obliged to the social goal (and make contributions—wherever possible) to the maintenance, creation and (appropriate) distribution of human, economically and environmentally compatible work.

Returning to the above-mentioned subjects of socially relevant aspects of the future of work, it is obvious that most of the topics are not in the microergonomic domain. These are—if accepted as such—tasks for macroergonomics.

As macroergonomics initially was labeled human factors in "organizational design and management" (ODAM), it could be helpful to look at the original disciplines of organization and management science to see what is going on there—and perhaps get some ideas for the development of macroergonomics.

DEVELOPMENTS IN ORGANIZATIONAL DESIGN AND MANAGEMENT

If we analyze organizational or, more generally, managerial concepts of recent years, we can realize some "overall" tendencies: While on the one hand, different approaches to "cost cutting" have been offered (entitled as lean XY or XY reengineering), we also can see progressively more approaches to overcome simple, partial concepts. One of the most prominent—which also is dealt with in the field of macroergonomics—is Total Quality Management (TQM) (see Zink, 1997). ISO 8402 describes it as:

> a management approach of an organization, centered on quality, based on the participation of all its members and aiming at long-term success through customer satisfaction, as well as benefits for all members of the organization and for society.

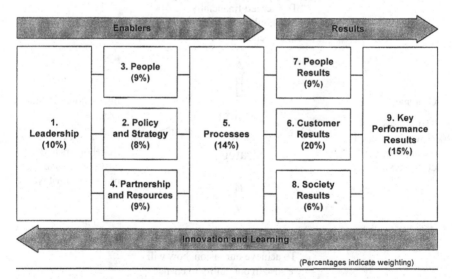

The EFQM Excellence Model

FIG. 18.1. Excellence model of the European foundation for quality management as the basis for the European Quality Award (EFQM, 200).

In this definition, we see a stakeholder orientation: customers, employees, society—but also shareholders. Another aspect of the definition in the context of holistic orientations is the goal of long-term success.

Though the term "Total Quality Management" might be at the end of its usefulness, the contents of this approach are not. As in the United States (Malcolm Baldrige National Quality Award) and in Australia (Australian Quality Award), Europe also has had a European Quality Award since the beginning of the 1990s. The assessment models behind these awards try to improve awareness of the need for more comprehensive evaluation concepts related to the overall quality of an organization.

A similar development can be seen in the "world of accounting": In 1987, Johnson and Kaplan wrote a book entitled *Relevance Lost—the Rise and Fall of Management Accounting*. Subsequently, Johnson proposed a bottom-up empowerment approach, which included activity-based cost management (Johnson, 1992) that has received considerable discussion in Europe. The most recent development in the accounting field focuses on a so-called "Balanced Scorecard" (Kaplan & Norton, 1996). "Balanced," in this context, can be interpreted as comprehensive or holistic, meaning that financial data alone are not enough to control an organization. Therefore (besides financial results), customers, processes, and learning and development (people) are included in the basic version of the Balanced Scorecard.

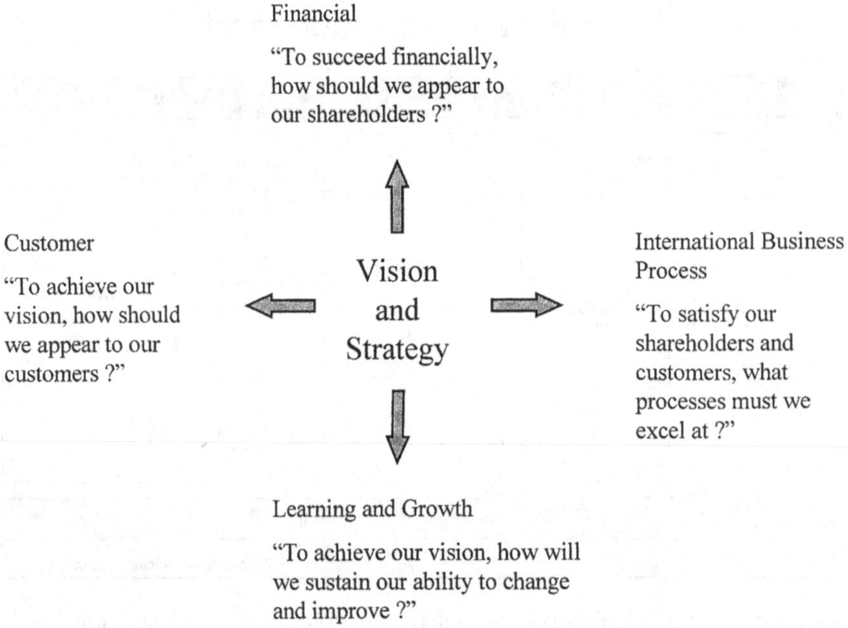

Financial

"To succeed financially, how should we appear to our shareholders ?"

Customer

"To achieve our vision, how should we appear to our customers ?"

Vision
and
Strategy

International Business Process

"To satisfy our shareholders and customers, what processes must we excel at ?"

Learning and Growth

"To achieve our vision, how will we sustain our ability to change and improve ?"

FIG. 18.2.   Balanced Scorecard (basic version).

There is a lot of similarity in these accounting concepts to the content of the so called Excellence Models as the basis for the (inter)national quality awards.

Lastly, looking at recent developments in organization theory, we find similar results to those in the recent management literature. According to Ulrich (Ulrich & Fluri, 1992, p. 19) there are three management dimensions to consider: *normative, strategic,* and *operative tasks.* However, their interdependence is explicitly emphasized. They are integrated via a management philosophy clarifying the theoretical basis of everyone's actions. It determines the vision, or final objective, and supports commitment of all participants.

Compared with the sociotechnical systems approach, the so-called St. Gallen model stresses the necessity of a commonly accepted vision as a striving force for the organization. It is deployed in a three-step process by different tools. In the sociotechnical concept, we have behavioral and structural dimensions, but in

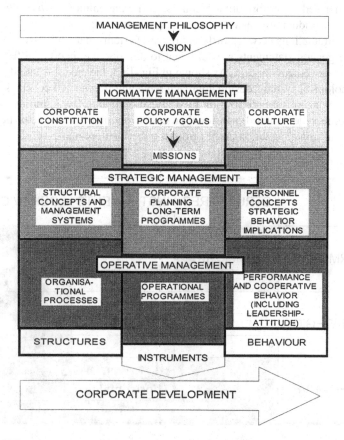

FIG. 18.3. Normative, strategic, and operative management (Zink (1997), according to Bleicher, (1996).

the St. Gallen model they are additionally structured in a normative, strategic, and operative dimension. The value of the model in creating a holistic approach results from its focus on harmonizing the different dimensions. According to Bleicher (1996), a holistic approach must integrate

- a basic fit,
- a vertical fit, and
- a horizontal fit.

Horizontal integration (fit) means, for example, that the corporate constitution is in harmony with the corporate culture and policy, and goals are based on both. *Vertical integration* includes the consistent transformation of normative elements into strategic ones, and strategic elements into operative ones. Management behavior and cooperation are determined by a strategic management concept derived from normative premises (corporate culture). A *basic fit* means that each individual dimension is consistent within itself; for example, that there are no contradictions between different reward systems in an organization.

These general statements are sufficient to describe the theoretical frame of a holistic management concept. As shown in Fig. 18.3, the model is based on a sociotechnological systems design approach replacing fragmented concepts and, as a result, replacing suboptimization by a global optimum. By definition, it must, therefore, include a multidimensional process to set objectives. The difficulty of its practical implementation comes from the fact that optimal conditions can be

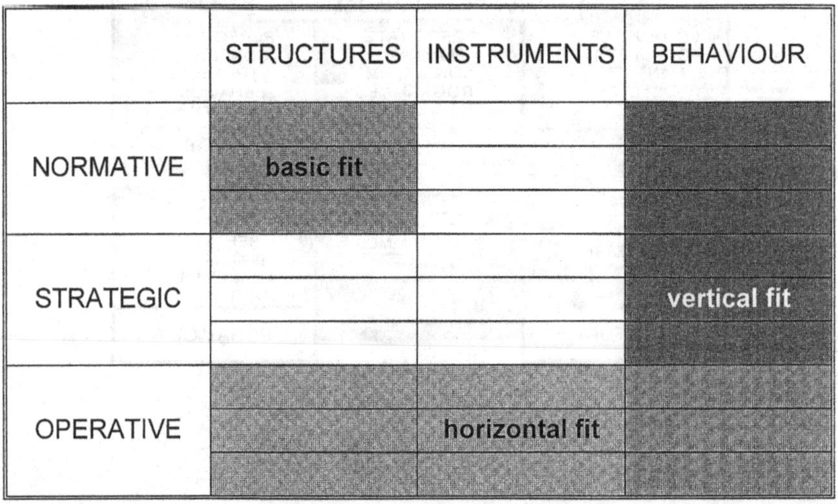

	STRUCTURES	INSTRUMENTS	BEHAVIOUR
NORMATIVE	basic fit		
STRATEGIC			vertical fit
OPERATIVE		horizontal fit	

FIG. 18.4 Harmonizing separate management dimensions (cf. Bleicher, 1996).

defined for technical or even economical subsystems. However, optimal conditions for social subsystems cannot, in many cases, be expressed in a general way—because of interindividual differences. Hence, all people affected must be involved in the design process.

Apart from this, it will become evident that problems will only be solved by an interdisciplinary and holistic approach (see Zink, 1984).

Knowing very well in ergonomics what interdisciplinary approaches mean, we must define the elements of "holistic" approaches before transferring our discussion to macroergonomics. Relevant aspects could be the following:

- stakeholder orientation
- integrative systems approach
- balance of respective demands
- long-term orientation (including sustainability)

Deploying holistic approaches therefore demands:

- interdisciplinary concepts for analysis, evaluation, and design
- participation
- process and result orientation
- critically analyzing so-called "systems limits"

Based on these recent developments, we now must ask, "What could be the role of macroergonomics?"

MACROERGONOMICS AS STAKEHOLDER-ORIENTED, HUMAN–SYSTEM INTERFACE DESIGN

As can be seen in many of the recent approaches, one important element of holistic or comprehensive concepts is a *stakeholder orientation*. This idea is not new (e.g., see Freeman, 1984), but the above-mentioned recent developments in some western countries (and in the broad definition of organizational ergonomics by the IEA) make it necessary to discuss its use in macroergonomics.

The advantages seem to be obvious:

1. Economic survival depends on customer orientation—and ergonomics can offer a lot to fulfil this demand.
2. Employee orientation is a traditional field of ergonomics.
3. As previously noted, the social challenges are progressively increasing and, thus, society is another important stakeholder.

4. Promoting the so far described stakeholders can also promote the share-holders (recent research is promising; e.g., see NIST, 1999).

How could a stakeholder-oriented system design be described and why could this be an upcoming task for macroergonomics? If we agree that customers, employees, shareholders, and society are the relevant stakeholders, we can find many examples in which ergonomics can support and optimize the human–system interface design:

Macroergonomics and Customers (Some Examples)

- *Kansei Engineering.* For years, Mitsuo Nagamachi has been showing how consumer-oriented product development can be successfully used to meet customer needs (e.g., see Nagamachi, 1994).
- *Usability testing.* Though Kansei Engineering demonstrates the better, more preventive approach, usability testing may help to correct faults in the product development process before negative reactions are received from the market.
- *Certifying the quality of use of ergonomic tools and procedures* in the product development and manufacturing process to give the customer the security that he or she is buying a product that includes all the ergonomic knowledge available at the moment. There are initial efforts within the IEA to build up a certification and accreditation body.
- *Combining ergonomics and ecology in product design, including aspects of sustainability.* New ways of an integrated product and process development may help customers to buy ergonomically and ecologically optimized products—products meeting the goal of sustainability for customer health as well as for society.

Macroergonomics and Employees (Some Examples)

- *All micro-ergonomic concepts.* The history of ergonomics is showing a huge mass of interventions, also described under the heading of Occupational Health and Safety, to make the work-place compatible with people's needs, abilities, and limitations.
- *All "traditional" macroergonomic approaches.* As described earlier, the still young approach of macroergonomics has a lot to offer to all employees. Taking participation as one example, these concepts are not restricted to physiological aspects but aim also at the realization of personal growth.

- *Macroergonomic interventions securing the survival of the organization through adequate change management.* As mentioned earlier, the complexity and dynamic of challenges caused by competition need both change management and an understanding of sociotechnological systems design. In the past, many workplaces lost their way because of too late or incorrect interventions. Using "whole picture" macroergonomics offers a lot of support.

Macroergonomics and Society (Some Examples)

- *Occupational health and safety.* It is in the interest of a society to ensure that working conditions do not lead to any health problems leading to early retirement causing—among others—financial problems. To this end, ergonomics should be supported by any society.
- *Security of complex systems.* Like using ergonomic principles to improve traffic, our power plant security should be in the interest of any community or state. Again, in addition to human suffering, these problems also have economic consequences.
- *Integrative health management systems.* Newer approaches in industry in the field of health promotion are using management concepts to improve acceptance (e.g., see Zink, 1999b). As health is influenced not only by work, approaches—such as those proposed by the World Health Organization—should include community or state activities.
- *Community ergonomics.* The activities started at the University of Madison-Wisconsin (see Chapter 14) show macroergonomics as a unique discipline with a true holistic approach that can help solve social problems in difficult areas. These ideas should be transferred to other regions—not only in the United States but all over the world.

Macroergonomics and Shareholders (Some Examples)

- *Total Quality Management and Change Management.* Many companies have not been able to implement Total Quality Management in a successful way. In most of the cases, TQM failed because of not realizing a sociotechnological approach. Here, macroergonomics can support organizations in introducing change management concepts in a comprehensive way.
- *Continuous improvement.* A core macroergonomic method is participation (see Chapter 2). In the early days of "Organizational Design and Management," quality circles and similar approaches (e.g., see Brown, 1986; and Zink, 1996) led to cost reductions and other savings. On the one hand, this

was, and is, a contribution to staff development; on the other, it has the potential to improve employee satisfaction and motivation.

- *Enlarging market share by selling customer-oriented products.* The use of ergonomic principles—as shown by Kansei Engineering and others—pays off. This potential is far from fully realized, and could be increased by combining ergonomics and environment-oriented design. In addition to positive economic effects, companies could also improve their public images.
- *Increasing shareholder value as a consequence of the above.* It is obvious that all the above-discussed examples must lead to an increase in shareholder value. This is not valuable per se, but in the context of positive results for the other stakeholders. In this sense, macroergonomic interventions often lead to a win-win situation.

Though these are only some first thoughts to discuss a stakeholder-oriented, human–system interface design, one can see that this combination of old and new approaches—classified in a new way—overcomes some limitations of "traditional" sociotechnical systems thinking (especially its "inside-orientation").

SOME CONCLUDING REMARKS

New paradigm shifts resulting in new forms of organization (e.g., virtual organizations), new forms of social challenge, and, last but not least, a growing shareholder value orientation in companies challenges macroergonomics to consider these tendencies for the future.

As the environments of organizations become progressively more complex and dynamic, we need comprehensive (i.e., holistic) problem-solving approaches. Ergonomics, by definition, can deliver a holistic concept based on its interdisciplinary nature. Regarding organizational as well as social developments, macroergonomics is the discipline challenged to develop answers. To broaden the scope, a stakeholder-oriented, human–systems interface design has been proposed in this chapter.

Though some authors fear that the stakeholder model is dead (see Beaver, 1999), showing as examples hostile takeovers, downsizing and mergers, and the influences of both institutional investors and executive compensation, it should be a normative demand for ergonomists to design a "human–(working) society interface" worth living for. If we want to keep a "human face" on capitalism, we must take part in activities to secure the survival of a society we would like to preserve for our children. Macroergonomics offers us a way.

References

Agarwal, A., Merrifield, J. & Tandon, R. (1985). *No place to run: Local realities and global issues of the Bhopal disaster.* New Market, TN: Highlander Research and Education Center.

Agricola, G. (1912). *De re metallica* (translated from the 1556 Latin edition by H. C. Hoover & L. H. Hoover). London: Salisbury House.

Alderfer, C. P. (1972). *Existence, relatedness and growth: Human needs in organizational settings.* New York: Free Press.

Andre, T., Kleiner, B. M. & Williges, R. C. (1998). A model for understanding computer-augmented distributed team decision making. *North Atlantic Treaty Organization (NATO) Research and Technology Agency/Human Factors and Medicine Panel 1st Symposium on Collaborative Crew Performance in Complex Operational Systems.* Neuilly-sur-Seine, France: NATO Research and Technology Agency.

Argyris, C. R. (1957). *Personality and organization.* New York: Harper & Row.

Argyris, C. (1971). *Management and organizational development.* New York: McGraw-Hill.

Azimi, H. (1991). *Madarhaaye toseah-nayafteghi in eqtesad-e-Iran* [Circles of underdevelopment in Iranian economy]. Teheran, Iran: Naey Publishing Company.

Babbage, C. (1832). *On the economy of machinery and manufacturers.* New York: Augustus M. Kelley Publishers. (reprinted in 1971)

Badham, R. & Schallock, B. (1991). Human factors in CIM: A human-centered perspective from Europe. *International Journal of Human Factors in Manufacturing, 1*(2), 121–141.

Bailey, R. W. (1989). *Human performance engineering: Using human factors/ergonomics to achieve computer system usability* (2nd ed.). Englewood Cliffs, NJ: Prentice Hall.

Bammer, G. (1987). VDUs and musculoskeletal problems at the Australian National University— A case study. In B. Knave & P. G. Wideback (Eds.), *Work with display units* (pp. 270–287). Amsterdam: North-Holland.

Bammer, G. (1990). Review of current knowledge—Musculoskeletal problems. In L. Berlinguet & D. Berthelette (Eds.), *Work with display units* (pp. 113–120). Amsterdam: North-Holland.

Bammer, G. (1993). Work-related neck and upper limb disorders—Social, organisational, biomechanical and medical aspects. In A. Gontijo & J. de Suza (Eds.), *Secundo congresso Lation-Americano e sexto seminario Brasileiro de ergonomia* (pp. 23–38). Florianopolis, Brazil: Ministerio de Trabalho.

Barnard, C. (1938). *Functions of the executive*. Cambridge, MA: Harvard University Press.

Barnatt, C. (1997). *Challenging reality*. Chichester, NY: John Wiley & Sons.

Baron, R. A. & Greenberg, J. (1990). *Behavior in organizations: Understanding and managing the human side of work* (3rd ed.). Boston: Allyn & Bacon.

Baum, A. (1988). Disasters, natural and otherwise. *Psychology Today*, April, 57–60.

Baum, A., Gatchel, R. & Schaeffer, M. (1983). Emotional, behavioral, and physiological effects of chronic stress at Three Mile Island. *Journal of Consulting and Clinical Psychology, 51*, 656–672.

Beardon, C. & Whitehouse, D. (Eds.). (1993). *Computers and society*. Oxford: Intellect Books.

Beaver, W. (1999). Is the stakeholder model dead? *Business Horizons, 42*(2), March–April, 8–12.

Becker-Lausen, E., Norman, S. & Pariante, G. (1987). *Human error in aviation: Information sources, research obstacles and potential*. Moffit AFB, CA: NASA Ames Research Center.

Bedeian, A. G. & Zammuto, R. F. (1991). *Organizations: Theory and design*. Chicago: Dryden.

Beekun, R. I., Glick, W. H. & Carsrud, A. L. (1984). User system analysis: Organizational design foundations of a management support system. In O. Brown, Jr. & H. W. Hendrick (Eds.), *Human factors in organizational design and management—II* (pp. 311–314). Amsterdam: North-Holland.

Bennis, W. G. (1969). Post bureaucratic leadership. *Transaction*, July–August, p. 45.

Benveniste, G. (1989). *Mastering the politics of planning: Crafting credible plans and policies that make a difference*. San Francisco: Jossey-Bass.

Blake, R. P. (Ed.). (1943). *Industrial safety*. New York: Prentice-Hall.

Blake, R. P. (1956). The accident causation model–88:10:2: con. *Journal of the American Society of Safety Engineers, 1*(19), 21–22.

Blanco, M. & Duggar, M. (1998). Macroergonomics class unpublished technical report. Blacksburgh, VA: Virginia Polytechnic Institute and State University.

Bleicher, K. (1996). *Das konzept integriertes management* [The concept of integrated management] (4th ed.). Frankfurt/Main: Campus.

Boeing Commercial Aircraft Group. (1993). Crew factor accidents: Regional perspective. *Proceedings of the 22nd Technical Conference of the International Air Transport Association (IATA) on Human Factors in Aviation* (pp. 45–61). Montreal, Canada: IATA.

Bongers, P. M., De Winter, C. R., Kompier, M. J. & Hildebrandt, V. H. (1993). Psychosocial factors and work and musculoskeletal disease. *Scandinavian Journal of Work, Environment and Health, 19*(5), 297–312.

Bordewich, F. (1987). The lessons of Bhopal. *The Atlantic*, March, 30–33.

Borowka, H. (1986). Management and safety. *Occupational Safety, 48*(7), 51–54.

Borowka, H. (1988). Managers, management, and the safety function. *Occupational Hazards, 50*(10), 137–140.

Bowonder, B., Kasperson, J. X. & Kasperson, R. E. (1985). Avoiding future Bhopals. *Environment, 27*(7), 6–37.

Bradley, G. (1977). *Computer technology, work life, and communication*. Stockholm: Liber. (In Swedish)

Bradley, G. (1979). Computerization and some psychosocial factors in the work environment. *Proceedings of the Conference on Reducing Occupational Stress* (pp. 30–40). (NIOSH Publication No. 78-140). Washington, DC: Department of Health, Education, and Welfare.

Bradley, G. (1986). *Stress and office automation*. Stockholm: Teldok. (In Swedish)

Bradley, G. (1989). *Computers and the psychosocial work environment*. London: Taylor & Francis.

Bradley, G. (1993). Psychosocial environment and the information age. In C. Beardon & D. Whitehouse (Eds.), *Computers and society*. London: Intellect Books.

Bradley, G. (2000). The information and communication society: How people will live and work in the new millennium. *Ergonomics, 43*.

Bradley, G. (Ed.). (2001). *Humans on the net*. Stockholm: Prevent.

Bradley, G., Bergström, C. & Lindeberg, S. (1988). *The role of engineers and the future computer technology*. Stockholm: Carlssons.

Bradley, G., Bergström, C. & Sundberg, L. (1984). *The role of secretaries and word processing—Office work in change.* Stockholm: Stockholm University.

Bradley G. & Bradley W. (1998). Towards global villages—Networks in action. In P. Vink, E. A. P. Konigsveld & S. Dhondt (Eds.), *Human factors in organizational design and management—VI* (pp. 33–37). Amsterdam: Elsevier Science.

Bradley, G. & Hendrick, H. W. (Eds.). (1994). *Human factors in organizational design and management—IV.* Amsterdam: North Holland.

Bradley, G. & Holm, P. (1991). *Knowledge based systems—Organizational and psychological aspects. Early experiences from a Swedish commercial bank.* Stockholm: Stockholm University, Institute of International Education.

Bradley, G. & Robertson, M. (1991). *Computers, psychosocial work environment, and stress. A comparative theoretical analysis of organisations and action strategies.* Presentation at the 11th Triennial Congress of the International Ergonomics Association (IEA), Paris.

Bradley, L., Andersson, N. & Bradley, G. (2000). *Home of the future—Information and communication technology (ICT)—Changes in society and human behaviour patterns in the net era.* Report R-00-1. Sweden: Fibre Science and Communication Network.

Bradley, L. & Bradley, G. (2000). Home of the future—Integration of professional and private roles. *Proceedings of the XIV Triennial Congress of the International Ergonomics Association and 44th Annual meeting of the Human Factors and Ergonomics Society, 2* (pp. 559–562). London: Taylor & Francis.

Brady, L. (1984). User system analysis. In H. W. Hendrick & O. Brown, Jr. (Eds.), *Human factors in organizational design and management* (pp. 173–177). Amsterdam: North-Holland.

Brödner, P. (1986). Skill based manufacturing vs. "unmanned factory"—Which is superior? *International Journal of Industrial Ergonomics, 1,* 145–153.

Brödner, P. (Ed.). (1987). Skill based automated manufacturing. In *Proceedings of the IFAC Workshop* (pp. 25–50). Oxford: Pergamon Press.

Brown, O., Jr. (1986). Participatory ergonomics: Historical perspectives, trends, and effectiveness of QWL programs. In O. Brown, Jr. & H. W. Hendrick (Eds.), *Human factors in organizational design and management—II* (pp. 433–437). Amsterdam: North-Holland.

Brown, O., Jr. (1990). Marketing participatory ergonomics: Current trends and methods to enhance organizational effectiveness. *Ergonomics, 33*(5), 601–604.

Brown, O., Jr. (1991a). Contemporary issues in participatory ergonomics. *Japanese Journal of Ergonomics, 27*(6), 291–294.

Brown, O., Jr. (1991b). Toward participatory ergonomics: A redefinition of the managerial role in organizational change. In Y. Queinnec & F. Daniellou (Eds.), *Designing for everyone* (pp. 1679–1681). London: Taylor & Francis.

Brown, O., Jr. (1993a). On the relationship between participatory ergonomics, performance, and productivity in organizational systems. In W. S. Marras, W. Karwowski, J. L. Smith & L. Pacholski (Eds.), *The ergonomics of manual work* (pp. 495–498). London: Taylor & Francis.

Brown, O., Jr. (1993b). Participatory ergonomics and participation: A review. In *Proceedings of the Scientific Conference Ergonomics in Russia, the Other Independent States, and Around the World, 2* (pp. 57–65). St. Petersburg: Russian Ergonomics Association.

Brown, O., Jr. (1994a). The evolution and development of participatory ergonomics. In *Proceedings of the XIIth Triennial Congress of the International Ergonomics Association, 6* (pp. 764–768). Toronto: Human Factors Association of Canada/ACE.

Brown, O., Jr. (1994b). High involvement ergonomics: A new approach to participation. In *Proceedings of the Human Factors and Ergonomics Society 38th Annual Meeting* (pp. 764–768). Santa Monica, CA: Human Factors and Ergonomics Society.

Brown, O., Jr. (1995a). Organizational issues in the implementation of high involvement ergonomics. In *Proceedings of the Human Factors and Ergonomics Society 39th Annual Meeting* (pp. 839–843). Santa Monica, CA: Human Factors and Ergonomics Society.

Brown, O., Jr. (1995b). The development and domain of participatory ergonomics. In *Proceedings of the IEA World Conference 1995* (pp. 28–31). Rio de Janeiro: Brazilian Ergonomics Association.

Brown, O., Jr. (1996). Participatory ergonomics: From participation research to high involvement ergonomics. In O. Brown, Jr. & H. W. Hendrick (Eds.), *Human factors in organizational design and management—V* (pp. 187–192). Amsterdam: North-Holland.

Brown, O., Jr. (2000). Participatory approaches to work systems and organizational design. In *Proceedings of the XIVth Triennial Congress of the International Ergonomics Association and 44th Annual Meeting of the Human Factors and Ergonomics Society, 2* (pp. 535–538). Santa Monica, CA: Human Factors and Ergonomics Society.

Brown, O., Jr. & Hendrick, H. W. (Eds.). (1986). *Human factors in organizational design and management—II*. Amsterdam: North-Holland.

Brown, O., Jr. & Hendrick, H. W. (Eds.). (1996). *Human factors in organizational design and management—V*. Amsterdam: North-Holland.

Bruening, J. C. (1989). Incentives strengthen safety awareness. *Occupational Hazards, 51*(11), 49–52.

Bruening, J. C. (1990a). Employee participation strengthens incentive programs. *Occupational Hazards, 52*(3), 45–47.

Bruening, J. C. (1990b). Shaping workers = attitudes toward safety. *Occupational Hazards, 52*(3), 49–51.

Bryce, G. K. (1981). *Joint labour-management occupational health committees: An example of worker participation in work site health and safety programs*. Unpublished master's thesis, University of Ottawa, Ottawa, Ontario, Canada.

Bureau of National Affairs. (1976). *Occupational safety and health reporter*. Washington, DC: Author.

Burns, T. & Stalker, G. M. (1961). *The management of innovation*. London: Tavistock.

Cakir, A., Hart, D. J. & Stewart, D. F. M. (1978). *Research into the effects of video display workplaces on the physical and psychological function of person*. Bonn, West Germany: Federal Ministry for Work and Social Order.

Calder, J. (1899). *The prevention of accidents*. New York: Longmans Green.

Caplan, R. D., Cobb, S., French, J. R., Jr., van Harrison, R. & Pinneau, S. R.(1975). *Job demands and worker health*. Washington, DC: U.S. Government Printing Office.

Cannon-Bowers, J. A., Salas, E. & Converse, S. A. (1993). Shared mental models in expert team decision making. In N. J. Castellan, Jr. (Ed.), *Current issues in individual and group decision making* (pp. 221–246). Hillsdale, NJ: Erlbaum.

Caplan, R. D., Cobb, S., French, J. R., Jr., van Harrison, R. & Pinneau, S. R. (1975). *Job demands and worker health*. Washington DC: U.S. Government Printing Office.

Carayon, P., Coujard, J. L. & Tarbes, J. (1997). The use of quality management methods in developing a network of enterprises. In P. Seppala, T. Luopajarvi, C.-H. Nygard & M. Mattila (Eds.), *Proceedings of the 13th Triennial Congress of the International Ergonomics Association, 1, Organizational Design and Management* (pp. 150–152). Helsinki: Finnish Institute of Occupational Health.

Carsrud, A. L. (1984). Organizational analyses: Idiosyncratic methodologies searching for holistic solutions, or what's wrong with the way they are done. In H. W. Hendrick & O. Brown, Jr. (Eds.), *Human factors in organizational design and management* (pp. 155–159). Amsterdam: North-Holland.

Chapanis, A. (1965). *Man-machine engineering*. Belmont, CA: Wadsworth.

Chapanis, A. (1974). National and cultural variables in ergonomics. *Ergonomics, 17*(2), 153–175.

Chapanis, A. (Ed.). (1975). *Ethnic variables in human factors engineering*. Baltimore: The Johns Hopkins Press.

Chapanis, A. (1991). To communicate the human factors message, you have to know what the message is and how to communicate it. *Human Factors Society Bulletin, 34*(11), 1–4.

Chernobyl Basiland

The Chernobyl accident and its consequences (1988). London: Atomic Energy Authority.

Chernousenko, V. (1992). *Chernobyl: Insight from the inside.* New York: Springer-Verlag.

Cherns, A. B. & Davis, L. E. (1975). Goals for enhancing the quality of working life. In L. E. Davis & A. B. Cherns (Eds.), *The quality of working life* (pp. 55–62). New York: Free Press.

Chicago Tribune. (1986). *The American millstone: An examination of the nation's permanent underclass.* Chicago: Contemporary Books.

Christensen, J. M. (1987). The human factors profession. In G. Salvendy (Ed.), *Handbook of human factors* (pp. 3–17). New York: Wiley.

Clegg, C., Ravden, S., Corbertt, M. & Johnson, S. (1989). Allocating functions in computer integrated manufacturing: A review and new method. *Behavior and Information Technology, 8,* 175–190.

Cleveland, R. J. (1976). *Behavioral safety codes in select industries.* Madison, WI: Wisconsin Department of Industry, Labor and Human Relations.

Cohen, A. L. (1977). Factors in successful occupational safety programs. *Journal of Safety Research, 9*(4), 168–178.

Cohen, A. L. (1996). Worker participation. In A. Bhattacharya & J. D. McGlothlin (Eds.), *Occupational ergonomics* (pp. 235–258). New York: Marcel Dekker.

Cohen, A. L., Gjessing, C. C., Fine, L. J., Bernard, B. P. & McGlothlin, J. D. (1997). *Elements of ergonomics programs. A primer based on workplace evaluations of musculoskeletal disorders.* Cincinnati, OH: National Institute for Occupational Safety and Health.

Cohen, A. L., Smith, M. J. & Anger, W. K. (1979). Self-protective measures against workplace hazards. *Journal of Safety Research, 11*(3), 121–131.

Cohen, M. D., March, J. G. & Olson, J. P. (1972). A garbage can model of organizational choice. *Administrative Science Quarterly, 17,* March, 1–25.

Cohen, W. J. (1997). *Community ergonomics: Design practice and operational requirements.* Madison, WI: University of Wisconsin Library.

Cohen, W. J. & Smith, J. H. (1994). Community ergonomics: Past attempts and future prospects toward America's urban crisis. In *Proceedings of Human Factors and Ergonomics Society 38th Annual Meeting* (pp. 734–738). Santa Monica, CA: Human Factors and Ergonomics Society.

Coleman, P. J. & Sauter, S. L. (1978). *The worker as a key control component in accident prevention systems.* Presented at the 1978 Convention of the American Psychological Association, Toronto, Ontario, Canada.

Coleman, P. J. & Smith, K. U. (1976). *Hazard management: Preventive approaches to industrial injuries and illnesses.* Madison, WI: Wisconsin Department of Industry, Labor, and Human Relations.

Committee For Economic Development. (1995). *Rebuilding inner-city communities: A new approach to the nation's urban crisis.* Washington DC: Author.

The Comptroller General Report to the Congress of the United States. (1980). *Three Mile Island: The most studied nuclear accident in history.* Washington, DC: The United States General Accounting Office.

Cooley, M. J. E. (1986). Problems of automation. In T. Lupton (Ed.), *Human factors: Man, machine and new technology.* Berlin: IFS and Springer-Verlag.

Corbett, J. M. (1985). Prospective work design of a human-centered CNC lathe. *Behaviour and Information Technology, 4*(3), 201–214.

Corbett, J. M. (1988). Ergonomics in the development of human-centered HAS. *Applied Ergonomics, 19,* 35–39.

Corbett, J. M. (1990). Human centered advanced manufacturing systems: From theoretic to reality. *International Journal of Industrial Ergonomics, 5,* 83–90.

Cotton, J. L. (1993). *Employee involvement.* Newbury Park, CA: Sage.

Cotton, J. L., Vollrath, D. A., Froggatt, K. L., Lengnick-Hall, M. L. & Jennings, K. R. (1988). Employee participation: Diverse forms and different outcomes. *Academy of Management Journal, 13,* 8–22.

Cotton, J. L., Vollrath, D. A., Lengnick-Hall, M. L. & Froggatt, K. L. (1990). Fact: The form of participation does matter—A rebuttal to Leana, Locke, and Schweiger. *Academy of Management Journal, 15,* 147–153.

Creswell, T .J. (1988). Safety and the management function. *Occupational Hazards, 50*(12), 31–34.

Cutter, W. A. & Wilkenson, T. H. (1959). Toward the profession of safety program management. *The Journal of the American Society of Safety Engineers*, October, 96–97, 227–238.

Dachler, H. P. & Wilpert, B. (1978). Conceptual boundaries and dimensions of participation in organizations: A critical evaluation. *Administrative Science Quarterly, 23*, 1–39.

Davidson, L. & Baum, A. (1986). Chronic stress and posttraumatic stress disorder. *Journal of Consulting and Clinical Psychology, 54*, 303–308.

Davis, L. E. (1982). Organizational design. In G. Salvendy (Ed.), *Handbook of industrial engineering* (pp. 2.1.1–2.1.29). New York: Wiley.

Davis, L. E. (1966). The design of jobs. *Industrial Relations, 6*, 21–45.

Davis, L. E. & Taylor, J. C. (1972). *Design of jobs*. Middlesex, UK: Penguin Books.

Davis, L. E. & Werling, R. (1960). Job design factors. *Occupational Psychology, 34*(2), 109–132.

Davis, R. A. (1993, October). *Human factors in the global marketplace*. Presented at the Annual Meeting of the Human Factors and Ergonomics Society, Seattle, WA.

Deci, E. L. (1975). *Intrinsic motivation*. New York: Plenum Press.

DeGreene, K. B. (1973). *Sociotechnical systems*. Englewood Cliffs, NJ: Prentice-Hall.

DeGreene, K. B. (1990). Supplementary systems paradigms for different stages of societal solution with special reference to war and peace. *Journal of Systems Research, 7*, 77–89.

DeJoy, D. M. (1990). Toward a comprehensive human factors model of workplace accident causation. *Professional Safety*, May, 11–16.

DeJoy, D. M. (1994). Managing safety in the workplace: An attribution theory analysis and model. *Journal of Safety Research, 25*(1), 3–17.

De Keyser (1991). Work analysis in French language ergonomics: Origin and current research trends. *Ergonomics, 34*, 653–669.

De Lisi, P. S. (1990). Lesson from the steel axe: Culture, technology and organizational change. *Sloan Management Review l32*(1) 83–93.

Demel, G. & Meshkati, N. (1989). Requisite variety: A concept to analyze the effects of cultural context for technology transfer. *Proceedings of the 33rd Annual Meeting of the Human Factors Society* (pp. 765–769). Santa Monica: CA: Human Factors Society.

Deming, W. E. (1986). *Out of the crisis*. Cambridge, MA: MIT Center for Advanced Engineering Study.

Dess, G. G., Rasheed, A. M., McLaughlin, K. J. & Priem, R. L. (1995). The new corporate architecture. *Academy of Management Executive, 9*, 7–20.

Dipiero, D. A,. Davis, G. L. & Krause, T. R. (1993), Safety and quality: Success at Georgia Gulf. *Occupational Hazards, 55*(10), 51–53.

Dray, S. M. (1985). Macroergonomics in organizations: An introduction. In I. D. Brown, R. Goldsmith, K. Combes & M. Sinclair (Eds.), *Ergonomics international* (pp. 520–522). London: Taylor & Francis.

Dray, S. M. (1986). The new internationalism in macroergonomics. In O. Brown, Jr. & H. W. Hendrick (Eds.), *Human factors in organizational design and management—II* (pp. 499–503). Amsterdam: North-Holland.

Driscoll, D. (1996). Mainteinance resource management training program. *Faces and places*. Pittspurgh, PA:US Airways Quality Assurance.

Driscoll, D., Kleiser, T. & Ballough, J. (1997). *US Airways maintenance resource management, aviation safety action program (MRM-ASAP)*. Pittsburg, PA: US Airways Quality Assurance.

Drucker, P. (1988). The coming of the new organization. *Harvard Business Review, 88*, 47.

Drury, C. G. (1991). Ergonomics practice in manufacturing. *Ergonomics, 34*(6), 825–839.

Drury, C. (1993). Chapter eight: Training for visual inspection of aircraft structures. In G. S. Corporation (Eds.), *Human Factors in Aviation Maintenance—Phase Three, Volume 1 Progress Report* (DOT/FAA/AM-93/15) (pp. 133–154). Washington, DC: FAA Office of Aviation Medicine.

Duncan, R. B. (1972). Characteristics of organizational environments and perceived environmental uncertainty. *Administrative Science Quarterly, 17*, 313–327.

Dupont, G. (1997). The dirty dozen errors in maintenance. In *Proceedings of the Eleventh FAA Meeting on Human Factors Issues in Aircraft Maintenance and Inspection* (pp. 45–49). Washington, DC: FAA Office of Aviation Medicine.

Durbin, T. J. (1993). Safety and quality at PPG industries. *Occupational Hazards, 55*(5), 79–82.

Ellegard, K., Engstrom, T. & Nilsson, L. (1992). *Reforming industrial work—Principles and realities.* Arbetsmiljonforden.

Ellis, L. (1975). A review of research on efforts to promote occupational safety. *Journal of Safety Research, 7*(4), 180–189.

Emery, F. E. & Trist, E. L. (1960). Sociotechnical systems. In C. W. Churchman & M. Verhulst (Eds.), *Management sciences: Models and techniques* (pp. 83–97). Oxford: Pergamon.

Emery, F. E. & Trist, E. L. (1965). The causal texture of organizational environments. *Human Relations, 18,* 21–32.

Emery, F. E. & Trist, E. L. (1978). Analytical model for sociotechnical systems. In W. A. Pasmore & J. J. Sherwood (Eds.), *Sociotechnical systems: A sourcebook* (pp. 120–133). LaJolla, CA: University Associates, Inc.

Endsley, M. R. & Robertson, M. M. (1986). *Team situation awareness in aircraft maintenance.* Lubbock, TX: Texas Tech University.

Endsley, M. R. & Robertson, M. M. (2000). Situation awareness in aircraft maintenance teams. *International Journal of Industrial Ergonomics, 26,* pp. 301–325.

ESPRIT consortiun AMICE, (1989). *Open system architecture for CiM.* Berlin: Springer-Verlag.

European Foundation for Quality Management (Ed.). (2000). *The European quality award.* Brussels: EFQM.

Everest, L. (1985). *Beyond the poison cloud: Union Carbide's Bhopal massacre.* Chicago: Banner Press.

Faucett, J. & Rempel, D. (1994). VDT-related musculoskeletal symptoms: Interactions between work posture and psychosocial work factors. *American Journal of Industrial Medicine, 26,* 597–612.

Federal Aviation Administration. (1991). *The aviation human factors national plan (Draft).* Washington, DC: Author.

Federal Aviation Administration. (1996, June 18). *Federal Aviation Administration human factors team report on: The interfaces between flightcrews and modern flight deck systems.* Washington, DC: Author.

Feigenbaum, A. V. (1961). *Total quality control.* New York: McGraw-Hill.

Feynman, R. P. (1986). Personal observations on reliability of shuttle. *Report of the Presidential Commission on the Space Shuttle Challenger Accident, 2.* Washington, DC: U.S. Government Printing Office.

Fitts, P. M. (1951). Engineering psychology in equipment design. In S. S. Stevens (Ed.), *Handbook of experimental psychology* (pp. 365–379). New York: Wiley.

Flach, J. M. & Hancock, P. A. (1992). An ecological approach to human–machine systems. In *Proceedings of the Human Factors Society 36th Annual Meeting* (pp. 1056–1058). Santa Monica, CA: Human Factors Society.

Forsythe, C. (1997). Human factors in agile manufacturing. *Human Factors and Ergonomics in Manufacturing, 7,* 3–10.

Forsythe, C. & Ashby, M. R. (1994). *Developing communications requirements for agile product realization* (Report No. SAND-94-0481C). New Mexico: Sandia National Laboratories.

Forsythe, C. & Ashby, M. R. (1996). Human factors in agile manufacturing. In Proceedings of the Human Factors and Ergonomics Society 39th Annual Meeting (pp. 538–540). Santa Monica

Forsythe, C. & Karwowski, W. (1994). Human factors in agile manufacturing. In *Proceedings of the Human Factors and Ergonomics Society 39th Annual Meeting* (pp. 538–540). Santa Monica, CA: Human Factors and Ergonomics Society.

Frankel, R. L., Priest, W. C. & Ashford, N. A. (1980). Occupational safety and health: A report on worker participation. *Monthly Labor Review,* September, 11–14.

Freeman, R. E. (1984). *Strategic management: A stakeholder approach*. Boston: Pitman.

Gagne, R., Briggs, L. & Wagner, R. (1988). *Principles of instructional design* (3rd ed.). New York: Holt Rinehart and Winston.

Gale, R. P. & Hauser, T. (1988). *Final warning: The legacy of Chernobyl*. New York: Warner Books.

Gasser, L. & Majchrzak, A. W. (1994). ACTION integrates manufacturing strategy, design, and planning. In Kidd, P. T. & Karwowski, W. (Eds.), *Advances in agile manufacturing* (pp. 133–136). Amsterdam: IOS Press.

Gesellschaft für Arbeitswissenschaft e.V. (GfA)(ed.), (1999). *Selbstverständnis der gesellschaft für arbeitswissenschaft e.V.* [Position Paper of GfA]. Dortmund: GfA Press.

Gesellschaft für Arbeitswissenschaft e.V. (Ed.). (2000). *Die zukunft der arbeit erforschen: Ein memorandum der Gesellschft für Arbeitswissenschaft e.V. zum strukturwandel der arbeit* [Research for the future of work: A memorandum of GfA related to the change of work]. Dortmund: GfA Press.

Giarini, O. & Liedtke, P. M. (1997). *Wie wir arbeiten werden* [The Employment dilemma and the future of work]. Hamburg: Hoffmann und Campe.

Gill, J. & Martin, K. (1976). Safety management: Reconciling rules with reality. *Personnel Management, 8*(6), 36–39.

Gleick, J. (1987). *Chaos*. New York: Penguin.

Glick, W. H. & Beekun, R. I. (1984). A theoretical overview of user systems analysis. In H. W. Hendrick & O. Brown, Jr. (Eds.), *Human factors in organizational design and management* (pp. 161–166). Amsterdam: North-Holland.

Goldman, M. L. (1986, July). Keeping the cold war out of Chernobyl. *Technology Review*, 18–19.

Goldman, S. & Preiss, K. (1992). *21st century manufacturing enterprise strategy, 2—Infrastructure*. Bethlehem, PA: Lehigh University.

Goldman, S. L., Nagel, R. N. & Preiss, K. (1995). *Agile competitors and virtual organizations*. New York: Van Nostrand Reinhold.

Goldsmith, W. W. & Blakely, E. J. (1992). *Separate societies: Poverty and inequality in U.S. cities*. Philadelphia: Temple University Press.

Goldstein, I. L. (1993). *Training in organizations* (3rd ed.). Belmont, CA: Wadsworth.

Goodstein, L. P., Anderson, H. B. & Olsen, S. E. (Eds.). (1988). *Tasks, errors and mental models*. London: Taylor & Francis.

Gordon, S. (1994). *Systematic training program design: Maximizing and minimizing liability*. Englewood Cliffs, NJ: Prentice Hall.

Gottlieb, M. S. (1976). *Worker's awareness of industrial hazards: An analysis of hazard survey results from the paper mill industry*. Madison, WI: Wisconsin Department of Industry, Labor and Human Relations.

Gottlieb, M. S. & Coleman, P. J. (1977). *Inspection impact on injury and illness totals*. Madison, WI: Wisconsin Department of Industry, Labor and Human Relations.

Gould, J. D. (1990). How to design usable systems. In M. Helander (Ed.), *Handbook of human-computer interaction* (pp. 757–789). Amsterdam: Elsevier.

Gould, S. J. (1981). *The mismeasure of man*. New York: Norton.

Graeber, R. C. (1994, May). *Integrating human factors knowledge into automated flight deck design*. Presented at the International Civil Aviation Organization (ICAO) flight safety and human factors seminar, Amsterdam.

Greiss, H. A. (1993). American industrial dominance will depend upon agility to manage change. *Focus*, June, 1.

Grenville, N. D. (1997). *A sociotechnical approach to evaluating the effect of managerial time allotment on department performance*. Unpublished master's thesis, Virginia Polytechnic Institute and State University, Blacksburg, VA.

Grenville, N. D. (2001). *Developing heuristics to optimize the configuration of the video-mediated environment*. Unpublished doctoral dissertation, Virginia Polytechnic Institute and State University, Blacksburg, VA.

Grenville, N. D. & Kleiner, B. (1997a). Sociotechnical systems approach to time allotment in manufacturing supervision. In *Proceedings of the 6th Annual Industrial Engineering Research Conference* (pp. 709–713). Norcross, GA: Institute of Industrial Engineers.

Grenville, N. D. & Kleiner, B. (1997b). Relationship between sociotechnical joint optimization and perceived department performance in manufacturing organizations. In *Proceeding of the Human Factors and Ergonomics Society 41st Annual Meeting* (pp. 772–776). Santa Monica, CA: Human Factors and Ergonomics Society.

Griffin, R. W. (1988). Consequences of quality circles in an industrial setting: A longitudinal assessment. *Academy of Management Journal, 31*, 338–358.

Grimaldi, J. V. (1970). The measurement of safety engineering performance. *Journal of Safety Research, 2*(3), 137–159.

Grimaldi, J. V. & Simonds, R. H. (1989). *Safety management* (5th ed.). Boston, MA: Irwin.

Grimston, M. C. (1977, April). *Comparison of Chernobyl and Bhopal accidents.* Presented at the meeting of the Society for Radiological Protection, London, England.

Grimston, M. C. (1997). Chernobyl and Bhopal Ten Years On — Comparisons and Contrasts. In Lewins (Ed.), *Advances in Nuclear Science and Technology*, 24, Plenum Press, New York, 1–45.

Grimston, M. C. (1977). Chernobyl and Bhopal ten years on—Comparisons and contrasts. In Lewins (Ed.), *Advances in nuclear science and technology, 24.* New York: Plenum Press.

Groesbeck, R., Sienknecht, T. & Merida, O. (1998). Macroergonomics class unpublished technical report, Virginia Polytechnic Institute and State University, Blacksburgh, VA.

Gross, C. M. & Fuchs, A. (1990). Reduce musculoskeletal injuries with corporate ergonomics program. *Occupational Health and Safety, 1*, 29–33.

Grugle, N. (2001). *An investigation into the effects of chemical protective clothing on team process performance.* Unpublished master's thesis, Virginia Polytechnic Institute and State University, Blacksburg, VA.

Guarnieri, M. (1992). Landmarks in the history of safety. *Journal of Safety Research, 23*, 151–158.

Guastello, S. J. (1993). Do we really know how well our occupational accident prevention programs work? *Safety Science, 16*, 445–463.

Guastello, S. J. (1995). *Chaos, catastrophe and human affairs.* Hillsdale, NJ: Lawrence Erlbaum.

Hackman, J. R. (Ed.). (1989). *Groups that work (and those that don't): Creating conditions for effective teamwork.* San Francisco: Jossey-Bass.

Hackman, J. R. (1990). *Groups that work.* San Francisco: Jossey-Bass.

Hackman, J. R. & Lawler, E. E. (1971). Employee reactions to job characteristics. *Journal of Applied Psychology Monograph, 55*, 259–286.

Hackman, J. R. & Oldham, G. R. (1975). Development of the job diagnostic survey. *Journal of Applied Psychology, 59*, 159–170.

Hackman, J. R. & Oldham, G. R. (1976). Motivation through the design of work: Test of a theory. *Organizational and Human Performance, 16*, 250–279.

Hackman, J. R. & Oldham, G. R. (1980). *Work redesign.* Reading, MA: Addisson-Wesley.

Hagberg, M., Silverstein, B., Wells, R., Smith, M. J., Hendrick, H. W., Carayon, P., Perusse, M., Kuorinka, I. & Forçier, L. (1995). *Work related musculoskeletal disorders (WMSDs): A reference book for prevention.* London: Taylor & Francis.

Hage, J. & Aiken, M. (1969). Routine technology, social structure, and organizational goals. *Administrative Science Quarterly, 14*, September, 72–91.

Haims, M. C. & Carayon, P. (1996). Implementation of an "in-house" participatory ergonomics program: A case study in a public service organization. In O. Brown, Jr., & H. W. Hendrick (Eds.), *Human factors in organizational design and management—V* (pp. 175–180). Amsterdam: North-Holland.

Haims, M. C. & Carayon, P. (1998). Theory and practice for the implementation of "in house", continuous improvement, participatory ergonomic programs. *Applied Ergonomics, 29*(6): 461–472.

Haines, H. M. & Wilson, J. R. (1998). *Development of a framework for participatory ergonomics.* Sudbury, Suffolk: Health and Safety Executive.

Hall, A. D. (1969). A three dimensional morphology of systems engineering. *IEEE Transaction System Science Cybernetics*, V *SCC-5*, 156–160.

Hall, R. H., Haas, J. E. & Johnson, N. J. (1967). Organizational size, complexity and formalization. *Administrative Science Quarterly, 12*, June, 303.

Hammer, W. (1972). *Handbook of systems and product safety*. Englewood Cliffs, NJ: Prentice-Hall.

Hammer, W. (1976). *Occupational safety management and engineering*. Englewood Cliffs, NJ: Prentice-Hall.

Handy, C. (1994). *The empty raincoat*. London: Hutchinson.

Haney, L. N., Reece, W. J., Wilhelmsen, C. J. & Romero, H. A. (1994). Modeling cognitive aspects of human error. In P.T. Kidd & W. Karwowski (Eds.), *Advances in agile manufacturing: Integrating technology, organization and people* (pp. 335–338). Amsterdam: IOS Press.

Hannum, W. & Hansen, C. (1992). *Instructional systems development in large organizations*. Englewood Cliffs, NJ: Educational Technology Publications.

Harrington, M. (1962). *The other America: Poverty in the United States*. New York: MacMillan.

Harvey, E. (1968). Technology and the structure of organizations. *American Sociological Review, 33*, April, 247–259.

Harvey, O. J. (1963). System structure, flexibility and creativity. In O. J. Harvey (Ed.), *Experience, structure and adaptability* (pp. 39–65). New York: Springer.

Harvey, O. J., Hunt, D. E. & Schroder, H. M. (1961). *Conceptual systems and personality organization*. New York: Wiley.

Hayes, R. H., Wheelwright, S. C. & Clark, K. B. (1980). *Dynamic manufacturing*. New York: The Free Press.

Hawkins, F. (1993). *Human factors in flight* (2nd ed.). Aldershot, England: Gower Technical Press.

Heinrich, H. W. (1931). *Industrial accident prevention: A scientific approach* (1st ed.). New York: McGraw-Hill.

Heinrich, H. W. (1956). The accident causation model. *Journal of the American Society of Safety Engineers, 1*(18), 20–22.

Heinrich, H. W. (1959). *Industrial accident prevention. A scientific approach* (4th ed.). New York: McGraw-Hill.

Helali, F. & Shahnavaz, H. (1996). Ergonomic intervention in industries of the industrially developing countries. Case study: Glucosan-Iran. In O. Brown, Jr. & H. W. Hendrick (Eds.), *Human factors in organizational design and management—V* (pp. 141–146). Amsterdam: North-Holland.

Helali, F. & Shahnavaz, H. (1998). Adapting macroergonomic approach for identification of workplace problems and development of low-cost/no-cost solutions in industrially developing countries. Case study: Glucosan-Iran. In P. Vink, E. A. P. Koningsveld & S. Dhondt (Eds.), *Human factors in organizational design and management—VI* (pp. 585–590). Amsterdam: North-Holland.

Helmreich, R. L. (1994a). Flightcrew perspective on automation: A cross-cultural perspective. In *Report of the Seventh ICAO Flight Safety and Human Factors Regional Seminar* (pp. 442–453). Montreal, Canada: International Civil Aviation Organization (ICAO).

Helmreich, R. L. (1994b). Anatomy of a system accident: Avianca Flight 052. *International Journal of Aviation Psychology, 4*(3), 265–284.

Helmreich, R. L. & Merritt, A. (1998). *Culture at work in aviation and medicine: National, organizational, and professional influences*. Brookfield, VT: Ashgate.

Hendrick, H. W. (1979). Differences in group problem solving behavior and effectiveness as a function of abstractness. *Journal of Applied Psychology, 64*, 518–525.

Hendrick, H. W. (1980). *Human factors in management*. Presented at the symposium on professional planning, 1980–2000 at the Human Factors Society 24th Annual Meeting, Los Angeles, CA.

Hendrick, H. W. (1981). Abstractness, conceptual systems, and the functioning of complex organizations. In G. England, A. Negandhi & B. Wilpert (Eds.), *The functioning of complex organizations* (pp. 25–50). Cambridge, MA: Oelgeschalger, Gunn and Hain.

Hendrick, H. W. (1984). Wagging the tail with the dog: Organizational design considerations in ergonomics. In *Proceedings of the Human Factors Society 28th Annual Meeting* (pp. 899–903). Santa Monica, CA: Human Factors Society.

Hendrick, H. W. (1986a). Macroergonomics: A conceptual model for integrating human factors with organizational design. In O. Brown, Jr. & H. W. Hendrick (Eds.), *Human factors in organizational design and management—II* (pp. 467–478). Amsterdam: North-Holland.

Hendrick, H. W. (1986b). Macroergonomics: A concept whose time has come. *Human Factors Society Bulletin, 30*(2), 1–3.

Hendrick, H. (1987). Organizational design. In G. Salvendy (Ed.), *Handbook of human factors* (pp. 470–494). New York: Wiley.

Hendrick, H. W. (1988). A macroergonomic approach to designing a university college. In *Proceedings of the Human Factors Society 32nd Annual Meeting* (pp. 780–784). Santa Monica, CA: Human Factors Society.

Hendrick, H. W. (1990). Perceptual accuracy of self and others and leadership status as functions of cognitive complexity. In K. E. Clark & M. B. Clark (Eds.), *Measures of leadership* (pp. 511–520). West Orange, NJ: Leadership Library of America.

Hendrick, H. W. (1991a). Human factors in organizational design and management. *Ergonomics, 34,* 743–756.

Hendrick, H. W. (1991b). Macroergonomics: A new concept leading to higher productivity. *Japanese Journal of Ergonomics, 27*(6), 297–300.

Hendrick, H. W. (1994a). Future directions in macroergonomics. In *Proceedings of the 12th Triennial Congress of the International Ergonomics Association, 1* (pp. 41–43). Toronto: Human Factors Association of Canada/ACE.

Hendrick, H. W. (1994b). Macroergonomics as a preventative strategy in occupational health: An organizational level approach. In G. E. Bradley & H. W. Hendrick (Eds.), *Human factors in organizational design and management—IV* (pp. 713–718). Amsterdam: North-Holland.

Hendrick, H. W. (1995a). Future directions in macroergonomics. *Ergonomics, 38,* 1617–1624.

Hendrick, H. W. (1995b). Humanizing re-engineering for true organizational effectiveness: A macroergonomic approach. In *Proceedings of the Human Factors and Ergonomics Society 39th Annual Meeting* (pp. 761–765). Santa Monica, CA: Human Factors and Ergonomics Society.

Hendrick, H. W. (1996). *Good ergonomics is good economics.* Santa Monica, CA: Human Factors and Ergonomics Society.

Hendrick, H. W. (1997). Organizational design and macroergonomics. In G. Salvendy (Ed.), *Handbook of human factors and ergonomics* (2nd ed.) (pp. 594–636). New York: Wiley.

Hendrick, H. W. (1998). Macroergonomics: A systems approach for dramatically improving occupational health and safety. In S. Kumar (Ed.), *Advances in occupational ergonomics and safety, 2* (pp. 26–34). Amsterdam: IOS Press.

Hendrick, H. W. & Brown, O., Jr. (Eds.). (1984). *Human factors in organizational design and management.* Amsterdam: North-Holland.

Hendrick, H. W. & Kleiner, B. M. (2001). *Macroergonomics: An introduction to work system design.* Santa Monica, CA: Human Factors and Ergonomics Society.

Herzberg, F. (1966). *Work and the nature of man.* New York: Brace & World, Harcourt.

Hickson, D., Pugh, D. & Pheysey, D. (1969). Operations technology and organizational structure: An empirical reappraisal. *Administrative Science Quarterly, 26,* 349–377.

Hirst, P. & Thompson, G. (1996). *Globalisation in question.* Cambridge: Polity Press.

Hofstede, G. (1980a). *Culture's consequences.* Beverly Hills, CA: Sage.

Hofstede, G. (1980b). Motivation, leadership, and organization: Do American theories apply abroad? *Organizational Dynamics,* Summer, 42–63.

Human Factors and Ergonomics Society. (1998). HFES strategic plan. *Human Factors and Ergonomics Society Directory and Yearbook, 1998–1999* (p. 388). Santa Monica, CA: Author.

Hollnagel, E. & Woods, D. (1983). Cognitive systems engineering: New wine in new bottles. *International Journal of Man-Machine Studies, 18*, 583–600.

Iacocca Institute (1991). *21st century manufacturing enterprise strategy. An industry-led view, 1 & 2.* Bethlehem, PA: Iacocca Institute.

Ignatenko, E. I., Voznyak, V. Y., Kopvalenko, A. P. & Troitskii, S. N. (1989). *Chernobyl—Events and lessons (questions and answers)*. Moscow: Political Literature Publishing House.

Illesh, S. (1987). *Chernobyl*. New York: Richardson and Steirman, Inc.

Imada, A. S. (1988). Participatory ergonomics. In A. S. Adams, R. R. Hall, B. J. McPhee & M. S. Oxenburgh (Eds.), *Ergonomics International 88: Proceedings of the Tenth Congress of the International Ergonomics Association, 2* (pp. 711–713). Sydney: Ergonomics Society of Australia.

Imada, A. S. (1991a). The rationale and tools of participatory ergonomics. In K. Noro & A. S. Imada (Eds.), *Participatory ergonomics* (pp. 30–49). London: Taylor & Francis.

Imada, A. S. (1991b). Participatory ergonomics, organizational complexity and chaos. In Y. Queinnec & F. Daniellou (Eds.), *Designing for everyone* (pp. 1676–1678). London: Taylor & Francis.

Imada, A. S. & Feiglstok, D. M. (1990). An organizational design and management approach for improving safety. In K. Noro & O. Brown Jr. (Eds.), *Human factors in organizational design and management—III* (pp. 479–482). Amsterdam: North-Holland.

Imada A. S. & Nagamachi, M. (1990). Improving occupational safety and health: Non-traditional organizational design and management approaches. In K. Noro & O. Brown, Jr. (Eds.), *Human factors in organizational design and management—III* (pp. 483–486). Amsterdam: North-Holland.

Imada, A. S. & Stawowy, G. (1996). The effects of a participatory ergonomics redesign of food service stands on speed of service in a professional baseball stadium. In O. Brown, Jr. & H. W. Hendrick (Eds.), *Human factors in organizational design and management—V* (pp. 203–208). Amsterdam: North-Holland.

International Atomic Energy Agency. (1986). Summary report on the post-accident review meeting on the Chernobyl accident. (INSAG-1) Vienna: Author.

International Atomic Energy Agency. (1987). *Nuclear safety report for 1987.* Vienna: Author.

International Atomic Energy Agency. (1991). *Safety culture.* (Safety Series No. 75-INSAG-4). Vienna: Author.

International Atomic Energy Agency (1992). *The Chernobyl accident: Updating of INSAG-1* (INSAG-7). Vienna: Author.

International Confederation of Free Trade Unions (ICFTU) (1985). *The Trade Union Report on Bhopal.* Bruxelles, Belgium: ICFTU.

International Ergonomics Association (Ed.). (2000). *The discipline of ergonomics—Definitions.* http://www.iea.cc/ergonomics/

Itoh, S. (1991). Customer-oriented manufacturing. *International Journal of Human Factors in Manufacturing, 1*(4), 365–370.

Jackson, S. E. (1992). *Diversity in the work place.* New York: Guilford Press.

Jaikumar, R. (1986). Post-industrial manufacturing. *Harvard Business Review*, November–December, 69–76.

Janis, I. L. & Feshback, S. (1953). Effects of fear-arousing communications. *Journal of Abnormal and Social Psychology, 48*, 78–92.

Jenkins, R. (1990, July 18–31). Bhopal: Five years latter. *In These Times*, 12–13.

Johnston, A. N. (1993). CRM: Cross-cultural perspectives. In E. L. Wiener, B. G. Kanki & R. L. Helmreich (Eds.), *Cockpit resource management* (pp. 367–397). San Diego, CA: Academic Press.

Jones, D. F. (1973). *Occupational safety programs—are they worth it?* Toronto: Labour Safety Council of Ontario, Ontario Ministry of Labour.

Jones, D. T. (1992). Beyond the Toyota production system: The era of lean production. In C.A. Voss (Ed.), *Manufacturing strategy, process, content* (pp. 189–210). London: Chapman & Hall.

Johnson, H. T. (1992). *Relevance regained: From top down control to bottom-up empowerment.* New York: The Free Press.

Johnson, H. T. & Kaplan, R. S. (1987). *The rise and fall of management accounting.* Boston: Harvard Business School Press.

Juran, J. M. (1964). *Managerial breakthrough. A new concept of the manager's job.* New York: McGraw-Hill.

Juran, J. M. (1995). *Managerial breakthrough. The classic book on improving Management performance* (revised ed.). New York: McGraw-Hill.

Kantola, J. & Karwowski, W. (1999). The CIMOP system for integration of technology, organization, and people. In P. Mondelo, M. Mathila & W. Karwowski (Eds), *Proceedings of the International Conference on Computer-Aided Ergonomics and Safety,* Barcelona, Spain, May.

Kaplan, M. C. & Coleman, P. J. (1976). *County highway department hazards: A comparative analysis of inspection and worker detected hazards.* Madison, WI: Wisconsin Department of Industry, Labor and Human Relations.

Kaplan, M. C., Knutson, S. & Coleman, P. J. (1976). *A new approach to hazard management in a highway department.* Madison, WI: Wisconsin Department of Industry, Labor and Human Relations.

Kaplan, R. S. & Norton D. P. (1996). *The balanced scorecard: Translating strategy into action.* Boston: Harvard Business School Press.

Karwowski, W. & Salvendy, G. (Eds.). (1994). *Organization and management of advanced manufacturing.* New York: Wiley.

Karwowski, W., Salvendy, G., Badham, R., Brodner, P., Clegg, C., Hwang, L., Iwasawa, J., Kidd, P. T., Kobayashi, N., Koubek, R., Lamarsh, J., Nagamachi, M., Naniwada, M., Salzman, H., Seppälä, P., Schallock, B., Sheridan, T. & Warschat, J. (1994). Integrating people, organization and technology in advance manufacturing. *Human Factors and Ergonomics in Manufacturing, 4*(1).

Kasarda, J. D. (1993). Cities as places where people live and work: Urban change and neighborhood distress. In H. G. Cisneros (Ed.), *Interwoven destinies: cities and the nation* (pp. 81–124). New York: W. W. Norton.

Katz, D. & Kahn, R. L. (1966a). Common characteristics of open systems. In D. Katz & R. L. Kahn (Eds.), *The social psychology of organizations* (pp. 14–29). New York: Wiley.

Katz, D. & Kahn, R. L. (1966b). *The social psychology of organizations.* New York: Wiley.

Kawakami, M. & Tange, T. (1980). An experimental comparison of two different assembly lines. *Japanese Journal of Industrial Engineering, 31*(2), 181–187.

Kedia, B. L. & Bhagat, R. S. (1988). Cultural constraints on transfer of technology across nations: Implications for research in international and comparative management. *Academy of Management Review, 13*(4), 559–571.

Keidel, R. W. (1994). Rethinking organizational design. *Academy of Management Executive, 8*(4), 12–30.

Kember, P. & Murray, H. (1988). Towards socio-technical prototyping of work systems. *International Journal of Production Research, 26*(1), 133–142.

Kemeny, J. (1980). Saving the American democracy: The lessons of Three Mile Island. *Technology Review, 83*(7), June/July, 65–75.

Kendall, R. M. (1986). Incentive programs with a competitive edge. *Occupational Hazards, 48*(3), 41–45.

Kendall, R. M. (1987). Recognition sparks safety excellence. *Occupational Hazards, 49*(3), 41–46.

Ketchum, L.(1984). Sociotechnical design in a third world country: The railway maintenance depot at Sennar in Sudan. *Human Relations, 37,* 135–154.

Kidd, P. T. (1987). The social shaping of technology: The case of a CNC lathe. *Behaviour and Information Technology, 7*(2), 193–204.

Kidd, P. T. (1990a). Human factors, CIM-Europe and the ESPRIT research programme. *International Journal of Industrial Ergonomics, 5,* 105–112.

Kidd, P. T. (1990b). Organisation, people and technology: Towards continuing improvement in manufacturing. In *Computer integrated manufacturing, Proceedings 6th CIM—European Conference.* London: Springer-Verlag.

Kidd, P. T. (1990c). Information technology: Design for human involvement or human intervention? In W. Karwowski & M. Rahimi (Eds.), *Ergonomics of hybrid automated systems II* (pp. 417–424). Amsterdam: Elsevier.

Kidd, P. T. (1991). Human and computer integrated manufacturing: A manufacturing strategy based on organization, people and technology. *International Journal of Human Factors in Manufacturing, 1*(1), 17–32.

Kidd, P. T. (1992). Interdisciplinary design of skill-based computer-aided technologies: Interfacing in depth. *International Journal of Human Factors in Manufacturing, 2*(3), 209–228.

Kidd, P. T. (1994). *Agile manufacturing: Forging new frontiers*. Reading, MA: Addison Wesley.

Kidd, P. T. & Karwowski, W. (1994). *Advances in agile manufacturing: Integrating technology, organization and people*. Amsterdam: IOS Press.

King, N. & Majchrzak, A. (1995). Concurrent engineering tools: Are the human issues being ignored. *IEEE Transactions on Engineering Management, 43*(2), 189–201.

King, N. & Majchrzak, A. (1996). Concurrent engineering tools: Are the human issues being ignored. *IEEE Transactions on Engineering Management, 43*(2), 189–201.

Kirkpatrick, D. (1975). Techniques for evaluating training programs. In D. L. Kirkpatrick (Ed.), *Evaluating training programs* (pp. 1–17). Madison, WI: American Society for Training and Development.

Kirkpatrick, D. (1979). Techniques for evaluating training programs. *Training and Development Journal, 31*(11), 9–12.

Kleindorfer, P. R. & Kunreuther, H. C. (Eds.). (1987). *Insuring and managing industrial risks: From Seveso to Bhopal and beyond*. New York: Springer.

Kleiner B. M. (1996). Macroergonomics lessons learned from large-scale change efforts in industry, government and academia. In O. Brown, Jr. & H. W. Hendrick (Eds.), *Human factors in organizational design and management—V* (pp. 483–487). Amsterdam: North-Holland.

Kleiner, B. M. (1997). An integrative framework for measuring and evaluating information management performance. *International Journal of Computers and Industrial Engineering, 32*, 545–555.

Kleiner, B. M. (1998). Macroergonomics directions in function allocation. In P. Vink, E. A. P. Koningsveld and S. Dhondt(Eds.) *Human factors in organizational design and management VI*, 635–640. Amsterdam: North-Holland.

Kleiner, B. M. & Drury, C. G. (1999). Large-scale regional economic development: Macroergonomics in theory and practice. *Human Factors and Ergonomics in Manufacturing, 9*(2), 151–163.

Kleiner, B. M. & Hertweck, B. (1996). By which method? Total quality management, reengineering or deengineering. *Engineering Management Journal, 8*(2), 13–18.

Knirk, F. G. & Gustafson, K. L. (1986). *Instructional technology: A systematic approach to education*. New York: Holt, Rinehart and Winston.

Knoke, D. & Kuklinski, J. (1982). *Network analysis*. Beverly Hills, CA: Sage.

Kofler, V. L. & Meshkati, N. (1987). Factors of success in transfer of technology: A case study in cross-cultural training. In M. Marquardt (Ed.), *Corporate culture: International HRD perspectives* (pp. 70–85). Alexandria, VA: American Society for Training and Development.

Kohn, A. (1993). Why incentive plans cannot work. *Harvard Business Review*, September–October, 54–63.

Konz, S. (1995). *Work design. Industrial ergonomics* (4th edition). Scottsdale, AZ: Publishing Horizons.

Kornhauser, A. (1965). *Mental health of the industrial worker: A Detroit study*. New York: John Wiley & Sons.

Kovac, F. J. (1993). Goodyear agility focuses on collaboration, technology and employee empowerment. *Focus, 2*. Brussels: FAST.

Krause, T. R. (1993). Safety and quality: Two sides of the same coin. *Occupational Hazards, 55*(4), 47–50.

Krause, T. R., Hidley, J. H. & Hodson, S. J. (1990). Broad-based changes in behavior key to improving safety culture. *Occupational Health & Safety, 59*(7), 31–37, 50.

Krishnan, U. (1987, June). *Ergonomics/human factors: Engineering applications—a review*. Presented at a lecture session at the Center for Ergonomics of Developing Countries (CEDC), Lulea University of Technology, Sweden. (Citation is by the author's permission.)

Kukkonen, R., Luopajarvi, T. & Riihimaki, V. (1983). Prevention of fatigue amongst data entry operators. In T. O. Kvalseth (Ed.), *Ergonomics of workstation design*. London: Butterworth.

Kuorinka, I. & Forcier, L. (Eds.) (1995). *Work related musculoskeletal disorders (WMSDs): A reference book for prevention*. London: Taylor & Francis.

Kuorinka, I. & Patry, L. (1995). Participation as a means of promoting occupational health. *International Journal of Industrial Ergonomics, 15*, 365–370.

LaBar, G. (1989). Incentives: Focus on safety performance. *Occupational Hazards, 51*(3), 30–33.

LaBar, G. (1990). What if your workers had the "right-to-act"? *Occupational Hazards, 52*(2), 49–53.

Landis, D., Triandis, H. C. & Adamopoulos, J. (1978). Habit and behavioral intensions as predictors of social behavior. *The Journal of Social Psychology, 106*, 227–237.

Lawler, E. E., III. (1969). Job design and employee motivation. *Personnel Psychology, 22*, Winter, 426–435.

Lawler, E. E., III. (1986). *High-involvement management*. San Francisco: Jossey Bass.

Lawler, E. E., III. (1991). *Managing employee involvement*. Los Angeles: University of Southern California.

Lawler, E. E., III. (1992). *The ultimate advantage: Creating the high-involvement organization*. San Francisco: Jossey-Bass.

Lawler, E. E., III. (1996). *From the ground up: Six principles for building the new logic organization*. San Francisco: Jossey-Bass.

Lawrence, P. R. & Lorsch, J. W. (1969). *Organization and environment*. Homewood, IL: Irwin.

Leana, C. R., Locke, E. A. & Schweiger, D. M. (1990). Fact and fiction in analyzing research on participative decision making: A critique of Cotton, Vollrath, Froggatt, Lengnick-Hall, and Jennings. *Academy of Management Review, 15*, 137–146.

Lemann, N. (1991). *The promised land: The great black migration and how it changed America*. New York: Alfred A. Knopf, Inc.

Lemann, N. (1994). The myth of community development. *New York Times Sunday Magazine*, January 9.

Leveson, N. (1995). *Safeware: System safety and computers*. New York: Addison Wesley.

Levine, D. I. & Tyson, L. D. (1990). Participation, productivity, and the firm's environment. In A. S. Blinder (Ed.), *Paying for productivity* (pp. 183–237). Washington: Brookings Institution.

Lewin, K. (1951). *Field theory in social science*. New York: Harper & Row.

Licklider, J. C. R. (1960). Man-computer symbiosis. *IRE Transactions on Human Factors in Electronics*, 4–10.

Liker, J. K., Fleischer, M. & Arnsdorf, D. (1992). Fulfilling the promises of CAD. *Sloan Management Review, 33*(3), Spring, 74–86.

Liker, J. K., Joseph, B. S. & Ulin, S. S. (1991). Participatory ergonomics in two US automotive plants. In K. Noro & A. S. Imada (Eds.), *Participatory ergonomics* (pp. 97–138). London: Taylor & Francis.

Liker, J., Majchrzak, A. N. & Choi, T. (1993). Impacts of programmable manufacturing technology: A review of recent studies and contingency formulation. *Journal of Engineering and Technology Management, 10*, 229–244.

Liker, J. K., Nagamachi, M. & Lifshitz, Y. R. (1988). *A comparative analysis of participatory program in U.S. and Japan manufacturing plants*. Ann Arbor, MI: The University of Michigan, Center for Ergonomics, Industrial and Operational Engineering.

Likert, R. (1961). *New patterns of management*. New York: McGraw-Hill.

Lillrank, B. & Kano, N. (1989). *Continuous improvement—Quality control circles in Japanese industries*. Ann Arbor, MI: The University of Michigan Center for Japanese Studies.

Lock, M. & Strutt, P. (1981). *Reliability of in-service inspection of transport aircraft structures* (CAA Paper 5013). London: Civil Aviation Authority.

Locke, E. A. & Schweiger, D. M. (1979). Participation in decision-making: One more look. In L. L. Cummings & B. Staw (Eds.), *Research in organizational behavior, 1* (pp. 265–339). Greenwich, CT: JAI Press.

Lodge, G. C. & Glass, W. R. (1982). The desperate plight of the underclass: What a business-government partnership can do about our disintegrated urban communities. *Harvard Business Review, 60,* 60–71.

Luczak, H. (1993). Problems and topics of ergonomics (Arbeitswissenschaft) in Germany. In O. Brown, Jr. (Ed.), *Ergonomics in Russia, the other independent states, and around the world: Past, present and future, 2* (pp. 40–59). St. Petersburg, Russia: Russian Ergonomics Association.

Luopajarvi, T. (1987). Workers' education. *Ergonomics, 30*(2), 305–311.

MacKenzie, D. (1985). Design failings that caused Bhopal disaster. *New Scientist,* March, 3–4.

MacKenzie, D. & Wajcman, J. (Eds.). (1985). *The social shaping of technology.* Philadelphia: Open University Press.

Macy, B. A. (1993). *Research in organizational change and development.* New York: JAI Press.

Mager, R. F. (1984). *Preparing instructional objectives* (2nd ed.). Belmont, CA: David S. Lake.

Magnusen, K. (1970). *Technology and organizational differentiation: A field study of manufacturing corporations.* Unpublished doctoral dissertation, University of Wisconsin, Madison, WI.

Mahoney, T. A. & Frost, P. J. (1974). The pole of technology in models of organizational effectiveness. *Organizational Behavior and Human Performance,* 122–138.

Majchrzak, A. (1988). *The human side of factory automation: Managerial and human resource strategies for making automation succeed.* San Francisco: Jossey Bass.

Majchrzak, A. (1990). Effects of CAD on the jobs of drafters and engineers: A quantitative case study. *International Journal of Man-Machine Studies, 32,* 245–262.

Majchrzak, A. (1995). Tools for technology integration. In G. H. Gaynor (Ed.), *Handbook of technology management* (pp. 11.3–11.9). New York: McGraw-Hill.

Majchrzak, A. & Gasser, L. (1992). Towards a conceptual framework for specifying manufacturing workgroups congruent with technological change. *International Journal of Computer-Integrated Manufacturing, 5,* 118–131.

Majchrzak, A. & Paris, M. L. (1995). High-performing organizations match technology and management strategies: Results of a survey. *International Journal of Industrial Ergonomics, 16*(4–6), 309–326.

Majchrzak, A. & Wang, Q. (1994), The human dimension of manufacturing: Results of a survey of electronics manufacturers. *The Journal of Applied Manufacturing Systems, 7*(1), 5–15.

Manufacturing Studies Board. (1986a). *Toward a new era in U.S. manufacturing: The need for national vision.* Washington, DC: National Academy Press.

Manufacturing Studies Board. (1986b). *Human resource practices for implementing advanced manufacturing technology.* Washington, DC: National Academy Press.

Marcum, C. E. (1984). Safety function management fundamentals. *Occupational Hazards, 46*(4), 73–76.

Marx, D. A. & Graeber, R. C. (1994). Human error in aircraft maintenance. In N. Johnston, N. McDonald & R. Fuller (Eds.), *Aviation psychology in practice* (pp. 87–104). Aldershot, UK: Avebury.

Maslow, A. H. (1943). A theory of human motivation. *Psychological Review, 50,* 370–396.

Maurino, D. (1994). Cross-cultural perspectives in human factors training: The lessons from the ICAO human factors programme. *The International Journal of Aviation Psychology, 4*(2), 173–182.

Maurino, D., Reason, J., Johnston, N. & Lee, R., (1995). *Beyond aviation human factors.* Brookfield, VT: Ashgate.

McAfee, R. B. & Winn, A. R. (1989). The use of incentives/feedback to enhance work place safety: A critique of the literature. *Journal of Safety Research, 20*(1), 7–19.

McCormick, E. J. (1957). *Human engineering.* New York: McGraw-Hill.

McCormick, E. J. (1970). *Human factors engineering* (3rd ed.). New York: McGraw-Hill.

McGinnies, E. (1970). *Social behavior: A functional analysis.* Boston: Houghton Mifflin.

McGregor, D. (1960). *The human side of enterprise.* New York: McGraw-Hill.

McWhinney, W. (1990). The power of myth in planning and organizational change. In *IEEE technics, culture and consequences* (pp. 179–184). Los Angeles, CA: IEEE.

Medvedev, G. (1991). *The truth about Chernobyl.* New York: Basic Books.

Meister, D. (1989). *Conceptual aspects of human factors.* Baltimore: Johns Hopkins University Press.

Meredith, J. (1997). *Empirical investigation of sociotechnical issues in engineering design.* Unpublished doctoral dissertation, Virginia Polytechnic Institute and State University, Blacksburg, VA.

Merritt, A. (1994, May). *Cross-cultural issues in CRM training.* Presented at the International Civil Aviation Organization (ICAO) flight safety and human factors seminar, Amsterdam.

Meshkati, N. (1986). Major human factors consideration in technology transfer to industrially developing countries: An analysis and proposed model. In O. Brown, Jr. & H. W. Hendrick (Eds.), *Human factors in organizational design and management—II* (pp. 351–363). Amsterdam: North- Holland.

Meshkati, N. (1988, October). *An integrative model for designing reliable technological organizations: The role of cultural variables.* Presented at the World Bank Workshop on Safety Control and Risk Management in Large-Scale Technological Operations, World Bank, Washington, DC.

Meshkati, N. (1989a). An etiological investigation of micro- and macroergonomic factors in the Bhopal disaster: Lessons for industries of both industrialized and developing countries. *International Journal of Industrial Ergonomics, 4,* 161–175.

Meshkati, N. (1989b, November). *Critical issues in the safe transfer of large-scale technological systems to the third world: An analysis and agenda for research.* Presented at the World Bank Workshop in Risk Management (in Large-Scale Technological Operations), Karlstad, Sweden.

Meshkati, N. (1989c). Technology transfer to developing countries: A tripartite micro- and macroergonomic analysis of human-organization-technology interfaces. *International Journal of Industrial Ergonomics, 4,* 101–115.

Meshkati, N. (1989d, November). *Self-organization, requisite variety, and cultural environment: Three links of a safety chain to harness complex technological systems.* Presented at the World Bank Workshop in Risk Management (in Large-Scale Technological Operations), Karlstad, Sweden.

Meshkati, N. (1991). Human factors in large-scale technological system's accidents: Three Mile Island, Bhopal and Chernobyl. *Industrial Crisis Quarterly, 5,* 133–154.

Meshkati, N. (1992). Ergonomics of large-scale technological systems. Impact of science on society. *The United Nations Educational, Scientific and Cultural Organization (UNESCO) Journal, 165,* 87–97.

Meshkati, N. (1994a). The critical role of ergonomics in ensuring the safety of nuclear power plants around the world. *Proceedings of the 12th Triennial Congress of the International Ergonomics Association, 5, Ergonomics and the Workplace* (pp. 1–3). Toronto: Human Factors Association of Canada/ACE.

Meshkati, N. (1994b). Cross-cultural issues in the transfer of technology: Implications for aviation safety. In *Report of the flight safety and human factors regional seminar and workshop, Amsterdam, the Netherlands* (pp. 116–137). Montreal: International Civil Aviation Organization (ICAO).

Meshkati, N. (1994). Cross-Cultural Issues in the Transfer of Technology: Implications for Aviation Safety. Invited presentation and paper in the *Report of the Flight Safety and Human Factors Regional Seminar and Workshop* (Amsterdam, the Netherlands, 16–19 May). Montreal, Canada: International Civil Aviation Organization (ICAO), 116–137.

Meshkati, N. (1996, October). The role of cultural factors in design and operation of aviation systems. *International Civil Aviation Organization (ICAO) Journal, 51*(8), 17–28.

Meshkati, N. (1996). The role of cultural factors in design and operation of aviation systems. *International Civil Aviation Organization (ICAO) Journal, 51*(8), 17–28.

Meshkati, N. & Robertson, M. M. (1986). The effects of human factors on the success of technology transfer projects to industrially developing countries: A review of representative case studies. In

O. Brown, Jr. & H. W. Hendrick (Eds.), *Human factors in organizational design and management—II* (pp. 343–350). Amsterdam: North-Holland.

Mileti, D. S., Gillespie, D. S. & Haas, J. E. (1977). Size and structure in complex organizations. *Social Forces, 56,* 208–217.

Millar, J. D. (1993). Valuing, empowering employees vital to quality health & safety management. *Occupational Health & Safety, 62*(9), 100–101.

Miller, J. G. & Miller, J. (1990). *The nature of living systems,* pp. 157–163.

Miller, J. M. (1982). The management of occupational safety. In G. Salvendy (Ed.), *Handbook of industrial engineering* (pp. 6.14.1–6.14.18). New York: Wiley.

Minter, S. G. (1991). Quality and safety: Unocal's winning combination. *Occupational Hazards, 53*(10), 47–50.

Mitchell, C. M. (1987). GT-MSOCC: A domain for research on human-computer interaction and decision aiding in supervisory control systems. *IEEE Transactions on Systems, Man, and Cybernetics, 17*(1), 553–572.

Mitroff, I. I. (1988). Crisis management: Cutting through the confusion. *Sloan Management Review,* Winter, 15–20.

Mohrman, S. A. (1982). *Employee participation programs: Implications for productivity improvement.* Los Angeles: University of Southern California.

Montanari, J. R. (1976). *An expanded theory of structural determinism: An empirical investigation of the impact of managerial discretion on organizational structure.* Unpublished doctoral dissertation, University of Colorado, Boulder, CO.

Montgomery, W. E. (1956). Machine guarding. *The Journal of the American Society of Safety Engineers,* November, 57–60.

Moon, S. D. & Sauter, S. L. (Eds.). (1996). *Beyond biomechanics: Psychosocial aspects of musculoskeletal disorders in office work.* London: Taylor & Francis.

Moretz, S. (1988). Incentives provide a competitive edge. *Occupational Hazards, 50*(3), 31–35.

Morris, J. (1984). Practical approaches to safety management—The future is now. *Occupational Hazards, 46*(5), 91–94.

Mosard, G. (1982). A generalized framework and methodology for systems analysis. *IEEE Transactions on Engineering Management, EM-29,* 81–87.

Munipov, V. (1990). Human engineering analysis of the Chernobyl accident. In M. Kumashiro & E. D. Megaw (Eds.), *Towards human work: Solutions and problems in occupational health and safety* (pp. 380–386). London: Taylor & Francis.

Munipov, V. M. (1992). Chernobyl operators: criminals or victims? *Applied Ergonomics, 23*(5), 337–342.

Murphy, S. (1989). The ESPRIT Project. In H. H. Rosenbrock (Ed.), *Designing human-centered technology* (pp. 145–168). New York: Springer-Verlag.

Nadler, G. (1981). *The planning and design approach.* Metheun: GOAL/QPC.

Nadler, G. (1992). The role and scope of industrial engineering. In G. Salvendy (Ed.), *Handbook of industrial engineering* (pp. 3–27). New York: Wiley.

Nadler, G. & Hibino, S. (1994). *Breakthrough thinking: The seven principles of creative problem solving* (2nd ed.). Rocklin, CA: Prima.

Nadler, G., Hibino, S. & Farrell, J. (1995). *Creative solution finding: The triumph of full-spectrum creativity over conventional thinking.* Rocklin, CA: Prima.

Nadler, G., Moran, J. W. Hoffherr, G. D. (1994). Breakthrough thinking in total quality management. *The Total Quality Review, 4*(2), 33–40.

Nagamachi, M. (1971a). Human factors study of monotonous work. *Industrial Engineering, 13*(6), 55–61.

Nagamachi, M. (1971b). Psychological and physiological study of monotony in a vigilance task. *Industrial Engineering, 6,* 50–65.

Nagamachi, M. (1973a). *Design of job enrichment.* Tokyo: Diamond.

Nagamachi, M. (1973b). A new job design and human factors. *Japanese Journal of Ergonomics, 9*(1), 187–196.

Nagamachi, M. (Ed.). (1975). *Theory and practice of job design.* Tokyo: Japan Management Association.

Nagamachi, M. (Ed.). (1985). *Textbook-job redesign.* Elderly Employment Development Association.

Nagamachi, M. (1977). Review of QWL-(2). *Japan Employer Association Times,* April 7.

Nagamachi, M. (1984). *Scientific knowledge about safety management.* Tokyo: Daily Engineering Newspaper Company.

Nagamachi, M. (1987). *Psychology of quality circle.* Tokyo: Kaibundo.

Nagamachi, M. (1991). Application of participatory ergonomics through quality circle activities. In K. Noro & A. S. Imada (Eds.), *Participatory ergonomics* (pp. 139–164). London: Taylor & Francis.

Nagamachi, M. (1993). Macroergonomics, job design and productivity. *The Ergonomics of Manual Works,* 471–476.

Nagamachi, M. (1994). Kansei engineering: A consumer-oriented technology. In G. E. Bradley & H. W. Hendrick (Eds.), *Human factors in organizational design and management—IV* (pp. 467–472). Amsterdam: North-Holland.

Nagamachi, M. (1995). Requisites and practices of participatory ergonomics. *International Journal of Industrial Ergonomics, 15*(5), 371–379.

Nagamachi, M. (1996). *Ergofactory—Challenge to Comfortable factory.* Japan Plant Maintenance Association.

Nagamachi, M. (1997). Challenge to comfortable factory: Concept of ergofactory and participation. In *Proceedings of the 13th Triennial Congress of the International Ergonomics Association, 2* (pp. 211–213).

Nagamachi, M. (1998). Japanese style of production system after lean production. *Ergonomics for Global Quality and Productivity,* 109–112.

Nagamachi, M. & Imada, A. S. (1992). A macroergonomic approach for improving safety and work design. In *Proceedings of the 36th Annual Meeting of the Human Factors and Ergonomics Society* (pp. 859–861). Santa Monica, CA: Human Factors and Ergonomics Society.

Nagamachi, M. Kaneda, K. & Matsubara, Y. (1993). A study of extended garbage can model on self-organization. *Japanese Journal of Industrial Management Association, 44*(3), 191–199.

Nagamachi, M., Matsubara, Y., Nomura, J. & Sawada, K. (1996). Virtual kansei environment and approach to business. In O. Brown, Jr. & H. W. Hendrick (Eds.), *Human factors in organizational design and management—V* (pp. 3–6). Amsterdam: North-Holland.

Nagamachi, M. & Yamada, Y. (1992). Design for manufacturing through participatory ergonomics. In M. Helander & M. Nagamachi (Eds.), *Design for manufacturing.* London: Taylor & Francis.

Nagel, R. N. & Dove, R. (1992). *21st century manufacturing enterprise strategy.* Bethlehem, PA: Iacocca Institute, Lehigh University.

Naisbitt, J. (1984). *Megatrends: Ten new directions transforming our lives.* New York: Warner.

Naschi, G. (1987). Engineering aspects of severe accidents, with reference to the Seveso, Mexico City, and Bhopal cases. In P. R. Kleindorfer & H. C. Kunreuther (Eds.), *Insuring and managing hazardous risks: From Seveso to Bhopal and beyond.* New York: Springer-Verlag.

Nathan, R. P. (1992). *A new agenda for cities.* Columbus, OH: Ohio Municipal League Educational and Research Fund.

National Transportation Safety Board. (1989). *Aircraft accidents report, Eastern Air Lines, Inc., L-1011, Miami, Florida, May 5, 1983.* Washington, DC: Author.

National Transportation Safety Board. (1991). *Aircraft accident report: Avianca, the airline of Columbia, Boeing 707-321B, HK 2016 fuel exhaustion, Cove Neck, New York, January 25, 1990* (Report No. NTSB-AAR-91-04). Washington, DC: Author.

Negandhi, A. R. T. (1975). *Modern organizational behavior.* Kent, OH: The Kent University Press.

Negandhi, A. R. T. (1977). A model for analyzing organization in cross cultural settings: A conceptual scheme and some research findings. In A. R. T. Negandhi, G. W. England & B. Wilpert (Eds.), *Modern organizational theory* (pp. 285–312). Kent State, OH: University Press.

Neumann, J. (1989). Why people don't participate when given the chance. *Industrial Participation, 601,* 6–8.

Newman, L. & Carayon, P. (1994). Community ergonomics: Data collection methods and analysis of human characteristics. In *Proceedings of the Human Factors and Ergonomics Society 38th Annual Meeting* (pp. 739–731). Santa Monica, CA: Human Factors and Ergonomics Society.

New York Times

The New York Times (2000). Pilots just say non to English-only. May 3, Section A, p. 4.

Nickerson, R. S. (1992). Looking ahead: Human factors challenges in a changing world. Hillsdale, NJ. Erlbaum.

NIOSH (1997). *Musculoskeletal disorders and workplace factors: A critical review of epidemiological evidence for work-related musculoskeletal disorders of the neck, upper extremity and low back* (DHHS Publication Number 97-14). Cincinnati, OH: Author.

Nisbett, R. & Ross, L. (1980). *Human inference: Strategies and shortcomings of social judgment.* Englewood Cliffs, NJ: Prentice-Hall.

Noro, K. & Brown, O., Jr. (1990). *Human factors in organizational design and management—III.* Amsterdam: North-Holland.

Noro, K. & Imada, A. (1991). *Participatory ergonomics.* London: Taylor & Francis.

Nuernberger, G. F. (1956). Principles of accident prevention and controls. *The Journal of the American Society of Safety Engineers,* November, 53–60.

Nuclear Safety Review for 1987. Vienna, Austria: International Atomic Energy Agency.

Nuclear Safety Review for 1989. Vienna, Austria: International Atomic Energy Agency.

Oberg, J. E. (1988). *Uncovering Soviet disasters: Exploring the limits of glasnost.* New York: Random House.

Occupational Safety and Health Administration. (1993). *Ergonomics program management guidelines for meatpacking plants* (OSHA 3123). Washington, DC: Author.

Occupational safety and health reporter. (1976). Washington, DC: Bureau of National Affairs.

Olphert, C. W. & Harker, S. D. P. (1994). The ORDIT method for organizational requirements definition. In G. E. Bradley & H. W. Hendrick (Eds.), *Human factors in organizational design and management—IV* (pp. 421–426). Amsterdam: North-Holland.

O'Neill, M. J. (1998). *Ergonomic design for organizational effectiveness.* New York: Lewis.

Organ, D. W. & Bateman, T. S. (1991). *Organizational behavior* (4th/ed.). Homewood, IL: Irwin.

Pasmore, W. A. (1988). *Designing effective organizations: The sociotechnical systems perspective.* New York: Wiley.

Perrow, C. (1967). A framework for the comparative analysis of organizations. *American Sociological Review, 32* 194–208.

Perrow, C. (1981). Normal accident at Three Mile Island. *Society,* July/August, 17–26.

Perrow, C. (1984). *Normal accidents.* New York: Basic Books.

Perrow, C. (1986a). *Complex organizations: A critical essay* (3rd ed.). New York: Random House.

Perrow, C. (1986b). The habit of courting disaster. *The Nation,* October 11.

Petersen, D. (1971). *Techniques of safety management.* New York: McGraw-Hill.

Pierce, B. F. (1963). *The ethnic factor in the man-machine relationship.* San Diego, CA: General Dynamics Corporation.

Pope, W. C. (1981). The strategy of change and safety management. *Occupational Hazards, 43*(11), 61–64.

Pope, W. C. (1990). *Managing for performance perfection. The changing emphasis.* Weaverville, NC: Bonnie Brae Publications.

Pope, W. C. & Cresswell, T. J. (1965). A new approach to safety programs management. *American Society of Safety Engineers Journal, X*(8), 11–16.

Porter, M. E. (1995). The competitive advantage of the inner city. *Harvard Business Review*, May–June, 55–71.

President's Commission on the Accident at Three Mile Island. (1979). *The need for change: The legacy of TMI*. Washington, DC: Author.

Prince-Embury, S. & Rooney, J. (1988, December). Psychological symptoms of residents in the aftermath of the Three Mile Island nuclear accident and restart. *The Journal of Social Psychology, 128*(6), 779–790.

Putz-Anderson, V. (1988). *Cumulative trauma disorders: A manual of musculoskeletal diseases of upper limbs*. Cincinnati: National Institute of Occupational Safety and Health; London: Taylor & Francis.

RAM - Reports no 1 -12. ISSN 0349 - 2015. Stockholm: Stockholm University.

Ramberg, B. (1987). Learning from Chernobyl. *Foreign Affairs*, Winter, 304–328.

Rasmussen, J. (1983). Skills, rules, and knowledge: Signals, signs, and symbols, and other distinctions in human performance models. *IEEE Transactions on Systems, Man and Cybernetics, SMC, 13*(3), 257–266.

Ray, P. S., Bishop, P. A. & Wang, M. Q. (1997). Efficacy of the components of a behavioral safety program. *International Journal of Industrial Ergonomics, 19*, 19–29.

Read, P. P. (1993). *Ablaze: The story of the heroes and victims of Chernobyl*. New York: Random House.

Reason, J. (1988, October). *Resident pathogens and risk management*. Presented at the World Bank Workshop on Safety Control and Risk Management, Washington, DC.

Reason, J. (1990). *Human error*. New York: Cambridge University Press.

Reid, R. (1987). Workers turn on to safety training. *Occupational Hazards, 49*(6), 43–47.

Richardson, V. L. (1973). *Hazard surveys at select employers*. Madison, WI: Wisconsin Department of Industry, Labor and Human Relations.

Robbins, S. R. (1983). *Organization theory: The structure and design of organizations*. Englewood Cliffs, NJ: Prentice-Hall.

Robertson, M. M. (1998). Maintenance resource management. In M. Maddox (Ed.), *Human factors guide for aviation maintenance* (pp. 16/1–16/92). Washington, DC: FAA Office of Aviation Medicine.

Robertson, M. M. (1999). Office ergonomics interventions. In H. J. Bullinger & J. Ziegler (Eds.), *Human computer interaction: Ergonomics and user interfaces, 1* (pp. 205–210). Mahwah, NJ: Lawrence Erlbaum.

Robertson, M .M. & Courtney, T. C. (2001). Office ergonomics: Analyzing the problem and creating solutions. *Professional Safety, 46*(4), 25–31.

Robertson, M. M. & Endsley, M. R. (1995). The role of crew resource management (CRM) in achieving situation awareness in aviation settings. In R. Fuller, N. Johnston & N. McDonald (Eds.), *Human factors in aviation operations* (pp. 281–286). Brookfield, VT: Ashgate.

Robertson, M. M. & Endsley, M. R. (1997). Creation of team situation awareness training for maintenance technicians. In *Human factors in aviation maintenance—phase seven, 1, progress report* (pp. 173–197). Washington, DC: FAA Office of Aviation Medicine.

Robertson, M. M. & O'Neill, M. J. (1994). A systems analysis for integrating macroergonomic research into office and organizational planning. In A. Grieco, G. Molteni, E. Occhipitini, B. Piccoli. (Eds.), *Work with display units, selected proceeding for WWDU'94*, (Elsevier, North-Holland), A6–A7.

Robertson, M. M. & O'Neill, M. J. (1999). Effects of environmental control on stress, performance and group effectiveness. *Proceedings of the Human Factors and Ergonomics Society 43rd Annual Meeting* (pp. 552–556). Santa Monica, CA: Human Factors and Ergonomics Society.

Robertson, M. M. & Rahimi, M. (1990). A systems analysis for implementing video display terminals. *IEEE Transactions on Engineering Management, 37*(1), 55–61.

Robertson, M. M. & Robinson, M. (1995). Enhancing user control of VDT work environments: Training as the vehicle. *Proceedings of the Human Factors and Ergonomics Society 39th Annual Meeting* (pp. 417–421). Santa Monica, CA: Human Factors and Ergonomics Society.

Robertson, M. M., Robinson, M., O'Neill, M. J. & Sless, J. (1998). Measuring the impact of work environment change programs: A systems approach. *Proceedings of the Human Factors and Ergonomics Society 42nd Annual Meeting* (pp. 984–988). Santa Monica, CA: Human Factors and Ergonomics Society.

Robertson, M. M. & Taylor, J. C. (1996). Team training in an aviation maintenance setting: A systematic evaluation. In B. Hayward & A. Lowe (Eds.), *Applied aviation psychology: Achievement, change and challenge* (pp. 373–383). Sydney, Australia: Avebury.

Robertson, M. M., Taylor, J. C., Stelly, J. W. & Wagner, R. H. (1994). Maintenance CRM training: Assertiveness attitudes and maintenance performance in a matched sample. In N. Johnston, N. McDonald & R. Fuller (Eds.), *Aviation psychology in practice*. Sidney, Australia: Avebury.

Robinson, R. D. (1988). *The international transfer of technology: Theory, issues, and practice.* Cambridge, MA: Balliner Publishing Company.

Rogers, A. G. (1991). Organizational factors in the enhancement of aviation maintenance. In *Proceedings of the Fourth Conference on Human Factors Issues in Aircraft Maintenance and Inspection* (pp. 45–59). Washington, DC: FAA Office of Aviation Medicine.

Rogovin, M. (1980). *Three Mile Island: A report to the commission and to the public, 1.* Washington, DC: U.S. Nuclear Regulatory Commission.

Rooney, E. F., Morency, R. R. & Herrick, D. R. (1993). Macroergonomics and total quality management at L. L. Bean: A case study. In N. R. Neilson & K. Jorgensen (Eds.), *Advances in industrial ergonomics and safety V* (pp. 493–498). London: Taylor & Francis.

Rosenbrock, H. H. (1983). Designing automated systems—Need skill be lost. *Science and Public Policy, 10,* 247–277.

Rosenbrock, H. H. (Ed.). (1989). *Designing human-centered technology.* New York: Springer-Verlag.

Rousseau, D. (1989). Psychological and implied contracts in organizations. *Employee Responsibilities and Rights Journal,* 121–139.

Ruffner, J. W. (1990). *A survey of human factors methodologies and models for improving the maintainability of emerging army aviation systems.* Alexandria, VA: U.S. Army Research Institute for the Behavioral and Social Sciences.

Rühl, G. (1974). *Menschengerechte arbeitsplätze durch soziotechnologische* systemgestaltung [Human oriented work design through sociotechnological systems design].

Rusk, D. (1993). *Cities without suburbs.* Washington, DC: Woodrow Wilson Center Press.

Sainfort, F. & Smith, M. J. (1996). Community quality. In J. O. Brown & H. W. Hendrick (Eds.), *Human factors in organizational design and management—V* (pp. 541–546). Amsterdam: North-Holland.

Saito, H. (1969). A study of monotonous work (I). *Labor Science, 45,* 679–739.

Saito, H. (1971). A study of monotonous work (II). *Labor Science, 47,* 243–286.

Salas, E., Dickinson, T. L., Converse, S. & Tannenbaum, S. I. (1992). Toward an understanding of team performance and training. In R. W. Swezey & E. Salas (Eds.), *Teams: Their training and performance* (pp. 3–29). Norwood, NJ: Ablex.

Salazar, N. (1989). Applying the Deming philosophy to the safety system. *Professional Safety,* December, 22–27.

Salvendy, G. & Karwowski, W. (Eds.). (1994). *Design of work and development of personnel in advanced manufacturing.* New York: John Wiley & Sons.

Sanderson, P. M. (1989). *The human planning and scheduling role in advanced manufacturing system* (unpublished report). Champaign-Urbana, IL: Department of Mechanical and Industrial Engineering, University of Illinois at Champaign-Urbana.

Sauter, S. & Smith, K. U. (1971). Social feedback: Quantitative division of labor in social interactions. *Journal of Cybernetics, 1*(2), 80–93.

Sauter, S. L., Dainoff, M. J. & Smith, M. J. (1990). *Promoting health and productivity in the computerized office*. London: Taylor & Francis

Sauter, S. L., Gottlieb, M. S., Rohrer, K. M. & Dodson, V. N. (1983). *The well-being of video display terminal users*. Madison: University of Wisconsin, Department of Preventive Medicine.

Savage, C. M. (1991). Fifth generation management: Integrating enterprises through human networking. Bedford, MA: Digital Press.

Schiller, B. R. (1989). *The economics of poverty & discrimination* (5th ed.). Englewood Cliffs: Prentice Hall.

Schlueter, C. F. (1965). Why should safety be integrated with production? *National Safety News*, February, 26–27, 76.

Schmidt, R. A. & Lee, T. D. (1998). *Motor control and learning, a behavioral emphasis*. Champaign, IL: Human Kinetics.

Scholtes, P. R. (1990). Joiner associates. In W. E. Demming (Ed.), *Quality, productivity and competitive position*. Los Angeles, CA: Quality Enhancement Seminars, Inc.

Schriber J. B. & Gutek, B. A. (1987). Some time dimensions of work measurement of an underlying aspect of organization culture. *Journal of Applied Psychology, 72*, 642–650.

Schwab, D. P. & Cummings, L. L. (1976). A theoretical analysis of the impact of task scope on employee performance. *Academy of Management Review, 1*, 23–35.

Schwarz, J. E. & Volgy, T. J. (1992). *The forgotten Americans: Thirty million working poor in the land of opportunity*. New York: W. W. Norton & Company.

Scott, W. F. (1991). Safety—total-quality-improvement view. *Engineering & Mining Journal, 192*(12), 19–21.

Senders, J. (1980). Is there a cure for human error? *Psychology Today*, April, 52–62.

Shahnavaz, H. (1987). Workplace injuries in the developing countries. *Ergonomics, 30*(2), 397–404.

Shahnavaz, H. (1989). Ergonomics: An emerging concept in industrially developing countries. *International Journal of Industrial Ergonomics, 4*, 91–100.

Shahnavaz, H. (1998). Cultural differences. In *Encyclopedia of occupational health and safety* (4th ed.), (pp. 29, 79). Geneva: International Labor Office.

Shannon, H. S. & Guastello, S. (1997). Workplace safety measures—What can we say about what works? In P. Seppälä, T. Luopajärvi, C. H, Nygärd & M. Mattila (Eds.), *Proceedings of the 13th Triennial Congress of the International Ergonomics Association, 1: Organizational design and management* (pp. 109–111). Helsinki, Finland: Finnish Institute of Occupational Health.

Sharif, N. M. (1988). Basis for techno-economic policy analysis. *Science and Public Policy, 15*(4), 217–229.

Shepherd, W. T., Johnson, W. B., Drury, C. G., Taylor, J. C. & Berninger, D. (1991). *Human factors in aviation maintenance phase 1: Progress report*, Chapter 6. Washington, DC: FAA/OAM.

Sheridan, T. B. (1991). New realities of human factors. *Human Factors Society Bulletin, 34*(2), 1–3.

Sherman, P. J., Helmreich, R. L. & Merritt, A. C. (1997). National culture and flightdeck automation: Results of a multi-nation survey. *International Journal of Aviation Psychology, 7*(4), 311–329.

Shiflett, S., Eisner, E. J., Price, S. J. & Schemmer, M. F. (1985). *The definition and measurement of small military unit team functions* (Report). Ft. Benning, GA: U.S. Army Research Institute for the Behavioral and Social Sciences.

Shrivastava, P. (1987). *Bhopal: Anatomy of a crisis*. Cambridge, MA: Ballinger Publishing Company.

Shrivastava, P., Mitroff, I. I., Miller, D. & Milan, A. (1988). Understanding industrial crises. *Journal of Management Studies, 25*(4), 285–303.

Sinclair, M. A., Siemienluch, C. E., Cooper, K. A. & Waddell, N. (1995). A discussion of simultaneous engineering and the manufacturing supply chain, from an ergonomic perspective. *International Journal of Industrial Ergonomic 16*(4–6), 263–281.

Sink, D. S. & Morris (1995). *By what method*. Norcross, GA: Industrial Engineering and Management Press.

Sink, D. S. & Tuttle, T. C. (1989). *Planning and measurement in your organization of the future.* Norcross, GA: Industrial Engineering and Management Press.

Slovic, P. (1987). Perception of risk. *Science, 236,* 280–285.

Smith, A. (1970). *The wealth of nations.* London: Penguin. (Originally published 1876)

Smith, J. H., Block, R. T. & Smith, M. J. (1996a). Cumulative social trauma and human stress disorders in the age of technology. In A. Derjani (Ed.), *Proceedings of the American Society for Ergonomic Systems Engineering First Ten Year Conference.* Madison, WI: Dept. of Industrial Engineering.

Smith, J. H., Cohen, W., Conway, F. & Smith, M. J. (1996b). Human centered community ergonomic design. In O. Brown, Jr. & H. W. Hendrick (Eds.), *Human factors in organizational design and management—V* (pp. 529–534). Amsterdam: North-Holland.

Smith, J. H., Conway, F. T. & Smith, M. J. (1996). Methods for the conduct of community ergonomic research: Findings from a community setting. In O. Brown Jr. & H. W. Hendrick (Eds.), *Human factors in organizational design and management—V* (pp. 535–540). Amsterdam: North-Holland.

Smith, J. H. & Smith, M. J. (1994). Community ergonomics: An emerging theory and engineering practice. In *Proceedings of Human Factors and Ergonomics Society, 38th Annual Meeting* (pp. 729–733). Santa Monica, CA: Human Factors and Ergonomics Society.

Smith, K. U. (1966). Cybernetic theory and analysis of learning. In E. Bilodeau (Ed.), *Acquisition of skill* (pp. 425–482). New York: Academic Press.

Smith, K. U. (1972). Cybernetic psychology. In R. N. Singer (Ed.), *The psychomotor domain* (pp. 285–348). New York: Lea and Febiger.

Smith, K. U. (1973). Performance safety codes and standards for industry: The cybernetic basis of the systems approach to accident prevention. In J. T. Widner (Ed.), *Selected readings in safety* (pp. 356–370). Macon, GA: Academy Press.

Smith, K. U. (1975a). *Hazard management: Behavioral practices in risk management, industrial safety and workers' compensation.* Madison, WI: Occupational Safety and Health Research Unit, Wisconsin Department of Industry, Labor and Human Relations.

Smith, K. U. (1975b). *Social and human factors design of safety and health programming of risk management of workers' compensation.* Madison, WI: Occupational Safety and Health Research Unit, Wisconsin Department of Industry, Labor and Human Relations.

Smith, K. U. (1979). *Human-factors and systems principles for occupational safety and health.* Cincinnati, OH: NIOSH, Division of Training and Manpower Development.

Smith, K. U. (1988). Human factors in hazard control. In P. Rentos (Ed.), *Evaluation and control of the occupational environment* (pp. 1–7). Cincinnati, OH: NIOSH, Division of Training and Manpower Development.

Smith, K. U. (1990). Hazard management: Principles, applications and evaluation. In *Proceedings of the Human Factors and Ergonomics Society 34th Annual Meeting* (pp. 1020–1024). Santa Monica, CA: Human Factors and Ergonomics Society.

Smith, K. U. & Kao, H. (1971). Social feedback: Determination of social learning. *The Journal of Nervous and Mental Disease, 152*(4), 289–297.

Smith, K. U. & Smith, W. M. (1962). *Perception and motion: An analysis of space structured behavior.* Philadelphia: Saunders.

Smith, K. U. & Smith (1966). *Cybernetic principles of learning and educational design.* New York: Holt, Rinehart and Winston.

Smith, M. J. (1994). Employee participation and preventing occupational diseases caused by new technologies. In G. E. Bradley & H. W. Hendrick (Eds.), *Human factors in organizational design and management—IV* (pp. 719–724). Amsterdam: North-Holland.

Smith, M. J., Bauman, R. D., Kaplan, R. P., Cleveland, R., Derks, S., Sydow, M. & Coleman, P. J. (1971). *Inspection effectiveness.* Washington, DC: Occupational Safety and Health Administration.

Smith, M. J. & Beringer, D. B. (1987). Human factors in occupational injury evaluation and control. In G. Salvendy (Ed.), *Handbook of human factors* (pp. 767–789). New York: Wiley.

Smith, M. J., Carayon, P., Smith, J. H., Cohen, W. & Upton, J. (1994). Community ergonomics: A theoretical model for rebuilding the inner city. In *Proceedings of Human Factors and Ergonomics Society 38th Annual Meeting* (pp. 724–728). Santa Monica, CA: The Human Factors and Ergonomics Society.

Smith, M. J., Cohen, H. H., Cohen, A. & Cleveland, R. J. (1978). Characteristics of successful safety programs. *Journal of Safety Research, 10*(1), 5–15.

Smith, M. J., Cohen, B. G., Stammerjohn, L. W. & Happ, A. (1981). An investigation of health complaints and job stress in video display operations. *Human Factors, 23*(4), 387–400.

Smith, M. J. & Sainfort, P. (1989). A balance theory of job design for stress reduction. *International Journal of Industrial Ergonomics, 4*, 67–79.

Smith, M. J., Sainfort, F., Sainfort, P. C. & Fung, C. (1989). Efforts to solve quality problems. In *Investing in people: A strategy to address America's workforce crisis* (pp. 1949–2002). Washington, DC: U.S. Department of Labor.

Smith, P. C. & Kendall, L. M. (1963). Retranslation of expectations: An approach to the construction of unambiguous anchors for rating scales. *Journal of Applied Psychology, 47*, 149–155.

Smith, P. M. (1992). Some implications of CRM/FDM for flight crew management. *Flight Deck, 5*, Autumn. Hounslow, Middlesex, UK: British Airways Safety Services.

Smith, T. J. (1993). The scientific basis of human factors—A behavioral cybernetic perspective. In *Proceedings of the Human Factors and Ergonomics Society 37th Annual Meeting* (pp. 534–538). Santa Monica, CA: Human Factors and Ergonomics Society.

Smith, T. J. (1994). Core principles of human factors science. In *Proceedings of the Human Factors and Ergonomics Society 38th Annual Meeting* (pp. 536–540). Santa Monica, CA: Human Factors and Ergonomics Society.

Smith, T. J. (1997). Ergonomics and quality—A behavioral cybernetic analysis. In P. Seppälä, T. Luopajärvi, C-H. Nygård & M. Mattila (Eds.), *Proceedings of the 13th Triennial Congress of the International Ergonomics Association, 1, Organizational Design and Management* (pp. 262–264). Helsinki: Finnish Institute of Occupational Health.

Smith, T. J. (1998). Context specificity in performance—The defining problem for human factors/ergonomics. In *Proceedings of the Human Factors and Ergonomics Society 42nd Annual Meeting* (pp. 692–696). Santa Monica, CA: Human Factors and Ergonomics Society.

Smith, T. J. (1999). Synergism of ergonomics, safety, and quality—A behavioral cybernetic analysis. *International Journal of Occupational Safety, 5*(2), 247–278.

Smith, T. J., Henning, R. A. & Smith, K. U. (1994). Sources of performance variability. In G. Salvendy & W. Karwowski (Eds.), *Design of work and development of personnel in advanced manufacturing* (pp. 273–330). New York: Wiley.

Smith, T. J., Henning, R. A. & Smith, K. U. (1995). Performance of hybrid automated systems—A social cybernetic analysis. *International Journal of Human Factors in Manufacturing, 5*(1), 29–51.

Smith, T. J. & Larson, T. L. (1991). Integrating quality management and hazard management: A behavioral cybernetic perspective. In *Proceedings of the Human Factors Society 35th Annual Meeting* (pp. 903–907). Santa Monica, CA: Human Factors Society.

Smith, T. J., Lockhart, R. W. & Smith, K. U. (1983). Safety cybernetics: Theory and practice of involving workers in hazard management programs. In Xth World Congress on the Prevention of Occupational Accidents and Diseases specialist day papers: *Analysis of the risk of accidents at work, methods and applications* (pp. 43–60). Ottawa-Hull, Canada: International Social Security Association for Research and Prevention of Occupational Risks.

Smith, T. J. & Smith, K. U. (1983). Behavioral cybernetic systems principles of hazard management. In *Proceedings of the Xth World Congress on the Prevention of Occupational Accidents and Diseases* (pp. 218–221). Ottawa-Hull, Canada: International Social Security Association for Research and Prevention of Occupational Risks.

Smith, T. J. & Smith, K. U. (1987). Feedback-control mechanisms of human behavior. In G. Salvendy (Ed.), *Handbook of human factors* (pp. 251–293). New York: Wiley.

Soviet Report on the Chernobyl Accident (1986, August 17). Data prepared for the International Atomic Energy Agency Expert Conference, 25–29 August 1986, Vienna, Austria (translated from the Russian). Washington, DC: Department of Energy, NE-40.

Speth, J. G. (1984). *The implications of the industrial disaster in Bhopal, India.* Transcripts of the Hearing Before the Subcommittee on Asian and Pacific Affairs of the Committee on Foreign Affairs, House of Representatives, Ninety-Eighth Congress, December 12, 1984. Washington, DC: U.S. Government Printing Office.

Stevenson, W. B. (1993). Organizational design. In R. T. Golembiewski (Ed.), *Handbook of organizational behavior* (pp. 141–168). New York: Marcel Dekker.

Summary report on the post-accident review meeting on the Chernobyl accident (Safety Series # 75-INSAG-1) (1986). Vienna, Austria: International Atomic Energy Agency.

Swain, A. D. (1973). An error-cause removal program for industry. *Human Factors, 15*(3), 207–221.

Swain, A. D. (1974). *The human element in systems safety: A guide for modern management.* London: Industrial and Commercial Techniques.

Swanson, N. G. & Sauter, S. L. (1999). Psychosocial factors and musculoskeletal disorders in computer work. In W. Karwowski & W. S. Marras (Eds.), *The occupational ergonomics handbook* (pp. 1813–1819). New York: McGraw-Hill.

Sydow, J. (1985). *Der soziotechnische ansatz der arbeits—und organisationsgestaltung: Darstellung, kritik, weiterentwicklung* [The sociotechnical approach to design of work and organization: Description, criticism, further development.]. Frankfurt/Main; New York: Campus.

Szilagyi, A. D., Jr. & Wallace, M. J., Jr. (1990). *Organizational behavior and performance* (5th ed.). Glenview, IL: Scott Foresman.

Taggart, W. (1990). Introducing CRM into maintenance training. *Proceedings of the Third International Symposium on Human Factors in Aircraft Maintenance and Inspection* (pp. 93–110). Washington, DC: Federal Aviation Administration.

Tapscott, D. (1996). *The digital economy.* New York: McGraw-Hill.

TAT (The Technology Atlas Team) (1987). Components of technology for Resource transformation. *Technological Forecasting and Social Change, 32*(1), 19–35.

Taylor, F. W. (1911). *Principles of scientific management.* New York: Harper.

Taylor, J. C. (1991). Maintenance organization. In W. Shepherd, W. Johnson, C. Drury, J. Taylor & D. Berninger (Eds.), *Human factors in aviation maintenance phase 1: Progress report* (pp.?). Washington, DC: FAA Office of Aviation Medicine.

Taylor, J. C. & Felton, D. F. (1993). *Performance by design.* Englewood Cliffs, NJ: Prentice-Hall.

Taylor, J. C. & Robertson, M. M. (1995). *The effects of crew resource management (CRM) training in airline maintenance: Results following three years' experience, contractor's report.* Moffett Field, CA: NASA Ames Research Center, Office of Life and Microgravity Sciences and Applications.

Taylor, J. C., Robertson, M. M. & Choi, S. (1997). Empirical results of maintenance resource management training for aviation maintenance technicians. *Proceedings of the Ninth International Symposium of Aviation Psychology* (pp. 1020–1025). Columbus, OH: Ohio State University.

Thomas, R. R., Jr. (1991). *Beyond race and gender.* New York: AMACOM.

Thompson, J. D. (1967). *Organizations in action.* New York: McGraw-Hill.

Thompson, T. (1964). Accidents and destructive tests. In T. J. Thompson (Ed.), *The technology of nuclear reactor safety.* Cambridge, MA: MIT Press.

Ting, T., Smith, M. & Smith, K. U. (1972). Social feedback factors in rehabilitative processes and learning. *American Journal of Physical Medicine, 51*(2), 86–101.

Topf, M. & Preston, R. (1991). Behavior modification can heighten safety awareness, curtail accidents. *Occupational Health & Safety, 60*(2), 43–49.

Trist, E. L. (1981). The sociotechnical perspective. In A. H. Van de Ven & W. F. Joyce (Eds.), *Perspectives on organization design and behavior* (pp. 19–75). New York: John Wiley.

Trist, E. L. & Bamforth, K. W. (1951). Some social and psychological consequences of the longwall method of coal-getting. *Human Relations, 4,* 3–38.

Trist, E. L., Higgin, G. W., Murray, H. & Pollock, A. B. (1963). *Organizational choice.* London: Tavistock.

Ulrick, P., Fluri, E. Management, 6 Aufl. (6th edition), Bern/stuttgart 1992.

Union Carbide Corporation (1985). *Bhopal methyl isocyanate incident investigation team report.* Danbury, CT: Union Carbide Corporation.

United Kingdom Atomic Energy Authority (1988). *The Chernobyl accident and its consequences.* London: U.K.

United States Department of Commerce (1994). *Empowerment zones enterprise communities guidebook: Federal programs, application guide, and strategic planning.* Washington, DC: United States Government Printing Office.

United States Department of Labor (1989). *The role of labor-management committees in safeguarding worker safety and health.* Washington, DC: U.S. Department of Labor, Bureau of Labor-Management Relations and Cooperative Programs.

USA Today (1989). *Toxic disaster is possible here; few cities are ready.* Arlington, Virginia.

Uth, H. J. (1988, November). *Can Bhopal happen in the Federal Republic of Germany? (Some aspects on managing high risk industries).* Presented at the International Conference on Industrial Risk Management and Clean Technologies, organized by the United Nations Industrial Development Organization (UNIDO) in cooperation with the International Association for Clean Technology (IACT), Vienna, Austria.

Van de Van, A. H. & Delbecq, A. L. (1979). A task contingent model of work-unit structure. *Administrative Science Quarterly,* June, 183–197.

Van Harrison, R. (1978). Person-environment fit and job stress. In C. L. Cooper & R. Payne (Eds.), *Stress at work* (pp. 175–205). New York: Wiley.

Veltri, A. (1991). Management principles for the safety function. *Journal of Safety Research, 22*(1), 1–10.

Verbeek, J. (1991). The use of adjustable furniture: Evaluation of an instruction program for office workers. *Applied Ergonomics, 22*(3), 179–184.

Vink, P., Koningsveld, E. A. P. & Dhondt, S. (Eds.). (1998). *Human factors in organizational design and management—VI.* Amsterdam: North-Holland.

von Bertalanffy, E. (1968). *General systems theory: Foundations, development, applications.* New York: Braziller.

Walker, C. R. & Guest, R. H. (1952). *The man on the assembly line.* Boston: Harvard University Press.

Wall, T., Corbett, J. M., Martin, R., Clegg, C. & Jackson, P. (1990). Advanced manufacturing technology, work design, and performance: A change study. *Journal of Applied Psychology, 75*(6), 691–697.

Wall, T. D., Clegg, C. W. & Kemp, N. J. (1987). *The human side of advanced manufacturing.* New York: John Wiley & Sons.

Walsham, G. (1993). Of IS in developing countries: Power to the people. *Journal of Information Technology, 8,* 74–81.

Walton, M. (1986). *The Deming management method.* New York: Putnam.

Wangler, R. B. (1986). Future safety and health management: A perspective on what lies ahead. *Occupational Hazards, 48*(5), 89–92.

Warnecke, H. J. (1993). *The fractal company—A revolution in corporate culture.* New York: Springer.

Weber, M. (1946). *Essays on sociology* (trans. H. H. Grath & C. W. Mills). New York: Oxford.

Weick, K. E. (1987). Organizational culture as a source of high reliability. *California Management Review, 24,* 112–127.

Weick, K. E. (1988). Enacted sensemaking in crisis situations. *Journal of Management Studies, 25,* 305–317.

Weick, K. E. (1995). *Sensemaking in organizations.* Thousand Oaks, CA: Sage.

Weir, D. (1987). *The Bhopal syndrome.* San Francisco: Sierra Club Books.

Westgaard, R. H. & Winkel, J. (1997). Ergonomic intervention research for improved musculoskeletal health: A critical review. *International Journal of Industrial Ergonomics, 20,* 463–500.

Weiger, W. & Rosman, D., (1989). *National plan to enhance aviation safety through human factors improvements* (p. 7). Washington, DC: Air Transport Association of America.

Wilson, J. R. (1991). Participation—A framework and foundation for ergonomics. *Journal of Occupational Psychology, 64,* 67–80.

Wilson, J. R. (1995). Ergonomics and participation. In J. R. Wilson & E. N. Corlett (Eds.), *Evaluation of human work* (pp. 1071–1096). London: Taylor & Francis.

Wilson, J. R. & Haines, H. M. (1997). Participatory ergonomics. In G. Salvendy (Ed.), *Handbook of human factors and ergonomics* (2nd ed.) (pp. 490–513). New York: Wiley.

Wilson, R. (1987). A visit to Chernobyl. *Science, 236,* 1636–1640.

Wilson, W. J. (1987). *The truly disadvantaged: The inner city, the underclass, and public policy.* Chicago: University of Chicago Press.

Winn, G. L. (1994). Total quality? The "new" paradigm seems out of reach for safety managers. *Occupational Health & Safety, 63*(10), 53–54.

Wisner, A. (1976). *Ergonomics in the engineering of a factory for exportion.* Presented at the Sixth Triennial Congress of the International Erogonomics Association, College Park, MD.

Wisner, A. (1984). Ergonomics or anthropotechnology, a limited or wider approach to working conditions in technology transfer. In H. Shahnavaz (Ed.), *Ergonomics in developing countries.* Lulea, Sweden: Lulea University Press.

Wisner, A. (1995). Situated cognition and action: Implications for ergonomic work analysis and anthropotechnology. *Ergonomics, 38,* 1542–1557.

Wobbe, W. & Charles, T. (1994). Human roles in advanced manufacturing technology. In W. Karwowski & G. Salvendy (Eds.), *Organization and management of advanced manufacturing.* New York: John Wiley & Sons.

Wolf, H. J. & Pearson, J. C. (1992). Happy workers mean fewer injuries. *Safety.*

Womack, J., Jones, D. & Roos, D. (1990). *The machine that changed the world.* New York: Rawson Associates.

Woodward, J. (1965). *Industrial organization: Theory and practice.* London: Oxford University Press.

Yankelovich, D. (1979). *Work values and the new breed.* New York: Van Nostrand Reinhold.

Yankelovich, D. (1988). *Starting with the people.* Boston: Houghton Mifflin.

Zaidel, D. M. (1991). *Specification of a methodology for investigating the human factors of advanced driver information systems.* Ottawa: Transport Canada Publication No. TP 11199.

Zhang, Y. K. & Tyler, J. S. (1990). The establishment of a modern telephone cable production facility in a developing country. A case study. *International Wire & Cable Symposium Proceedings, 96–104.*

Zink, K. J. (1984). Zur notwendigkeit eines sozio-technologischen ansatzes [The necessity of a sociotechnological systems approach]. In K. J. Zink (Ed.), *Sozio-technologische systemgestaltung als zukunftsaufgabe* [Sociotechnological systems design as task of the future] (pp. 25–49). München: Hanser.

Zink, K. J. (1996). Continuous improvement through employee participation: Some experiences from a long-term study in Germany. In O. Brown, Jr. & H. W. Hendrick (Eds.), *Human factors in organizational design and management—V* (pp. 155–160). Amsterdam: North-Holland.

Zink, K. J. (1997). *Total quality management as a holistic management concept.* Heidelberg/ New York: Springer.

Zink, K. J. (1998). Self assessment—A holistic approach for the evaluation of health promotion projects. In P. A. Scott, R. S. Bridger & J. Charteris (Eds.), *Global ergonomics, proceedings of the ergonomics conference* (pp. 233–238). Amsterdam: Elsevier.

Zink, K. J. (1999a). Human factors and business excellence. In J. Axelson, B. Bergmann & J. Eklund (Eds.), *Proceedings of the international conference on TQM and human factors, 1* (pp. 9–27). Linköping, Sweden: Linköpings University. CMTO.SE-58183.

Zink, K. J. (1999b). Ergonomics and human factors in Germany: A discussion at the edge of the new century. In The Hellenic Ergonomics Society (ed.), *Proceedings of the symposium, strength & weaknesses, threats & opportunities of ergonomics in front of 2000* (pp. 83–87). Athens: The Hellenic Ergonomics Society.

Zink, K. J. (2000). Ergonomics in the past and the future: From a German perspective to an international one. *Ergonomics, 43*, 920–930.

Zohar, D. (1980). Safety climate in industrial organizations: Theoretical and applied implications. *Journal of Applied Psychology, 65*, 96–102.

Zwerman, W. L. (1970). *New perspectives on organization theory.* Westport, CT: Greenwood.

Glossary

Abstract functioning: Cognitively complex conceptual functioning characterized by a high degree of differentiation and integration.

Accident: An event that takes place without one's foresight or expectation; an event that proceeds from an unknown cause, or is an unusual effect of a known cause, and therefore not expected.

ACTION: An interactive software system and a methodology that embodies an extensive knowledge base about the relationships among technical, organizational, and strategic features of the manufacturing enterprises.

Action groups: An action group is individuals with similar interests about an issue that form an association to pursue action to promote the issue of interest.

Adhocracy: A rapidly changing adaptive work system organized around problems to be solved by groups of relative strangers with diverse professional skills.

Advanced information technologies (AIT): Emerging computer-based innovations in the early stages of their life cycle that make use of such capabilities as broadband and wireless technology.

Agile manufacturing: A manufacturing philosophy and methodology that enables a firm to respond successfully to changes in the marketplace.

Akamiso: Negative affect. Feelings or emotional states or temporal moods that can cause operators to underestimate the real risk of their actions. Understanding the sources of *akamiso* is an important macroergonomic step in improving systems safety and, often, productivity.

Amoeba-type organization. See Free-form design.

Behavioral cybernetics: The science that applies feedback principles to the study and improvement of human performance and learned behavior. The theoretical and empirical analysis of behavior as a closed-loop, self-governed control process.

Behavioral hazards: System performance hazards linked to variability in the behavior of those responsible for system operation.

Behavioral safety: Those attributes of the safety performance of a system influenced by the behavior of those responsible for system operation.

Bottom-up: A work system analysis and design approach that proceeds from the individual worker level up through the work system's subunits to the overall work system level.

Boundaries: Work system borders that separate domains of responsibility.

Breakthrough thinking: A conscious process of thinking and problem solving that applies seven principles of solution-finding to any problem faced by an individual, group, or company. The seven principles are: uniqueness, purposes, solution-after-next, systems, limited information collection, people design, and betterment timeline.

Centralization: The degree to which formal decision making is concentrated in a relatively few individuals, group, or level, usually high in the organization.

Change agent: A person whose role is to facilitate change within an organization to better enable it to meet its goals.

Checkpoint: Specific, standardized places in the work process of a work system.

CIM: Computer Integrated Manufacturing. A manufacturing strategy, incorporating computers, for linking existing technology and people to optimize business objectives.

CIMOP: Computer Integrated Manufacturing, Organization, and People. A methodology that evaluates the CIM system, organizational design, technology, people, and information system to determine overall system design quality.

Closed loop: A feedback system in which the output from a system is fed back to the system input to provide incremental improvement in the output.

CNC: Computer Numerical Control (CNC) machines comprised of small, inexpensive, microprocessors with large memories, and programmable computers that have such functions as program storage, tool offset and compensation, program-editing capability, various degrees of computation, and the ability to send and receive data from a variety of sources.

Cognitive complexity: A higher order structural personality trait; the extent to which people have developed differentiation and integration in their conceptual functioning; the extent to which persons are abstract, as opposed to concrete, in their conceptualizing.

Cognitive ergonomics: The aspect of ergonomics concerned with the design of the interfaces among human mental, perceptual, and information processing characteristics with other sociotechnical system elements—particularly software.

Collaboration: Group participation on a common task.

Community-based organization: An institution, organization, association, or group that has local ties to the community in which the organization is involved.

Community development group: An association of people with a primary interest in enhancing the economic, social, and/or safety aspects of a community.

Community-environment system: An extrapolation from the P:E Fit Theory which describes the relationships between the community and the environment in which the community operates. The community-environment system is one with multilateral and continuous interactions among populations, conditions, institutions, situations, and objects. These are linked through feedback, forming a unified whole leading to the vital functions of lifespan development. The community is surrounded by an environment comprised of other communities and institutions, which may have very different beliefs, values, and modes of behavior. This community-environment system may include institutions for education, financial transactions, government and politics, commerce and business, law enforcement, transportation, and housing.

Community ergonomics: The application of systems theory and ergonomic principles of fitting the environment to the capabilities of the community to produce improvements in the community.

Community self-regulation: The control of the environment by the community using technology, activities (tasks), and work organization processes.

Compatibility: The sociotechnical principle that for a work system to exhibit certain characteristics (e.g., human-centered), its design or redesign process must incorporate those same characteristics.

Complexity: The degree of *differentiation* and *integration* existing in a work system. Complexity also can refer to cognitive complexity (see definition above).

Computer-aided design (CAD): Designing products using computers; utilizes computer software programs specifically developed for aiding in design.

Computer-integrated manufacturing (CIM): Linking together through computer technology all the various departments in an industrial company so they operate smoothly as a single, integrated business system.

Computer-supported collaborative work technology: Information technology that supports human-to-human interaction for shared tasks.

Concrete functioning: Cognitively simple conceptual functioning; characterized by a relatively low degree of differentiation and integration.

Concurrent engineering: Engineering functions, such as design, analysis, and production that are performed simultaneously with a lot of cross-functional interaction.

Consensus: Final decision agreement/support within a group; does not necessarily mean that individual members agree personally.

Consumer feedback: Information from those who received products and/or services about their satisfaction with those products and/or services.

Consumers: Those who receive and usually pay for outputs (i.e., products and/or services) from a work system.

Context specificity: Variability in system performance attributable to context (design) features of the performance environment.

Continuous flow manufacturing: Similar to *just-in-time* and *flexible manufacturing*; the primary objective is to produce a high-quality product in the shortest possible production time at the lowest possible cost. Provides techniques to reduce product cycle times, minimize inventories, improve quality, and increase inventory turns.

Continuous improvement: The incremental betterment of performance over time. In the sociotechnical systems literature, it is sometimes called *incompletion*.

Control chart: A quality assurance tool that uses historical data to create upper and lower control limits (from standard deviations) to plot data and evaluate current performance of processes.

Coupling: Whether participative data are directly or remotely used, in which the former involves little or no filtering of participant input and the latter involves some filtering or translation, usually by managers or consultants.

Cross-functional team: A parallel or informal team composed of members from several functional units or departments. These members can represent different skills and ranks. The team usually has partial decision-making authority (i.e., recommendations) and often is involved in TQM (*see* Total quality management or other process improvement activities.

Cumulative social trauma (CST): Results from long-term exposure to extreme detrimental societal conditions leading to a vicious cycle of dependency, social isolation, and learned helplessness. CST is a chronic social condition of a community that leads to community deterioration. Symptoms include community flight, economic depression, housing decay, decreased physical and mental health, and reduced opportunity for a quality life.

Decision support technologies: Mechanisms that assist in converting data to information so decisions are easier to make or better.

Deming flow diagram: A type of organizational input/output diagram popularized by the late W. Edwards Deming, beginning with his involvement in Japan in the 1950s and extending through the TQM movement in the United States in the 1980s.

Departmentalization: Division of a work system's labor into groups of specialists.

Design specificity: *See* Context specificity.

Differentiation: The number of conceptual categories a person has developed for storing experiential information. In organizational design, the number of hierarchical levels and/or departments that comprise the structure of the work system.

Domain: An organization's range of products or services offered and market share.

Empowered: Having a sufficient degree of decision-making authority.

Engineering controls: Reliance on modification in the engineering features of a system for the purpose of hazard control.

Environmental uncertainty: The extent to which an organization's specific task environment is (a) complex, in terms of its number of components, and (b) dynamic versus stable over time.

External environment: Factors external to the organization that permeate it. Examples are materials sources, customers, government policies and regulations, and stockholders.

Facilitator: A person who manages the meeting process of a group; for example, managing the time and uniformity of participation.

Fitt's list: Original list of basic human versus machine capabilities and limitations to guide system design function allocation.

Flexibility: The capability to change in response to environmental change.

Flexible manufacturing: Ability for the internal manufacturing system to cope with changes dictated by a dynamic environment.

Focal role: The work system function that is expected to control the most significant key variances.

Focus groups: A temporary collection of selected participants to discuss specific issues or test-market specific ideas or products.

Formalization: The extent to which jobs within a work system are standardized, including use of formal rules and procedures and explicit job descriptions.

Free-form design: An adhocracy type of organizational design in which there is no functional departmentalization and the shape of the work system changes rapidly in response to its external environment in order to survive. Departmentalization is replaced by a profit center arrangement. Also referred to as an *amoeba-type organization*.

Full direct participation: Having all those affected by a decision or design become involved.

Function allocation: A methodology for assigning tasks and/or functions to humans and/or machines.

Function analysis modeling: A macroergonomic approach for identifying work system functions and related quantitative and qualitative personnel subsystem requirements.

Gap: A variance or deviation from what is expected or needed.

Garbage can model for organizational design: A modification of the garbage can model for organizational decision making for use as a tool to evaluate work system design alternatives.

Harmonized work system: A work system in which all subsystems and components are synchronized and behave as a single unit.

Hazard: A factor inherent to the design and/or human elements of a system that elevates the risk of decremental system performance.

Hazard management: That approach to system safety management that focuses on the detection, evaluation, and abatement of system hazards.

HCCIM: Human-Centered Computer Integrated Manufacturing. A methodology that focuses on the unification of planning, execution, and monitoring of work in order to minimize division of labor.

High involvement: On the upper end of the participation continuum; exemplifying significant design or decision-making involvement.

HITOP: High Integration of Technology, Organization, and People. A methodology that allows one to conduct an analysis of a technology-organization-people (TOP) system.

Horizontal differentiation: The degree of departmentalization and specialization within a work system.

Human-centered approach: An approach to human–machine function and task allocation that first considers the capabilities and limitations of the human and whether the function or task justifies the use of a human. Also called the *humanized task* approach.

Human-centered manufacturing: A manufacturing concept that emphasizes human skills; control over technology; unification of planning, execution, and monitoring; maximization of human operator knowledge; and consideration of ergonomic factors.

Human error: Error in system performance that is attributed to, or associated with, human behavior.

Human networking: One of the generic features of the agile manufacturing paradigm in which workers are networked together by interactions that occur among all people in the organization for generating expected productivity and quality results.

Human relations theory: In contrast to classical theory, a school of thought that centers on the assumption that workers' feelings and attitudes are important and can possibly have an impact on performance.

Human-system interface technology: The unique technology of the human factors/ergonomics discipline; it consists of empirically derived design principles, guidelines, specifications, tools, and methods to design human–organization, human–job, human–machine, human–software, and human– environment interfaces.

Humanized task approach: *See* Human-centered approach.

Ideal bureaucracy: The classical bureaucratic design developed by Max Weber.

Incompletion: *See* Continuous improvement.

Input variance: An unexpected or unwanted deviation of a resource from standard operating conditions, specifications, or norms.

Instructional objectives: Specify what the trainee can accomplish on completion of the training at a specified standard level.

Instructional strategy: A plan for assisting learners with their study efforts for each performance objective. Developing a strategy is completed before developing the training materials to outline how the instructional activities will relate to the accomplishment of the objectives. Instructional events are defined and the instructional designer directs the attention of the learner to the objectives, informs the learner of the objectives, presents the stimulus materials, and provides feedback on their learning performance.

Instructional systems design model (ISD): Uses a systematic approach to designing, developing, and evaluating training and instructional programs. It has five stages of activities beginning with a needs analysis phase, followed by a design phase in which instructional objectives and strategies are defined, then the development phase of the training materials, the implementation and delivery of the training, and finally, the evaluation phase to determine the effectiveness of the training.

Instrumental feedback: The response that the kinesthetic and tactile receptors receive from the person's interface with tools and technology. It often is called the "feel" of the tool.

Integration: The number of rules and combinations of rules a person has developed for integrating conceptual information. In organizational design, the number of mechanisms designed into the work system for ensuring communication, coordination, and control among the differentiated elements (e.g., standard operating rules and procedures, committees, task teams).

Internal control: Mechanisms and/or processes that serve as checks and balances on the internal processes of a work system.

Investment zone: A preferred area of the community that is given special tax advantages to attract investment.

Job: A formal position in an organization as documented and detailed by a formal job description.

Job enlargement: Tasks added at the same level of responsibility.

Job enrichment: Tasks added at a higher level of responsibility.

Joint design: Attending to both personnel and technical factors simultaneously in the design process.

Joint optimization: A sociotechnical systems design principle that states that the technological and personnel subsystems must be designed jointly in order to achieve the most effective functioning of the work system.

Kanzei engineering: Ergonomic technology of product development that translates a consumer's feelings about a new product into design requirements.

Key variance: A deviation from what is expected, needed, or wanted that affects key performance criteria significantly or a variance that has a multiplicative relationship with other variances.

Key variance control table: A tabular representation of which roles control which key variances and how. In addition, special technological support and training requirements needed to control key variances are identified.

Knowledge-based technology: A means of classifying technology based on task variability, or the number of exceptions or nonroutine problems created by the technology, and task analyzability, or the extent to which the problems created lend themselves to rational-logical, quantitative, and analytical thinking as opposed to reliance on the experience, judgment, and intuition of the problem-solver.

Lean manufacturing: A manufacturing concept that requires a system-level change for the organization, a change that affects every segment of the company, from accounting to shipping, that begins with the manufacturing system.

Learned helplessness: A human condition produced by cumulative exposure to failure and rejection that leads to ineffective behavior when challenged.

Learning hierarchy: Developing a learning hierarchy involves designing effective conditions for learning the defined skill and the proper sequence of prerequisite skills. The learning hierarchy is a result of the learning-task analysis in that a hierarchy of intellectual skill objectives is arranged into a pattern that shows the prerequisite relationships among them.

Learning outcomes: Defined by the instructional objectives and organized by five categories of learning domains: intellectual skills, cognitive strategies, verbal information, motor skills, and attitude. Understanding and identifying these learning outcomes help determine the appropriate instructional conditions for the trainees.

Learning task analysis: Conducted to determine the necessary objectives for the learner. Enabling objectives are objectives that the learner must accomplish first in the sequence of a learning hierarchy. Typically this analysis is conducted for intellectual skills.

Likert-type (survey): A questionnaire with a graded response to each statement, typically on a 5-point scale: strongly agree, agree, undecided, disagree, and strongly disagree.

Machine bureaucracy: The bureaucratic form that evolved from Max Weber's *ideal bureaucracy* and Fredrick W. Taylor's *scientific management*. It is characterized by narrowly defined jobs, routine and well-defined tasks, a well-defined hierarchy, high formalization, and centralized decision making.

Macroergonomic analysis and design (MEAD): A 10-step framework for conducting work system improvements.

Macroergonomic hazards: That class of system hazards related to defects in the organizational design and management of system safety.

Macro-ergonomic level: The overall work system level of ergonomic application.

Macroergonomics: The subdiscipline of ergonomics that focuses on the design of the overall work system. Conceptually, a top-down sociotechnical systems approach to the design of work systems and the carry-through of the overall work system design characteristics to the microergonomic design of human–job, human–machine, and human–software interfaces to ensure that the entire work system is fully harmonized.

Maintenance resource management (MRM): A human factors program designed to improve communication, effectiveness, and safety in airline maintenance operations. MRM also is used to change the "safety culture" of the organization by establishing a positive attitude towards safety among maintenance personnel.

Mass production: The mode of production in which the items produced are essentially the same and thus lend themselves to mass production techniques, such as assembly lines.

Matrix organization: An adhocracy type of organizational design that combines departmentalization by function with departmentalization by project or product line.

Mechanistic work systems: Work systems characterized by high vertical and horizontal differentiation, formalization, and centralization. They typically have routine tasks and programmed behaviors and can respond to change only slowly.

Mental model: Cognitive representations, based on past experience, that guide current perceptual activity.

Micro-ergonomics: Those aspects of ergonomics primarily focused on the design of the interfaces between the individual and other system elements, including human–job, human–machine, human–software, and human–environment interfaces.

Middle-out: A work system analysis and design approach that proceeds from an intermediate or sub-unit level of the work system both up to the overall work system level and down to the individual worker level.

Mission: The goal or purpose of a work system.

Modular form: A relatively new form of adhocracy that outsources nonvital functions while retaining full strategic control.

Monte Carlo technique: An empirical study of statistics using random numbers. It is applied to empirical studies of behavioral models or methods that the investigator wishes to explore. A way of randomly selecting a variety of items within a set so the selection is constrained by the correlations between the dimensions.

Open systems: Systems that are open to being influenced by, and influencing, their external environment, such as sociotechnical systems.

Operational hazards: That class of system hazards related to operational aspects of the system. Risk factors for injury or illness produced by the person's interaction with technology. They are the interaction of behavior and physical hazards that may potentiate the risk of injury. They often are associated with dynamic operational circumstances and/or with intermittent or transitory conditions of system performance for which no physical standards or standard operating procedures exist.

Organic work systems: Work systems characterized by relatively low vertical differentiation and formalization with decentralized tactical decision making, enabling them to be flexible and adapt quickly to change.

Organization: The planned coordination of two or more people who, functioning on a relatively continuous basis and through division of labor and a hierarchy of authority, seek to achieve a common goal or set of goals.

Organizational design: The design of a work system's structure and related processes to achieve the organization's goals.

Organizational requirements definition tools (ORDIT): A set of automated tools designed to assist in the specification of work system requirements for information technology systems using an integrated methodology. Developed by the HUSAT Research Institute at the Loughborough University of Technology, UK.

OSHA: U.S. Department of Labor Occupational Safety and Health Administration.

Pareto analysis: Any procedure that identifies which 20% of the variance causes 80% of an impact on performance.

Partial direct participation: Having a representative subset of participants involved, rather than the entire group; usually necessary because of economic considerations.

Participation: A general term for involvement of users or others in a task.

Participative management: A style of supervision that involves workers in decision making, at least at the level of providing recommendations and sometimes at a level of complete delegation.

Participatory ergonomics: The involvement of employees in the ergonomic analysis and design of their work environments and activities.

Participatory hazard management: Involving workers in the hazard management process by assigning responsibility to them for hazard recognition, evaluation, and abatement, as well as hazard management decision making.

Passive-aggressive: The acting out of anger or hostility by being passive, not doing things, or doing them slowly or inefficiently. In organizations, this often takes the form of doing the minimum to get by rather than what is really required to get the job done effectively.

P-E Fit Theory: Theory proposed by researchers at the Institute for Social Research (ISR) at the University of Michigan to define the relationship between the person and the environment in which the person operates. The theory proposed ways to improve the "fit" between people and the environment. These researchers included Robert Kahn, Jack French, Robert Caplan, Stan Seashore, and others. The core premise of the person–environment fit theory is that stress arises not from the person or environment separately, but rather by their fit or congruence with one another.

Performance standards: Standards of system safety performance associated with the performance of those responsible for managing system safety.

Personnel subsystem: One of the four basic elements of a sociotechnical system. It consists of the people who make up the organization's workforce.

Physical hazards: That class of system hazards related to defects in physical/environmental design features of the system.

"Picking the low-hanging fruit": Selecting projects for ergonomic intervention in which there is a high probability of showing improvement in productivity, health, safety, or other important organizational criteria in a relatively short time.

Presence support: Technological augmentation that attempts to simulate the perception or feeling of being in a real environment.

Principles: The guiding principles (for behavior) of a work system.

Process: A series of steps that convert inputs to outputs in a system.

Process production: The mode of production in which the production process is continuous, such as oil and chemical refineries.

Production technology: A means of classifying technology based on mode of production (e.g., unit, mass, or process).

Production type: The categories of production systems offered by various taxonomies, such as "craft," "unit," "mass," and "process" in the case of the production mode classification system.

Professional bureaucracy: A bureaucratic design that relies on a high degree of professionalism in the jobs that comprise the work system. It is characterized by more broadly defined and less routine jobs than found in a machine bureaucracy, relatively low formalization, and decentralization of tactical decision making.

Profit center: A structural characteristic of free-form adhocracies used in place of departmentalization. Profit centers consist of highly professionalized, results-oriented work teams.

Psychosocial: The reciprocal influence or interaction of the mental and emotional characteristics of the individual with the social characteristics of the group.

Quality circles: Permanent organizational teams focused on problem solving, popularized by the Japanese in the 1960s.

Quality management: Total quality management (TQM) is an approach for continuously improving the quality of goods and services delivered and, therefore, meeting or exceeding the needs and expectations of customers through the participation of all levels and functions of the organization.

Quasi-experiment: An empirical study in which independent variables are manipulated to evaluate the effect on dependent variable(s) but in which representative selection and/or assignment are not performed.

Reactive feedback: Reactive feedback is the response that the sensory receptors in the muscles and joints receive from motor activity.

Reengineering: The redesign of work processes to improve efficiency and productivity.

Relevant task environment: Those parts of an organization's external environment that can positively or negatively influence the organization's effectiveness.

Role analysis: The process of evaluating work roles. This can include the expectations others have for these roles as well as the perceptions held by role occupants.

Role network: Sometimes called *role set*, the conceptualization of various work roles and how they interact with one another to form a social subsystem.

Role set: Sometimes called *role network*, the conceptualization of various work roles and how they interact with one another to form a social subsystem.

Safety: Freedom/security from danger, injury, or damage.

Safety management: That organizational function or program with a general focus on safety and accident prevention.

Safety performance: The integrated performance of all organizational and individual participants in a system whose activities affect system safety.

Scanning: A term from sociotechnical systems theory referring to a broad-based evaluation or analysis, usually focused on the system or environment being studied or improved. A preliminary high-level analysis.

Scientific management: A method of work design, developed by Frederick W. Taylor at the beginning of the 20th century. The systematic observation of workers to determine the "one best way" of performing each task, and then, training workers to follow it.

Self-managed teams: In contrast to a cross-functional team, a permanent team with significant decision-making authority, usually including selection, assignment, scheduling, and work design. These teams are sometimes called *autonomous* or *high-performance* work teams.

Self-regulation: The control of the environment by the individual, organization, or community using technology, activities (tasks), and work organization processes.

Semistructured interview: An interview procedure in which a set of basic questions are developed for each interviewee, but then the interviewer improvises with additional questions as required to follow up on the interviewee's answers to the basic questions to gain additional information.

Sequential engineering: Engineering functions, such as design, analysis, and production, are performed independently with the output of one function serving as the input to the next.

SHELL model: Software, Hardware, Environment, Liveware model. This model represents how human factors is defined as a system and the various interactions that occur between the subsystems and the human operator as it relates to the aviation community.

Situational awareness: Perception of the immediate environment.

"Smart" products: System outputs with ingrained information-processing intelligence.

Social boundaries: The borders created by the formal organization chart's definition of jobs.

Social tracking (or social feedback control): A synonym for social feedback control. It is used to designate the fact that such control is continuous and not discrete, that it involves self-generated action and not S-R [Stimulus-Response] processes, and that it entails various modes, conditions, and parameters of closed-loop regulation of interpersonal and group activity.

Socio-informational view: Refers to an effort to develop new work structures having greater system transparency, higher system flexibility, and decentralization. This view emphasizes congruency and integration of people, technology, and organization.

Sociotechnical systems: Work systems composed of (a) a technological subsystem, (b) a personnel subsystem, (c) an external environment that interacts with the organization, and (d) an organizational design.

Span of control: The number of employees a given manager can directly supervise effectively. Span of control is affected by a number of work system design factors.

Spatial dispersion: The extent to which an organization's activities are performed in multiple locations. It is measured by (a) the number of geographic locations constituting the total work system, (b) the average distance of the separated locations from the organization's headquarters, and (c) the proportion of employees in these separated units relative to the number in the headquarters.

Specific task environment: The particular combination of relevant task environments for a given organization.

Stakeholders: Individuals or groups with a vested interest in a work system.

Steady state: A constant trend in performance.

Strategic decisions: Decisions that deal with the long-range vision and goals of the organization.

Strategic planning: Long-range planning for an organization, usually looking 5 to 10 or more years into the future. Strategic planning typically results in strategic planning documents and "roadmaps." Popularized by General Electric in the 1960s.

Stratified semistructured interview: A procedure in which a sample of interviewees is systematically selected by organizational level, department, and so on to ensure that it will be representative of the entire work system of interest. Each selected interviewee then goes through a semistructured interview process with an interviewer. This procedure is widely used in conducting organizational assessments.

Structural analysis: Analysis of the organizational structure of a work system.

Structural form: The type of organizational structure utilized by a given work system.

Subenvironments: Categorical subunits within the external environment of a work system.

Suppliers: Providers of resources or inputs to a work system.

Survey feedback method: A method in which the data gained from an organizational questionnaire survey are summarized and subgrouped statistically and by organizational level, department, project, and so on and then are fed back to the individual organizational units for interpretation and, where applicable, action to improve organizational functioning.

System harmonization: The work system condition achieved when all subsystems are synchronized and behaving as a single unit.

System inputs: Resources provided by suppliers to a work system.

Systematic organizational design methodology (SORD): Developed for designing U.S. Army organizational units, SORD is a step-by-step set of computer-assisted procedures for the comprehensive, systematic, integrative, and reliable design of work systems.

Tactical decisions: Decisions that deal with the day-to-day operation of the work system.

Task allocation: The process of assigning tasks to humans or machines in designing or modifying a work system; includes the allocation of tasks to specific work modules and jobs.

Technological complexity: A scale of production mode complexity, with unit production being the least technologically complex, mass production intermediate, and process production the most complex.

Technological imperative: The often-held view, unsupported by the research literature, that technology has a compelling influence on work system structure and should determine work system design.

Technological subsystem: One of the four basic elements of a sociotechnical system; it consists of the machines, tools, software, and other technological components of the organization.

Territorial boundaries: The borders around the physical space used for product conversion.

Throughput boundaries: The work system borders, from the input owned by the system to the output for distribution to consumers.

Throughput variance: An unexpected or unwanted deviation in a process from standard operating conditions, specifications, or norms.

Time boundaries: The temporal borders related to such characteristics as seasonality, and the number and timing of shifts.

Top-down: A work system analysis and design approach that proceeds from the overall work system level down through the work system's subunits to the individual worker level.

Total quality management (TQM): A management system and philosophy that uses cross-functional teams and tools to continuously improve business processes for the satisfaction of customers. Deming, Juran, and Crosby are some of the names associated with the movement. In the United States, the Malcolm Baldrige National Quality Award recognizes companies for achievement in this area.

Unit operations: Groupings of conversion steps from inputs to outputs that together form a complete or whole set of tasks and are separated from other steps by territorial, technological, or temporal boundaries.

Unit production: The mode of production in which each item produced is unique, rather than essentially the same; for this reason, items do not lend themselves to mass production techniques.

Usability: The extent to which a given hardware or software product can readily, effectively, and safely be operated or maintained (used) by people from the intended user population; often dependent on the extent to which the product has been well designed ergonomically.

User-centered design: A design philosophy that uses participative approaches to involve users in the design process.

User systems analysis: An approach for assessing work system needs for information-processing equipment and software and evaluating related task and work system design factors. Central to the approach is an analysis of the user's environment, functions, and tasks, and related user information needs.

Value: An attribute that guides behaviors and attitudes.

Variance: An unexpected or unwanted deviation from standard operating conditions, specifications, or norms.

Variance table: A graphic representation that positions variances along both axes and illustrates which variances are related to which other variances. Key variances are usually identified as well.

Vertical differentiation: The hierarchical structure of the work system. It is measured by the number of hierarchical levels separating the chief executive position from the jobs directly involved with the work system's output.

Virtual enterprise: The means by which development of an appropriate response to global market opportunities can be realized.

Virtual organization: A relatively new form of adhocracy that consists of a continually evolving network of independent companies.

Virtuality: The simulation of real environments or tasks through information technologies.

Vision: The long-term view or desire of what a work system is to become in the future.

"Walk the talk": Managers exhibiting the same behaviors and/or attitudes (driven by corporate values) that they expect of their subordinates.

Worker hazard survey: A hazard management approach involving surveying workers for purposes of hazard identification and evaluation.

Workers' compensation: A state-managed and employer-funded insurance system that provides indemnification for workers that experience job-related injury and/or illness.

Work flow integration: A scale for defining technology in both manufacturing and service organizations in terms of a combination of three factors: (a) degree of equipment automation, or extent to which work activities are performed by machines; (b) work flow rigidity, or the extent to which the sequence of activities is inflexible; and (c) specificity of evaluation, or the degree to which work activities can be assessed by specific, quantitative means.

Work-related musculoskeletal disorders (WMSDs): Musculoskeletal disorders that result from work that requires excessive lifting, repetition, awkward postures, or other stress factors in the work environment, including psychosocial factors.

Work role: The actual functions and tasks expected or required to be performed to control variances in a work system. These functions and tasks may or may not be consistent with the formal job description.

Work system: A system that involves two or more persons interacting with some form of (a) hardware and/or software, (b) internal organizational environment, (c) external environment, and (d) organizational design.

Author Index

401

Subject Index

Printed in the United States
by Baker & Taylor Publisher Services